Vehicular Cloud Computing for Traffic Management and Systems

Jyoti Grover
Manipal University Jaipur, India

P. Vinod
SCMS School of Engineering and Technology, India

Chhagan Lal
Manipal University Jaipur, India

A volume in the Advances in Computer and Electrical Engineering (ACEE) Book Series

IGI Global
DISSEMINATOR OF KNOWLEDGE

Published in the United States of America by
IGI Global
Engineering Science Reference (an imprint of IGI Global)
701 E. Chocolate Avenue
Hershey PA, USA 17033
Tel: 717-533-8845
Fax: 717-533-8661
E-mail: cust@igi-global.com
Web site: http://www.igi-global.com

Library of Congress Cataloging-in-Publication Data

Names: Grover, Jyoti, 1980- editor. | Vinod, P., 1979- editor. | Lal, Chhagan, 1985- editor.
Title: Vehicular cloud computing for traffic management and systems / Jyoti Grover, P. Vinod, and Chhagan Lal, editors.
Description: Hershey, PA : Engineering Science Reference, [2018] | Includes bibliographical references and index.
Identifiers: LCCN 2017025495| ISBN 9781522539810 (hardcover) | ISBN 9781522539827 (ebook)
Subjects: LCSH: Vehicle-infrastructure integration. | Vehicular ad hoc networks (Computer networks) | Cloud computing.
Classification: LCC TE228.3 .V465 2018 | DDC 388.3/12028546782--dc23 LC record available at https://lccn.loc.gov/2017025495

This book is published in the IGI Global book series Advances in Computer and Electrical Engineering (ACEE) (ISSN: 2327-039X; eISSN: 2327-0403)

British Cataloguing in Publication Data
A Cataloguing in Publication record for this book is available from the British Library.

All work contributed to this book is new, previously-unpublished material. The views expressed in this book are those of the authors, but not necessarily of the publisher.

For electronic access to this publication, please contact: eresources@igi-global.com.

Advances in Computer and Electrical Engineering (ACEE) Book Series

Srikanta Patnaik
SOA University, India

ISSN:2327-039X
EISSN:2327-0403

Mission

The fields of computer engineering and electrical engineering encompass a broad range of interdisciplinary topics allowing for expansive research developments across multiple fields. Research in these areas continues to develop and become increasingly important as computer and electrical systems have become an integral part of everyday life.

The **Advances in Computer and Electrical Engineering (ACEE) Book Series** aims to publish research on diverse topics pertaining to computer engineering and electrical engineering. **ACEE** encourages scholarly discourse on the latest applications, tools, and methodologies being implemented in the field for the design and development of computer and electrical systems.

Coverage

- Analog Electronics
- VLSI Fabrication
- Optical Electronics
- Sensor Technologies
- Microprocessor Design
- Power Electronics
- Computer Architecture
- Digital Electronics
- Chip Design
- Applied Electromagnetics

IGI Global is currently accepting manuscripts for publication within this series. To submit a proposal for a volume in this series, please contact our Acquisition Editors at Acquisitions@igi-global.com or visit: http://www.igi-global.com/publish/.

The Advances in Computer and Electrical Engineering (ACEE) Book Series (ISSN 2327-039X) is published by IGI Global, 701 E. Chocolate Avenue, Hershey, PA 17033-1240, USA, www.igi-global.com. This series is composed of titles available for purchase individually; each title is edited to be contextually exclusive from any other title within the series. For pricing and ordering information please visit http://www.igi-global.com/book-series/advances-computer-electrical-engineering/73675. Postmaster: Send all address changes to above address.

Titles in this Series

For a list of additional titles in this series, please visit: www.igi-global.com/book-series

EHT Transmission Performance Evaluation Emerging Research and Opportunities
K. Srinivas (Transmission Corporation of Andhra Pradesh Limited, India) and R.V.S. Satyanarayana (Sri Venkateswara University College of Engineering, India)
Engineering Science Reference • copyright 2018 • 160pp • H/C (ISBN: 9781522549413) • US $145.00 (our price)

Fuzzy Logic Dynamics and Machine Prediction for Failure Analysis
Tawanda Mushiri (University of Johannesburg, South Africa) and Charles Mbowhwa (University of Johannesburg, South Africa)
Engineering Science Reference • copyright 2018 • 301pp • H/C (ISBN: 9781522532446) • US $225.00 (our price)

Creativity in Load-Balance Schemes for Multi/Many-Core Heterogeneous Graph Computing Emerging Research and Opportunities
Alberto Garcia-Robledo (Center for Research and Advanced Studies of the National Polytechnic Institute (Cinvestav-Tamaulipas), Mexico) Arturo Diaz-Perez (Center for Research and Advanced Studies of the National Polytechnic Institute (Cinvestav-Tamaulipas), Mexico) and Guillermo Morales-Luna (Center for Research and Advanced Studies of the National Polytechnic Institute (Cinvestav-IPN), Mexico)
Engineering Science Reference • copyright 2018 • 217pp • H/C (ISBN: 9781522537991) • US $155.00 (our price)

Free and Open Source Software in Modern Data Science and Business Intelligence Emerging Research and Opportunities
K.G. Srinivasa (CBP Government Engineering College, India) Ganesh Chandra Deka (M. S. Ramaiah Institute of Technology, India) and Krishnaraj P.M. (M. S. Ramaiah Institute of Technology, India)
Engineering Science Reference • copyright 2018 • 189pp • H/C (ISBN: 9781522537076) • US $190.00 (our price)

Design Parameters of Electrical Network Grounding Systems
Osama El-Sayed Gouda (Cairo University, Egypt)
Engineering Science Reference • copyright 2018 • 316pp • H/C (ISBN: 9781522538530) • US $235.00 (our price)

Design and Use of Virtualization Technology in Cloud Computing
Prashanta Kumar Das (Government Industrial Training Institute Dhansiri, India) and Ganesh Chandra Deka (Government of India, India)
Engineering Science Reference • copyright 2018 • 315pp • H/C (ISBN: 9781522527855) • US $235.00 (our price)

IGI Global
DISSEMINATOR OF KNOWLEDGE

701 East Chocolate Avenue, Hershey, PA 17033, USA
Tel: 717-533-8845 x100 • Fax: 717-533-8661
E-Mail: cust@igi-global.com • www.igi-global.com

Table of Contents

Detailed Table of Contents

Vehicular clouds is an active area of research that has emerged at the nexus of conventional cloud computing and vehicular networks. The defining differences between conventional and vehicular clouds include the heterogeneity and volatility of compute resources and the bandwidth-challenged network fabric. A variety of new architectures and services for vehicular clouds have been proposed, mostly as incremental extensions of the VANET platform. As vehicular cloud research continues and expands, a careful eye should be kept on the restrictions that come with the mobility, limited network, and heterogeneity of resources. The first main contribution of this chapter is to survey recent work of VCs with an eye on the realistic and unrealistic. Our second main goal is to realign the VC community with a realistic vision for the future by spelling out a number of challenges faced by the VC research community.

Cloud computing provides a way to avail hardware and software to provide services over the networks to users, whereas in mobile cloud computing, mobile devices are a part of cloud users and service providers. Vehicular ad hoc networks (VANETs) are developed with the help of mobile ad hoc networks (MANET) with the difference that the data are exchanged among mobile vehicles to enhance the road safety and traffic efficiency. VANETs safety applications includes traffic signal violation, curve speed warning, emergency brake lights, pre-crash sensing, collision warning, left turn assist, lane change warning, and stop sign assist. But due to the requirement of smaller size, one of the significant challenges of the vehicular network is the limited resource with respect to memory, computation power, and bandwidth. It leads to slow data processing capability. However, the demanding requirement of emerging application is complex computation with high storage capacity. This necessity can be fulfilled by sharing the available resource among all the vehicles which are in near proximity. It can be achieved by integrating the vehicular network with cloud computing which is called as vehicular cloud computing (VCC). VCC is a new technological shift that can take the advantages of cloud computing to afford the services to drivers of VANET. But VCC is still in early stage and due to its unique features and applications it has become a significant emerging research area which needs to be explored further.

Vehicular ad-hoc networks (VANET) have become an important research area due to their ability to allow sharing resources among the users to carry out their application and provide services of transport and traffic management. VANET communication allows exchange of sensitive information among nearby vehicles such as condition of weather and road accidents in order to improve vehicle traffic efficiency through Intelligent Transportation Systems (ITS). Many technologies have been developed to enhance ITS. Recently, vehicular cloud computing (VCC) has been developed in order to overcome the drawbacks VANET. VCC technology provides low-cost services to vehicles and capable of managing road traffic efficiently by using the vehicular sources (such as internet) to make decisions and for storage. VCC is considered as the basis for improving and developing intelligent transportation systems. It plays a major role in people's lives due to its safety, security, trust, and comfort to passengers and drivers. This chapter investigates the vehicular cloud computing. The authors first concentrate on architectures. Then, they highlight applications and features provided by VCC. Additionally, they explain the challenges for VCC. Finally, the authors present opportunities and future for VCC.

Vehicular ad hoc networks are relied upon to essentially enhance movement security and transportation productivity. Accessible correspondence, stockpiling, and calculation assets of the associated vehicles are not well used to meet the administration necessities of canny transportation frameworks. Vehicular cloud computing (VCC) is a promising methodology that makes utilization of the preferences of distributed computing and applies them to vehicular systems. Vehicles have turned out to be prepared with different sensors and assets, making them competent to convey, to share assets, and to carry on helpfully. VC can be depicted from the point of view of uses and administrations offered by vehicles that have a place with a vehicular ad hoc network (VANET) due to the impediment of calculation, stockpiling, and data transfer capacity assets for complex in-vehicle sight and sound amusement applications.

There are rapid technological advancements in the field of transportation. Vehicles cooperate among themselves and establish an intelligent transportation system to fulfill the needs on traffic management. To make this possible, the vehicles are equipped with on board unit which shares information regarding the condition of roads, traffic, and management. Such information is stored in cloud server to be accessed by the vehicles at any time. During the message communication, the origin of the message is checked for its authenticity by the receiver. During signature verification, to preserve the privacy of the user, it is required that the ID of the vehicle should not be revealed. Security and privacy of the information stored in the cloud is a challenging task. The privacy of the vehicle user and the information should be secured from the threats. The solutions in the existing literature do not fulfill the requirements of security and privacy. In this chapter, a secured and dynamic privacy preserving scheme is proposed for vehicular cloud computing. Analysis on the proposed privacy scheme is done based on anonymity and un-linkability. The analysis shows that the proposed dynamic scheme provides better privacy compared to static schemes.

Integration of vehicular ad-hoc network (VANET) and cellular network is a promising architecture for future machine-to-machine applications. This integration helps the vehicles have steady internet connection through cellular network (i.e., LTE), and at same time communicate with other vehicles. However, dead spot areas and unsuccessful handoff processes due to the high speed of vehicles that can disrupt the implementation of this kind of architecture. In this chapter, simplified cluster-based gateway selection (SCGS) scheme for multi-hop relay in VANET network is proposed. The scheme is achieved by utilizing a new routing protocol called an enhanced hybrid wireless mesh protocol (E-HWMP). The simulations results show that SCGS scheme through E-HWMP protocol performed better than ad-hoc on demand distance vector (AODV) routing protocol. Furthermore, SCGS scheme through E-HWMP is compared with other cluster-based gateway selections used in the previous works; the result shows that SCGS scheme through E-HWMP protocol outperforms the other cluster-based gateway selections schemes.

In vehicular cloud computing systems (VCC), the overlapping transmission range of each vehicle ensures a unified and common channel for communication among the vehicles. The flexibility of VCC systems opens the door to myriad applications that contribute to the safety and comfort of the passengers. They are distributed, self-organizing communication networks built up by moving vehicles, and are thus characterized by very high node mobility and limited degrees of freedom in mobility patterns. Such particular features often make the standard networking protocol inefficient or unusable, hence the growing effort in the development of communication protocols, which are specific to vehicular networks. Routing protocols should be selected carefully after carrying out literature review. This chapter has investigated different on-demand routing protocols and focused to identify the efficient on-demand routing protocol that can give better performance in realistic environments of vehicular cloud computing systems.

A vehicular ad hoc network (VANET) is the network of mobile devices as well as stationary objects that can communicate with each other. This technology comprises of both vehicle-to-vehicle (V2V) and

Preface

Number of road accidents are increasing day by day all over the world due to impaired driving and traffic congestion. Recently, vehicular ad-hoc networks (VANETs) have gained a significant attention both from industrial and research communities to increase road safety and traffic management. In VANET, Vehicle to vehicle (V2V) and vehicle to infrastructure (V2I) communication is used to propagate safety information to nearby vehicles on time. Each vehicle is equipped with high computation, communication and storage unit in VANET. But most of the time, these resources are under-utilized. Vehicular Cloud Computing (VCC) is the paradigm emerged to utilize the VANET resources efficiently. In VCC, a group of vehicles cooperate with each other to dynamically share computing, sensing and communication resources for decision making on the road in order to improve traffic management and road safety. A vehicular cloud provides a convenient platform for vehicles and people to share sensing data and computing resources for ITS applications. Some examples of VCC applications are:

1. Local traffic condition can be collected from nearby vehices for route planning.
2. Current transportation systems can be improved by big data processing of traffic information by local traffic authorities.
3. Collaborative image of critical events can be reconstructed such as car accident, congestion on the road etc.

These emerging applications (such as vehicular social networking, in-vehicle multimedia entertainment) demand complex computation and large storage. However, it is very difficult for an individual vehicle to compute on big data efficiently. VCC is a very promising solution to share the computation and storage resources among nearby vehicles in order to implement these types of applications. This book will cover all the aspects of vehicular cloud computing as well as the existing techniques that utilizes cloud computing to improve VANET performance. It will also discuss the open issues and research directions related to VANET cloud computing.

An intelligent transportation system aims to improve the efficiency and safety of travelling by consolidating the operations of vehicles, managing vehicle traffic and help drivers with alert and safety messages. VANET is used to provide the updated and dynamic information about the conditions of road traffic so that number of road accidents can be minimized. Vehicles can collect, transmit and interpret the information through wide range of sensors, computing and communication resources, therefore assist in the computation and acquisition of data and facilitates drivers to take appropriate actions. However, most of the times, VANET resources are underutilized. Vehicular Cloud Computing is the paradigm that facilitates the optimization of these resources. It consists of a group of nearby vehicles and roadside infrastructure cooperating with each other in order to share the resources of computing, sensing and communication for decision making on the road to improve Intelligent Transportation System (ITS).

VCC is still in its infancy stage of research and need to be explored more. The objective of the book is to publish papers studying this emerging paradigm of Vehicular cloud computing. Contributed submissions to this book present novel ideas, models, system design and architecture, security and privacy, QoS issues, big data analysis and performance evaluation techniques etc. Chapters of the book deal with the following aspects:

1. Vehicular Cloud Computing models, architectures, platforms, challenges and opportunities
2. On-Demand Vehicular Cloud Computing Systems and Services
3. V2V and V2R communication protocols in Cloud assisted Vehicular Networks
4. Resource management in Vehicular Cloud Computing
5. Security and Privacy in Vehicular Cloud Computing
6. Storage and Data Processing for VCC
7. Big Data analysis for VCC
8. QoS management for VCC
9. Performance Evaluation of VCC architectures

This comprehensive and timely publication aims to be an essential reference source, building on the available literature in the field of VCC while providing for further research opportunities in this upcoming field. It is hoped that this text will provide the essential knowledge for efficient VCC architecture.

Research on Vehicular cloud computing is interdisciplinary that is very much supporting Intelligent Transportation System (ITS) applications. It involves several disciplines such as advances in wireless communication and VANET technology, sensor technology, cloud computing, big data analytics, artificial intelligence and many others. The concept of VCC is new and attaining growing interest from the research community. The objective of the book is to publish papers studying this emerging paradigm of Vehicular cloud computing. Contributed submissions to this book may present novel ideas, models, system design and architecture, security and privacy, QoS issues, big data analysis and performance evaluation techniques etc.

TARGET AUDIENCE

The book adopts multidisciplinary approach and present both theoretical and practical approaches in order to target mass audience. The book is useful to academicians, researchers, Under Graduate and Post Graduate students, technology developers, policy makers and government officials. This book provides the readers with the essential research exposure about pertinent topics in vehicular cloud computing by presenting chapters about concepts, models, technologies, approaches, methodologies and its practices. It also covers other issues of VCC e.g. security and privacy, big data analysis, Quality of service and performance evaluation etc. It can also be a reference for selection by the audience with multiple field backgrounds. Audience of this book also includes members of professional societies, such as Computer and Communication Society of IEEE, ACM, Springer, CSI and other related ones.

ORGANIZATION OF THE BOOK

The book is organized into nine chapters. A brief description of each chapter follows below:

Chapter 1 elaborates on Vehicular clouds. Vehicular Clouds is an active area of research that has emerged at the nexus of conventional cloud computing and vehicular networks. The defining differences between conventional and vehicular clouds include the heterogeneity and volatility of compute resources and the bandwidth-challenged network fabric. A variety of new architectures and services for vehicular clouds have been proposed, mostly as incremental extensions of the VANET platform. As Vehicular Cloud research continues and expands, a careful eye should be kept on the restrictions that come with the mobility, limited network, and heterogeneity of resources. The first main contribution of this chapter is to survey recent work of VCs with an eye on the realistic and unrealistic. Second main goal is to realign the VC community with a realistic vision for the future by spelling out a number of challenges faced by the VC research community.

Chapter 2 elaborates on the architecture and applications of Vehicular Cloud Computing. VANETs safety applications includes traffic signal violation, curve speed warning, emergency brake lights, pre-crash sensing, collision warning, left turn assist, lane change warning, and stop sign assist. But due to the requirement of smaller size, one of the significant challenges of the vehicular network is the limited resource with respect to memory, computation power and bandwidth. It leads to slow data processing capability. However the demanding requirement of emerging application is complex computation with high storage capacity. This necessity can be fulfilled by sharing the available resource among all the vehicles which are in near proximity. It can be achieved by integrating the vehicular network with cloud computing which is called as vehicular cloud computing (VCC). VCC is a new technological shift which can take the advantages of cloud computing to afford the services to drivers of VANET. The goal of this chapter is to facilitate readers to better understand the fundamental of emerging paradigm of vehicular cloud computing mechanisms to lead their research activities. Authors have introduced the background of vehicular cloud computing such as VANET, cloud computing, and Mobile cloud computing. It also gives about the overview of VCC and comparative study between the VANET, cloud computing, Mobile cloud computing and VCC. They also explained the VCC architecture, framework and its design principles. Specifically in detail about the integration of the Mobile cloud computing with VANET, cloud based middle ware for VANET, and cloud-assisted information dissemination system in VANET.

Chapter 3 discusses about the challenges and opportunities of Vehicular Cloud Computing. Vehicular ad-hoc networks (VANET) have become an important research area due to its ability to allow sharing resources among the users to carry out their application, and provide services of transport and traffic management. A VANET establishes communications between the vehicles to exchange sensitive information like a condition of weather and road accidents to improve vehicle traffic efficiency through Intelligent Transportation Systems (ITS). Many technologies have been developed to enhance ITS. Recently, Vehicular Cloud Computing (VCC) has been developed in order to overcome the drawbacks of VANET. VCC technology provides low-cost services to vehicles and capable of managing road traffic efficiently by using the vehicular sources (such as Internet) to make decisions and for storage purpose also. VCC is considered as the basis for improving and developing intelligent transportation systems. It plays a major role in peoples' lives due to its safety, security, trust, and comfort to passengers and drivers. This work investigates the vehicular cloud computing. Authors have explained the challenges for VCC. Finally, various opportunities and future for VCC is also presented.

Chapter 4 elaborates on the resource management in VCC. The distributed computing world view has empowered the misuse of overabundance registering power. The endless number of vehicles on avenues, roadways and parking areas will be dealt with as abundant and underutilized computational assets, which can be utilized for giving open administrations. Consistently, numerous vehicles, invest hours in a stopping carport, garage or parking area. The stopped vehicles are an unlimited unexploited asset, which is squandered basically. These elements make vehicles the ideal contender for hubs in a distributed computing system. A few vehicle proprietors may consent to lease overabundance on board assets, like colossal processing holders and storerooms who lease their abundance limit and advantage financially. The voyagers regularly stop their autos in air terminal parking spots while they are travelling. The airplane terminal expert determination the vehicles' processing assets and take into account on request access to this parking structure server farm. So also, the drivers stuck in movement blockage will concur give their on board figuring assets to help city activity specialists run complex recreations intended to expel clog by rescheduling the movement lights of the city. In this chapter, authors have discussed about vehicular resources and how communication with each other is performed. They also discussed how to exploit these resources properly to meet the service requirements and get best results.

Chapter 5 discusses about a secured dynamic privacy preserving scheme for VCC. Recently there are rapid technological advancements in the field of transportation. Vehicles cooperate among themselves and establish an Intelligent Transportation System to fulfill the needs on traffic management. To make this possible the vehicles are equipped with On Board Unit which shares information regarding the condition of roads, traffic and management. Such information is stored in cloud servers to be accessed by the vehicles at any time. During the message communication, the origin of the message is checked for its authenticity by the receiver. During signature verification, to preserve the privacy of the user, it is required that the ID of the vehicle should not be revealed. Security and privacy of the information stored in the cloud is a challenging task. The privacy of the vehicle user and the information should be secured from the threats. The solutions in the existing literatures do not fulfill the requirements of security and privacy. In this book chapter, a secured and dynamic privacy preserving scheme is proposed for vehicular cloud computing. Analysis on the proposed privacy scheme is done based on anonymity and un-link ability. The analysis shows that the proposed dynamic scheme provides better privacy compared to static schemes.

Chapter 6 presents V2V communication protocols in cloud assisted vehicular networks. Consequently, the main challenges facing VANET are to decide upon the routing protocol that should be used to control the process of forwarding packets through nodes in the network; determining how to select the next-hop node to use to forward packets to their final destination, particularly in a sparse environment, depending on the presence of the unique characteristics of VANET. Therefore, providing a robust routing protocol is considered to be the most crucial solution in VANET. This involves using new parameters to take the decisions regarding selecting the next-hop node to enhance the efficiency of the routing process and increase performance. As a result, it is anticipated that designing an efficient routing protocol will aid in accomplishing the task of delivering packets to their destinations in VANET via a more realistic method, which promises to apply road safety efficiently. The specific objectives of this chapter are to design end to end multi-hop scheme to extend the coverage of VANET network and develop a new cluster-based gateway (cluster head) selection algorithm for Vehicular communication. Authors have also evaluated the proposed scheme under multi-metric design parameters for overall throughput.

Chapter 7 presents on demand routing protocols in VCC. The flexibility of Vehicular Cloud Computing Systems opens the door to a myriad of applications that contribute to the safety and comfort of the passengers. VANETs represent a rapidly emerging, particularly challenging the class of mobile ad-hoc network (MANET). They are distributed, self-organizing communication networks built up by moving vehicles, and are thus characterized by very high node mobility and limited degrees of freedom in mobility patterns. Such particular features often make the standard networking protocol inefficient or unusable, hence the growing effort in the development of communication protocols, which are specific to vehicular networks. The high mobility of communicating vehicles necessitated that protocols be vastly revised or re-written from scratch. Routing protocols should be selected carefully after carrying out literature review. This Chapter has investigated about different on-demand routing protocols and focus to identify the efficient on-demand routing protocol that can give better performance in realistic environments of Vehicular Cloud Computing Systems. Finally the feasibility analysis with the help of on-demand protocols and different environment performance metrics will describe with analysis and conclusion are presented in this chapter.

Chapter 8 presents a survey report on "smart traffic control" system that incorporates traffic-related parameters to further augment the control and management of vehicular movements on the roadways. This can support efficient management of traffic in the city, and cops can communicate with each other, get real-time, accurate, status update of the traffic, track the vehicular movements, etc. In addition, jam control mechanism can be placed on heavy traffic days to optimize the routes. This system introduces Artificial Intelligence (AI) that can optimize deployment of cops and find alternate routes for the driver to reach the destination address without much fuel consumption.

Chapter 9 elaborates on Fog computing in vehicular networks. In recent years, there has been huge developments in automotive industry, computing and communication technologies. This has led Vehicular Cloud Computing (VCC) as a solution to satisfy the requirements of VANETs such as computing, storage and networking resources. Vehicular Fog computing (VFC) is a standard that comprehends cloud computing and related services to the proximity of a network. Since VANET applications have special mobility, low latency and location awareness requirements, fog computing plays a significant role in VANET applications and services. In this chapter, real time applications of VANET that can be implemented using VFC, are presented. Cloud computing is the central portal of information, does not have location awareness. Fog computing overcome this issue by providing localized services to specific deployment sites. Fog computing is a most suit-able approach that can control critical resources like energy, traffic and healthcare etc. It is shown that for a scenario where 25% of applications demand real-time low latency services, mean energy expenditure is fog computing is 40% less than conventional cloud computing model. This analysis motivates the research and academia to explore fog computing more extensively.

Chapter 1
Vehicular Clouds:
A View From Above

Ryan Florin
Old Dominion University, USA

Stephan Olariu
Old Dominion University, USA

ABSTRACT

Vehicular clouds is an active area of research that has emerged at the nexus of conventional cloud computing and vehicular networks. The defining differences between conventional and vehicular clouds include the heterogeneity and volatility of compute resources and the bandwidth-challenged network fabric. A variety of new architectures and services for vehicular clouds have been proposed, mostly as incremental extensions of the VANET platform. As vehicular cloud research continues and expands, a careful eye should be kept on the restrictions that come with the mobility, limited network, and heterogeneity of resources. The first main contribution of this chapter is to survey recent work of VCs with an eye on the realistic and unrealistic. Our second main goal is to realign the VC community with a realistic vision for the future by spelling out a number of challenges faced by the VC research community.

1. INTRODUCTION

The vision and promise of vehicular networks, populated by vehicles that can communicate to improve safety and provide traveler information, has captivated the networking research community over the past two decades. Applications and protocols have been proposed, standards were and are still being written, and hardware is being developed to support various applications. One of the turning points that we have witnessed in the past decade was the emergence of *self-driving* cars. Several manufacturers have already committed resources towards large-scale production of such self-driving cars. This trend will drive the production of vehicles equipped with advanced computational, communication, storage, and sensing capabilities. With close to seven million new cars purchased in the US alone each year, we will soon have many millions of these advanced vehicles on our roadways and city streets.

DOI: 10.4018/978-1-5225-3981-0.ch001

Many of these advanced vehicles will spend hours at a time in a parking garage, parking lot, or driveway. While parked, or while crawling in a traffic slowdown, the computing and storage resources of these vehicles will be untapped and the opportunity for their use will be wasted. Given the right *incentives*, the owner of a vehicle may decide to rent out excess on-board capabilities. Owners of parking garages may decide to provide such incentives for vehicle owners to allow the use of their parked vehicles. For example, future airports will power travelers' vehicles to harness their computing resources, allowing for on-demand access to parking lot data centers.

Likewise, vehicles crawling in a traffic slowdown will be able to donate their compute resources so that municipal traffic management centers can run complex simulations designed to help alleviate the effects of congestion by city-wide rescheduling of traffic lights. We also anticipate that in emergency evacuation situations, the combined compute power of the vehicles will allow evacuation management centers to run sophisticated simulations in close to real-time in order to effectively manage the evacuation event.

To enable such a future, Eltoweissy *et al.* (Eltoweissy, Olariu, & Younis, 2010) introduced and developed the concept of a Vehicular Cloud (VC) defined to be:

A dynamic group of vehicles whose excess computing, communication, storage, and sensing resources are coordinated and dynamically allocated to authorized users.

As it turns out, VCs are a non-trivial extension of conventional cloud computing. The distinguishing characteristic of VCs is mobility, which results in a dynamically changing amount of available resources. It is therefore clear that the VCs can only achieve their full potential if their basic architecture is tailored to offer a seamless integration of the cyber-physical resources of the participating vehicles. In particular, the architecture must adapt its managed vehicular resources allocated to an application according to dynamically changing requirements and systems conditions.

While some flavors of VCs may be run much like conventional clouds, their novelty is that they can also be used to supplement a municipal separtment of transportation's computing resources during extended periods of traffic congestion or even during emergency conditions, such as an evacuation (Abuelela & Olariu, 2010).

In such scenarios, the VC can provide on-demand solutions to traffic events that have occurred but whose timely resolution cannot be met reasonably with pre-assigned assets or in a proactive fashion. It is essential to investigate issues related to dynamic resource management in a VC -- a challenging, dynamically varying wireless environment. The results obtained for VCs could be applied in the development of other transformative mobile computing applications.

The first main goal of this chapter is to survey recent work of VCs with an eye on the realistic and unrealistic. Our second main goal is to realign the VC community with a realistic vision for the future by spelling out a number of challenges faced by the VC research community.

At this point, it is appropriate to give the reader a synopsis of the chapter. After this introductory section, Section 2 offers an overview of vehicular networks, one of the possible entities that are very likely to constitute the backbone of vehicular clouds. Section 3 provides a succinct overview of cloud computing and cloud services that has motivated the vision of vehicular clouds. Next, in Section 4 introduces vehicular clouds and discusses some of their distinguishing characteristics Specifically, Subsection 4.1 offers a historical perspective of VCs, while Subsection 4.2 points out the main differences between VC and conventional clouds. Section 5 discusses a typical architecture of a VC. Subsection 5.1 introduces the main components of such an architecture; Subsection 5.2 discusses setting up a VC; Subsection 5.3 offers a number of challenges related to the task of engineering a VC. Section 6 presents a taxonomy of VCs where we distinguish between static VCs involving vehicles in a parking lot and vehicles that are moving around. These flavors of VCs will be discussed in detail in Subsections 6.1 and 6.2. Section 7 of the chapter surveys recent VC literature. Subsection 7.1 looks at job scheduling and resource provisioning, Subsection 7.2 deals with VM migration in VCs, Subsection 7.3 surveys the important topic of service management, one of the backbones of VCs, Subsection 7.4 looks at various papers dealing with data dissemination and, finally, Subsection 7.5 surveys the important topic of data sharing. Section 8 enumerates a number of research challenges for VCs. Subsection 8.1 makes a case for identifying realistic applications for VCs; Subsection 8.2 proposes resource management as another challenges for VC research; Subsection 8.3 suggests VC federation and the creation of VC ecosystems as yet another worthwhile challenge; Subsection 8.4 suggests devoting attention to quantitative aspects of VC research; Subsection 8.5 makes a case for the importance of testing out ideas by providing large-scale implementations of VCs; Subsection 8.6 points out the importance of determining conditions under which VCs can support Big Data applications; Subsection 8.7 suggests that VCs will play an important role in the Smart Cities of the future. Finally, Section 9 offers concluding remarks.

2. VANET: VEHICULAR AD-HOC NETWORKS

The past two decades have witnessed a growing interest in vehicular networking and its host of potential applications. The initial vision that had fueled research in vehicular networking had originated in an altruistic impulse, namely that radio-equipped vehicles can somehow network together and, by exchanging and aggregating individual traffic views, can keep the drivers informed about potential traffic safety risks and can heighten their awareness of road conditions and other traffic-related events (Olariu & Weigle, 2009).

The unmistakable promise of vehicular networking has lead to a rapid convergence with Intelligent Transportation Systems leading to the emergence of Intelligent Vehicular Networks, expected to revolutionize the way we drive by creating a safe, secure, and robust environment that will eventually pervade our highways and city streets (USDOT, 2015).

In support of vehicular networking and, more generally, of traffic-related communications, the US Federal Communications Commission (FCC) has allocated 75MHz of spectrum in the 5.850 to 5.925 GHz band for the exclusive use of Dedicated Short Range Communications (DSRC). As pointed out by a number of workers including Rawat *et al.* (Rawat, Yan, Popescu, Weigle, & Olariu, 2009), the DSRC spectrum by far exceeds the needs of traffic safety applications. The availability of bandwidth has motivated the emergence of a host of third-party applications that can take advantage of the excess DSRC spectrum. Initially, these non-safely related applications included vehicular peer-to-peer networking and rudimentary multimedia content delivery. More recently, they were expanded to include location-specific services, on-line banking, as well as on-line gaming and other forms of mobile entertainment. In due time, we expect to see the emergence of a large array of commercial applications targeted at the traveling public and made possible by the excess DSRC bandwidth.

In view of the short communication range stipulated by DSRC, we expect third-party infrastructure providers to deploy various forms of road-side infrastructure as well as advanced in-vehicle resources such as embedded powerful computing and storage devices, cognitive radios, and multi-modal programmable sensor nodes. As a result, in the near future, vehicles equipped with computing, communication, and sensing capabilities will be organized into ubiquitous and pervasive networks with a significant Internet presence while on the move. This is apt to revolutionize the driving experience making it safer, more enjoyable and more environmentally friendly. As a pleasant side benefit, the unsightly billboards that flank our highways will disappear and will be replaced by in-vehicle advertising that the driver can filter according to their wants and needs. However, we believe that the impressive array of on-board capabilities present in our vehicles is likely to remain under-utilized by the above-mentioned applications.

The original goal of VANET was to provide drivers with notification of real-time traffic conditions. In the past decade a number of (mostly small-scale) prototypes have been deployed in Europe and Japan by consortia involving car manufacturers, government agencies, and various standardization bodies. American car manufacturers have recently jumped on the bandwagon, embracing the idea that Vehicle-to-Vehicle (V2V) and Vehicle-to-Infrastructure (V2I) communications could have a tremendous marketing potential. The most noteworthy US initiatives are the California Partners for Advanced Transit and Highways (PATH) that has investigated issues ranging from automated driving, to traffic modeling and collision warning systems, and the recently proposed Connected Vehicles Initiative of US-DOT that proposes to investigate the integration of Intelligent Transportation Systems with vehicular communications (Roess, Prassas, & McShane, 2011).

The original impetus for the interest in VANET was provided by the need to inform fellow drivers of actual or imminent road conditions (Olariu & Weigle, 2009; Rawat, Yan, Popescu, Weigle, & Olariu, 2009).

Therefore, most VANET applications focus on traffic status reports, collision avoidance, emergency alerts, cooperative driving, and other similar concerns. Almost across the board, the community of researchers and practitioners anticipate that advances in VANET, or other emerging vehicle-based computing and communications technology, are poised to have a huge societal impact.

The vision of VANET has fascinated the networking research community over the past two decades: a number of important applications have been identified and implemented over the years. The societal benefits and potential for rapid commercialization will continue to drive the production of vehicles equipped with advanced processing, communication, storage, and sensing capabilities. This huge vehicular fleet on our roadways and city streets is a valuable resource that should be put to good use: VCs are just one of the ways in which one could harness the potential of VANET – a potential killer app.

2.1. DSRC/Wave

As already mentioned, the US Federal Communication Commission has allocated 75MHz of spectrum between 5.850 GHz and 5.925GHz for use by ITS; known as Dedicated Short Range Communications (DSRC). Initially intended to be used for traffic safety applications only, the amount of bandwidth in the range has been shown to provide support for many other applications (Nakano & Olariu, 2000; Rawat, Yan, Popescu, Weigle & Olariu, 2009).

Due to the US government's Connected Vehicle program, DSRC/WAVE is becoming the standard method of communication for vehicular communications, allowing the research community to use it as a platform on which to base further networked vehicle research.

3. CLOUD COMPUTING

Cloud Computing (CC), a catchy metaphor for *utility computing*, implemented through the provisioning of various types of hosted services over the Internet, has seen a phenomenal growth in the past decade. The underlying business model of CC is the familiar pay-as-you-go model of metered services, where a user pays for whatever he/she uses and no more, and where additional demand for service can be met in real time.

This powerful idea has been suggested, at least in part, by the pervasive presence of relatively low-cost high-speed Internet, a good handle on virtualization and advances in distributed computing.

Utility computing, perceived through the prism of CC, is a paradigm shift adopted by a large number of infrastructure providers, both in the US and around the world, whose large installed infrastructure often goes under-utilized. The unmistakable appeal of utility computing is that it provides scalable access to seemingly unbounded computing resources and to a multitude of IT services that anyone with a valid credit card can use.

CC has arisen at the confluence of a number of enabling technologies, of the availability of a chronic excess of compute resources and of the need for on-demand scalable, cheap and secure compute power that can be provided in real time (Barroso, Calderas, & Holzle, 2013; Buyya, Vecchiola, & Selvi, 2013). One of the appealing consequences of CC is that instead of investing in infrastructure, many businesses and individual entrepreneurs find it useful to rent the infrastructure and, oftentimes, the software required to run their applications (Marinescu, 2013).

It is fair to say that the success of CC was due, to a large extent, to its ability of providing the illusion of scalable access to seemingly unbounded compute resources and to a mind-boggling number of IT services. However, in order to achieve almost unbounded scalability, reliability, and availability, CC requires substantial architectural support ranging from virtualization, to server consolidation, to file system support, and to memory hierarchy design (Hennessy & Patterson, 2012).

4. VEHICULAR CLOUDS

4.1. How It All Started

It was noticed that under present-day state of the art, the vehicles are mere spectators that witness traffic-related events, without being able to participate in the mitigation of their effect. Eltoweissy *et al.* (Eltoweissy, Olariu & Younis, 2010), and Abuelela and Olariu (Abuelela & Olariu, 2010) suggested that in such situations the vehicles themselves have the potential to cooperate with various authorities to solve problems that otherwise would either take an inordinate amount of time to solve (e.g. dissipating traffic jams) or could not be solved for lack for adequate resources that could be brought to bear. Specifically, they proposed to "take vehicular networks to the clouds" so that the vehicular fleet on our roadways and streets can be integrated with our productivity, comfort, safety, and economic prosperity. These were the ideas that have laid the foundations of what is not called Vehicular Clouds (VC).

In addition, VCs were suggested and inspired by conventional cloud computing (CC), since it was clear that most of the on-board resources in our vehicles are under-utilized. Historically, the first papers that introduced the concept of a VC were (Eltoweissy, Olariu & Younis, 2010; Abuelela & Olariu, 2010; Olariu, Khalil & Abuelela, 2011; Olariu, Eltoweissy & Younis, 2011). Inspired by the success of conventional clouds and realizing that most of the on-board resources of present-day vehicles are chronically under-utilized, these early papers combined the two ideas to form the concept of VCs. The vision was that putting the under-utilized vehicular resources to work in a meaningful way will, in all likelihood, have a significant societal impact. The visionaries have anticipated that in a VC the under-utilized vehicular resources including computing power, storage, and Internet connectivity can be shared between drivers or rented out over the Internet to various customers, very much as conventional cloud resources are. In VCs, the vehicles on the road share their resources much like that of servers in the conventional cloud. Since their inception in 2010, VCs have become a booming area of research with numerous papers describing new variants of VCs and all sorts of fancy cloud services.

The early papers mentioned above defined various possible variants of VCs, their possible applications and numerous research challenges. Specifically, *autonomous clouds*, a precursor of VCs, were proposed for the first time in (Olariu, Eltoweissy & Younis, 2011), where they also surveyed a number of important applications and research challenges.

It was suggested that, even under current technology, many non-trivial forms of VCs are technologically feasible and economically viable. Recently, Olariu *et al.* (Olariu, Hristov, & Yan, 2013) argued that, once adopted, VCs will turn out to be the next paradigm shift with a lasting technological and societal impact.

VCs are still in their infancy and we have barely scratched the surface of their potential applications. To realize their full potential a number of challenges have to be met. Some of these challenges will be discussed in the chapter.

4.2. How Are Vehicular Clouds Different?

The most obvious difference between CC and VC is that, unlike CCs where the compute resources (i.e. servers) are proprietary, in VC the ownership of the compute resources is distributed over a large driver population. In CC, servers leave the cloud only when they unexpectedly crash or when shut down for maintenance. In VCs, vehicles leave the cloud on the whim of the driver. If the VCs underlying network is wireless, it may be as simple as the driver taking a turn away from the VC coverage. In cases of a VC in a parking lot, it may be that driver leaves the parking lot. No matter the case, the mobility of vehicles must be considered. Another difference is that the storage, computing, and sensing capabilities of vehicles are not homogeneous. Each vehicle is owned by an individual and the amount of storage, processing power, and availability of sensors varies from vehicle to vehicle. Of course, an exception can be made for special cases; perhaps a municipality outfits a fleet of vehicles, buses, law enforcement vehicles, or ambulances, with a certain specification of equipment.

Third among the differences between VC and CC is the network fabric. VCs using vehicles in a parking lot may benefit from plugging into a wired network, but this is a special case. VCs using vehicles on the road must use wireless communication. The bandwidth of these VCs is far less than that of the CC.

The original intention of VCs was not to use them as if they were CCs on wheels; instead, it was to utilize the on-board resources of the participating vehicles in a way that benefits the vehicles on the road. Vehicles are unique in that they have plenty of power, storage, processing ability, and an array of sensing equipment. They are additionally unique in that they have some communication abilities, although much more limited than that of the CC.

5. VEHICULAR CLOUDS: THE BASIC ARCHITECTURE

If the VC concept is to achieve its full potential, its basic architecture must be engineered to offer a seamless integration of the excess cyber-physical resources of the participating vehicles. In particular, the architecture must dynamically adapt its managed vehicular resources allocated to an application according to dynamically changing requirements and systems conditions. In this section we offer a high-level overview of the architecture and challenges involved in engineering a VC.

5.1. Architectural Overview

At the logical level, a typical cloud computing architecture consists of a client-side API translator, a cloud controller, several cluster controllers, and even more node controllers. A client application on one side of the network communicates with a cloud controller that facilitates the exchange of data to and from multiple services hosted on the back-end. The cloud controller in turn communicates with multiple cluster controllers, which manage multiple node controllers. In the generic form of the cloud architecture, each of these node controllers is an application that runs on top of a server in the data center.

The conventional cloud architecture is built based on the hardware specifications of datacenter servers interconnected with a minimum of 1 Gbps Ethernet. The node controllers are hosted on multi-core, multi-processor machines, with high-memory density. The required specifications allow for virtualization of the node hosts to provide virtual machines (VMs) to run user services. A challenging aspect in a VC is the relatively low hardware specifications of embedded units. We assume that vehicles in the future will have enhanced capabilities, but in all likelihood they will remain below that of data center machines. The main challenges in developing a VC are these computational limits and the requirement, in many cases, of wireless communications. The current implementation of cloud computing requires virtualization, and thus VCs require a method of pooling the resources of multiple vehicles together to generate a single VM.

As illustrated in Figure 1, a VC involves certain cyber-physical components.

Figure 1. A high level view of the VC architecture

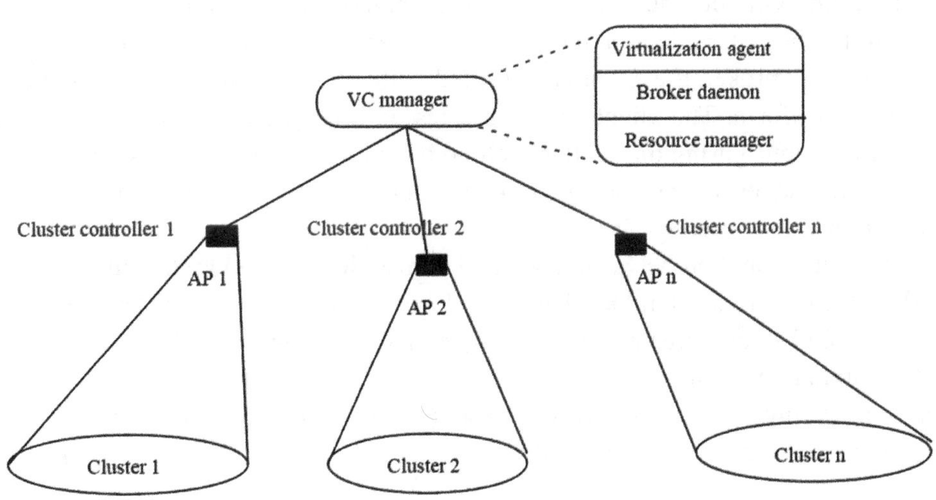

A *VC manager* that plays the role of the *cloud controller* to interface with the outside world. In turn, as shown in Figure 1, the VC manager consists of the following components:

- **A Broker Daemon:** In charge of negotiating cloud service requests on behalf of the VC from various potential clients. The broker has the authority to accept feasible requests and to commit cloud resources to meet their specifications;
- **A Virtualization Agent:** In charge of configuring the available cloud resources in ways that best match the commitments underwritten by the broker. It is important to note that in the VC, virtualization takes place here, rather than at individual nodes;
- **A Resource Manager:** In charge of discovering and managing the dynamically-changing cloud resources. This includes discovering new resources as they become available, managing available resources, predicting the availability of resources using a suitable stochastic model, migrating tasks from departing vehicles to available ones, and balancing the load on the existing clusters.

In addition to the VC manager, the VC architecture contains a number of *cluster controllers*, each in charge of a resource cluster. In most cases, clusters would be geographically organized and of the size, for example, of a typical parking lot on a university campus. The cluster controller is in charge of logical operations and communicates with the resource manager;

Referring, again, to Figure 1, each cluster controller interfaces with the vehicles in its cluster through a smart *access point* (AP), responsible for managing the wireless connection to the vehicles in the cluster. The smart AP is in charge of managing physical resources in the cluster;

Each vehicle participating in a cluster lends its excess on-board resources in return for some quantifiable benefit. The VC resources are the pooled resources of these individual vehicles.

5.2. Setting Up a Vehicular Cloud

We now describe the actions involved in setting up a cloud service in a VC. To begin, the broker daemon receives a request for cloud services and decides, in consultation with the resource manager and the virtualization agent, whether or not the currently available cloud resources suffice to meet the anticipated demands of the requested services. If the cloud service request is accepted, the broker daemon's job is to ensure that the client will not request more resources than originally stipulated.

Figure 2 illustrates the logical data-path between the client and the VC. Once a cloud service request has been accepted by the broker, the client is presented with a VM, essentially an API that provides the illusion of a virtual platform. The client communicates with the virtualization agent, which, in turn, specifies strategies for configuring the available cloud resources in ways that best matches the needs of the client application. The resource manager's job is to allocate physical resources that, together, provide the virtual environment requested by the client.

Note the important distinction between a VC and conventional cloud architecture. Virtualization is performed at the VC manager level, rather than at the level of individual nodes. This means that a single VM may be supported by resources provided by several vehicles in multiple clusters, rather than the resources of an individual vehicle.

In addition to allocating the physical resources, the resource manager, in coordination with the cluster controllers, performs the following tasks:

Figure 2. Illustrating an end-to-end data-path in a VC

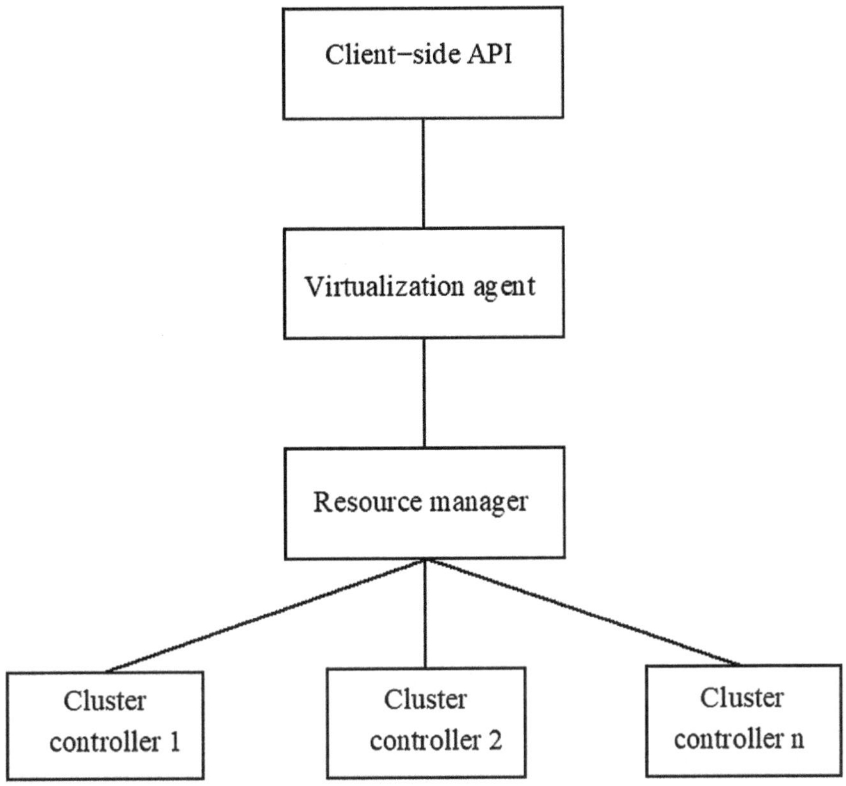

- Discovers new resources as soon as they become available;
- Keeps track of currently-available resources;
- Predicts the availability of resources using a suitable stochastic model;
- Migrates tasks from a vehicle leaving the VC to an available vehicle;
- Balances the loads of the existing clusters.

The mobility attribute of the VC, combined with the fact that the presence of vehicles in close proximity to an event is very often an un-planned process implies that the pooling of the resources of those vehicles that for a VC in support of mitigating the event must occur spontaneously by the common recognition of a need for which there are no pre-assigned or dedicated resources available. This agility of action does not exist in conventional clouds and turns out to be an important defining characteristic of VCs.

5.3. Engineering a Vehicular Cloud: Research Challenges

VCs are complex cyber-physical entities that must be designed to withstand structural stresses induced by the inherent instability of the operating environment. A VC is defined by its aggregated cyber and physical resources: their aggregation, coordination and control are non-trivial research challenges.

- **Architectural Challenges:** These challenges include issues related to the organization of the cyber structure of the VC and its interaction with the physical resources. There is a critical need to efficiently manage host mobility and heterogeneity (including computing, communication, and storage capabilities) and vehicle membership (change in interest or location, resource denial, failure, etc.).
- **Functional Challenges:** In order for the VC vision to become reality, the problems of assuring emergent trust and security in VC communication and information need to be addressed. Research is needed on developing a trustworthy base, negotiation, and strategy formulation methodology (e.g., game theory), efficient communication protocols, data processing, etc.
- **Security and Privacy:** These are two main concerns when allowing multiple users to share the same set of resources, and VCs are no exception. The security and privacy of both the vehicle's owner and the VC customers who rent the resources must be preserved. We recognize that there will be many challenges involved in making the VC vision reality, and security is a very important one. In VCs, issues such as security and privacy rely on solutions currently used by conventional CC systems, which mainly use the security features provided by virtualization. Yan *et al.* (Yan, Wen, Olariu, & Weigle, 2013) pointed out a number of security challenges for VCs; however, in this chapter the focus is on quantitative challenges.

6. A HIGH-LEVEL TAXONOMY OF VEHICULAR CLOUDS

In their seminal papers (Eltoweissy, Olariu & Younis, 2010; Abuelela & Olariu, 2010; Olariu, Khalil & Abuelela, 2011) have argued that, in the near future, vehicles will autonomously self-organize into VCs utilizing their corporate resources on-demand and largely in real-time for resolving critical problems that may occur unexpectedly. While static VCs may mimic the behavior of a conventional cloud, many vehicles spend a substantial amount of time on the road and may be involved, on a daily basis, in various dynamically changing situations, ranging from normal traffic to congestion, to accidents, to other similar traffic-related events.

VCs are still in their infancy and so are the myriad potential applications and significant research challenges facing the VC research community. While a good number of papers were written about various flavors of VCs, there are virtually no studies concerning the practical feasibility of VCs. As already mentioned, a number of recent papers proposed VCs built on top of moving vehicles interconnected by some form of wireless communication fabric; some other authors consider a mix of moving and stationary vehicles. However, based on their extensive simulations Florin *et al.* (Florin, Abolghasemi, Zadeh & Olariu, 2017) have shown that VCs based on wireless communications (as are, per force, employed by moving vehicles) do not appear to be able to support data-intensive applications effectively.

It is our hope that this negative result might serve as a wakeup call to the VC community. One of the significant research challenges in the realm of VCs is to identify conditions under which dynamic VCs can support data-intensive applications. Such applications, with stringent data processing requirements might not be efficiently supported by ephemeral VCs, that is to say, VCs where the residency time of vehicles in the cloud is too short for supporting VC setup and VM migration. Similarly, VC implementations with bandwidth-challenged interconnection topologies are not the perfect choice for applications that consume or produce a large amount of data and therefore need to be run for an extended amount of time.

6.1. Static Vehicular Clouds

In static VCs cars are parked or non-mobile for an extended period of time such as when parked in a parking lot (Arif, Olariu, Wang, Yan, Yang, & Khalil, 2012). In the most ideal of parking lots, cars will park and will plug into a power source and network essentially creating an environment similar to a conventional cloud. A group of parking spots may be organized into a cluster, with several clusters into a group, with several groups into a region, making up a hierarchal network fabric. This mimics the conventional CC setup where the vehicle is an analog to the server, the cluster is an analog to the rack, etc.

The obvious difference is the residency time of a vehicle in a parking lot is relatively short in comparison to the residency time of a server in a data center. In conventional clouds, residency time of the server depends on hardware malfunctions and maintenance schedules; whereas, the residency time of a vehicle in a parking lot depends on the whim of the driver. In certain circumstances, perhaps at an airport where the schedule of the driver is known, the residency time of the vehicle may be known before hand, but this is atypical. At best, statistics of the mean and variance of residency times may be known. Whatever information that is known about residency times of vehicles must be taken advantage of to get the best performance from the VC.

In some parking lots, a plug-in network will not be available, leaving wireless as the only means of communication between vehicles. In these VCs, the amount of data transferred must be low; otherwise, congestion of the network is almost certainly unavoidable, unless creative solutions are used. Furthermore, some parking lots may not provide plug in power. In these VCs, compute power must be as efficient and distributed properly as not to drain the batteries of vehicles.

The defining point of the static vehicular clouds is that there is essentially no mobility until the vehicle leaves. Redundancy must be considered because when a vehicle leaves the parking lot, its data will no longer be accessible.

6.2. Mobile Vehicular Clouds

In mobile VCs, vehicles are moving about. They may stop at a traffic light for some time, but are still considered mobile. Note, that at certain moments the relative change in distance between vehicles on the road may be very small, leading to a situation where a mobile VC can temporarily assume the properties of a static VC. In these situations, vehicles may work together and share information back and forth for long periods of time. In other situations the relative change in distance between cars may seem to be chaotic. Vehicles may still work together and share information back and forth, but they only do so for short periods of time.

When mobility is low, there is the chance for redundancy of computing, data, and an increase of shared resources between a given set of vehicles. When mobility is high, vehicles may be treated as data mules, disseminating data throughout the network quickly, but resource sharing becomes less likely.

In a mobile VC, vehicles use a combination of V2V and V2I communications. While V2V communications are "zero-infrastructure", V2I communication requires a roadside unit (RSU) that acts as a gateway that supports the communication between vehicles and, say, the Internet. It is well known that RSUs are expensive to deploy and maintain and so, not surprisingly, very few urban communities have deployed them (Roess, Prassas & McShane, 2011).

However, there are many possible ways of implementing the *functionality* of an RSU without physically deploying RSUs at great expense. For example, parked vehicles could play to role of RSUs and so are, to some extent, vehicles in the public transportation fleet, taxi cabs that crisscross our streets, or indeed, police cruisers. We expect many novel forms of RSUs to be proposed in the near future.

7. SURVEY OF VEHICULAR CLOUD LITERATURE

The main goal of this section is to offer a succinct review of the various trends the authors have noticed in the VC literature surveyed. Due to the space limitations imposed by the editor, only a fraction of the papers could be mentioned here. However, the authors trust they are fairly representative of the entire VC literature of the past seven years.

7.1. Job Scheduling and Resource Provisioning

Within conventional CC, it is essential to the perceived quality of the customer to only accept jobs when there is a very good chance of a success. Nowadays, one failed job may result in the customer changing providers, resulting in loss of income for the cloud service provider.

In VCs, the stakes are similar. If there is a perceived notion that the VC cannot successfully complete a job, then the user may not use the service. As such, there is a need to assign jobs to the VC only when there is a reasonable chance of success.

Zheng *et al.* suggest resources in the VC determine whether to accept a job or to send it to a remote conventional cloud using a semi-Markov decision process to maximize a long-term reward (Zheng, Meng, Chatzimisios, Lei, & Shen, 2015). They assume the job can be submitted to and split between up to three vehicles.

In a series of papers, Florin *et al.* require the job be submitted to two or three vehicles, but to mitigate system failure, the same computation is performed on both (Florin, Ghazizadeh, Ghazi Zadeh, & Olariu, 2017). The authors developed formulas for MTTF (Ghazizadeh, Florin, Ghazi Zadeh, & Olariu, 2016) and expected job completion time (Florin, Ghazizadeh, Ghazi Zadeh, & Olariu, 2017) for their J2 and J3 job scheduling. They find for short term residency of vehicles, a redundancy of computation is required to handle system failures.

Salahuddin *et al.* propose a dynamic resource allocation scheme based on a Markov decision process (Salahuddin, Al-Fuqaha, & Guizani, 2016). In static resource allocation, an expected demand is determined; if the actual demand deviates above the expected demand, then QoS suffers. On the other hand, if the actual demand is lower, then the VC will be underutilized. In dynamic resource allocation, the demand is estimated regularly decreasing the chances of QoS violations and VC underutilization. The authors propose a reinforcement-learning-based Markov decision process to more accurately match the demand of the VC.

Given the dynamic nature of the VC, jobs may be scheduled, but without knowing exactly when the vehicle will leave, it is difficult to determine when the host vehicle should accept a job. Much of the work in the literature looks at how to get the most out of the available resources with little regard to system failure. This is expected to lead to additional system failures when the host vehicle leaves the network. Instead, as in (Ghazizadeh, Florin, Ghazi Zadeh, & Olariu, 2016; Florin, Ghazizadeh, Ghazi Zadeh, & Olariu, 2017), redundancy may be introduced to lessen the occurrence of a job failures.

The quality of the job scheduling scheme should be judged based on job completion time, the span of time starting when the job is originally accepted into the VC until the time it finished successfully, and the number of jobs completed. Accepting more jobs may increase the number of jobs completed, but may also increase the job completion time if doing so incurs failures. The job scheduling scheme should be optimized to maximize both according to the needs of the requestor.

7.2. Virtual Machine Migration

In conventional clouds, servers do not leave the network unless they unexpectedly crash or scheduled maintenance is to be performed. In VCs, vehicles may leave the cloud on the whim of the driver. If the VCs network is wireless, it may be as simple as the driver taking a turn away from network coverage. In cases of a VC in a parking lot, it may be that the driver leaves the parking lot. In VCs, to handle the vehicles leaving the cloud, the Virtual Machine (VM) must be transferred to another vehicle; otherwise, the work done will be lost.

In work starting with Ghazizadeh *et al.*, VM migration occurs as a result of a vehicle leaving the parking lot (Ghazizadeh, Florin, Ghazi Zadeh, & Olariu, 2016; Florin, Ghazizadeh, Ghazi Zadeh, & Olariu, 2017). In the author's Jn job assignment strategy, there must always be n vehicles running the job, unless recruitment is occurring. The recruitment occurs when a vehicle leaves; a recruiter is chosen, then it copies the state of its VM and sends it to the recruited vehicles. When recruitment finishes, each job is restarted.

Refaat *et al.* look specifically at VM migration between mobile vehicles, proposing four such methods (Refaat, Kantarci, & Mouftah, 2016). In the first, branded VVMM-Uniform, any vehicle is considered as a new host for the migrated virtual machine, despite not having the proper time or space for the new VM. Essentially this is a benchmark used to compare the rest of the methods. The next method, VVMM-Least Workload, only considers vehicles that have the proper resources for the migrating VM. If there are several choices, the one with the least workload will be chosen. The Uniform and Least Workload methods may fail because there is not enough time for the migration to complete. The third method is VVMM-Mobility Aware, where only vehicles with enough time to migrate are considered. If there are multiple choices, the one with the least workload is then selected. The final method, Mobility and Destination Workload Aware Migration, ensures that the vehicle receiving the VM has enough time to receive it, as well as will have enough time to migrate it to another vehicle before leaving the network. This adds an interesting criterion to the VM migration that considers if the new host will have enough time to migrate the VM before it leaves the network.

VM migration is a network intensive process. Each time the VM must move from one host to another, there is a delay to the VM's job completion as well as the expense of using the limited bandwidth of the network. If the VM migration is essential to protect against a failure, then the network will need give it a priority in being sent to its new host. On the other hand, if there is redundancy in the job, as in (Ghazizadeh, Florin, Ghazi Zadeh, & Olariu, 2016; Florin, Ghazizadeh, Ghazi Zadeh, & Olariu, 2017), then the VM migration is not essential to the life of the VM. Several single VMs may fail, as long as at least one VM migration is successful.

It is critical to the VC that the VM migration policy is optimal. Migrating too soon may leave a vehicle unused until it waits to leave the VC; even worse, it may cause additional VM migrations than are truly needed. On the other hand, migrating too late result in a failed job if the vehicle leaves before the migration is complete. An optimal time to migration must be found unless redundancy of VMs is introduced.

7.3 Service Management

In service management, there are sets of vehicles that want to provide access to their available resources to be used as a service for a consumer vehicle. Additionally, a consumer vehicle requires more resources than it currently has. There is a need to match the two together based on some set of conditions. A vehicle that wishes to provide its additional resources is a provider vehicle. A vehicle that wishes to consume these resources is a consumer vehicle.

Brik *et al.* propose that public buses are known as a trusted source of information, its location and route is easily determined, and, presumably, there are enough to provide adequate coverage (Brik, Lagraa, Ghamri-Doudane & Lakas, 2016). The provider vehicle registers with a bus that is within communication range. The consumer vehicle chooses a bus within range and sends a request for a service packet containing service and quality required. The bus then provides any available provider vehicles that meet the requirements to the consumer vehicle. Upon choosing the best *provider vehicle*, the *consumer vehicle* sends a request for resource packet using only buses to route the packets. Finally, the *provider vehicle* returns data using the buses to route packets back to the *consumer vehicle*.

Aloqaily *et al.* propose a framework for matching service providers to service consumers that is based on game theory (Aloqaily, Kantarci, & Mouftah, 2016). Each consumer has preferences for price, delay, and privacy to which they are matched to a service provider fairly. Three models are proposed, the base multiagent/multiobjective interaction game system (MMIGS), an additional model that considers negotiations (MMIGS-WN), and a third model that includes QoE-awareness and a trusted third party. Interestingly they have found that the number of stages to reach equilibrium does not rely on the number of vehicles.

Meneguette *et al.* provides a model for managing and identifying resources without roadside infrastructure (Meneguette, Boukerche, & de Grande, 2016). It uses a peer to peer protocol based on concepts from Gnutella. A requestor sends a message to a controller, which is the vehicle farthest from the requestor. This is done to maximize the size of the network. The controller attempts to fulfill the request using resources near it. If not, it will send a message to the next controller, the vehicle farthest from it.

An important aspect of VCs, is that vehicles can be highly mobile. In a mobile VC, in the absence of a fully connected network, provider and consumer vehicles need to be matched such that the mobility patterns keep them within wireless range throughout the length of the service. This may be ideal if vehicles are within a platoon and are known to be nearby for a long time.

In a fully connected network, where the network fabric allows a consumer and provider to be connected even though far apart physically, then the length of service is easier to guarantee, as long as both stay within the confines of the connected network.

VCs should conceptually work no different than a conventional CC, in that service management should be handled by the VC itself. A job or service request is submitted and the VC handles the rest. Using buses or other non-traditional types of infrastructure, assuming the coverage is adequate, may work. In the presence of a disconnected network a job or request may succeed, fail, or may just currently be unavailable. It will not be possible to determine the difference between a job or shared resource that has failed or is currently unavailable.

7.4. Data Dissemination

In content delivery, a large file or data stream is being sent to a target vehicle. Typically, the vehicle will pass by a RSU and will collect the file when within range of the RSU. To be most efficient, the route of the vehicle is known, and the content is split into pieces that can be delivered within the time the vehicle is in range of the RSU.

Gong *et al.* suggest that instead of using a static RSU, cars parked on the side of the road are used (Gong, Yu, Liu & Zhang, 2016). The authors claim a car is a valid stand in for the RSU. Huang and Xu suggest additional cloud-like resources, a fog, are attached to each RSU to give additional resources to be used for mobile content distribution (Huang & Xu, 2016). The idea is further extended by Vigneri *et al.* where the fog resources are pushed down further to the vehicle level (Vigneri, Spyropoulos & Barakat, 2016). The bandwidth of the RSU is seen as a limited resource, so the authors suggested using vehicles as mobile distributors of content.

Kim *et al.* propose that content is also saved at the RSU level to be distributed the vehicles that request it (Kim, Lim, & Krishnamachari, 2016). The authors look at choosing the proper amount of data to put on each RSU to maximize the chance of getting to the target vehicle, assuming a limit to the storage space and bandwidth on the RSU.

Baron *et al.* suggest certain bus stops are used as file storage repositories (Baron, Spathis, de Amorim, & Ammar, 2016). As buses move from stop to stop, they act as data mules to transfer data between repositories. They look into the problem of locating where repositories should be placed within the network.

The bandwidth in the VC is a limited resource. The expectation that real-time data or large files can be delivered is naïve to the amount of bandwidth that is available. In content delivery, the content to be delivered must be carefully chosen to ensure the quality expected by the user. If data is to be transferred to individual vehicles, the chunk size of each piece delivered at the RSU is very small in relation to the total bandwidth.

Additionally, this type of model presupposes the route and future location of the vehicle is known. If the driver takes a different route or is faster than expected, then the content may not be at the RSU at the proper time. To be able to handle such occurrences, additional network bandwidth must be used to handle all the vehicles possible locations. Such a disconnected network may be called a VC in name only, since keeping track of jobs and resources in a disconnected network is difficult.

7.5. Data Sharing

Vehicles are equipped with a wide array of sensors that help the driver make better decisions. One of the great improvements that V2V will bring is the sharing of this sensor data with other vehicles. Today's vehicles must have a particular sensor available to benefit from it. In the future, VCs will allow this data to be shared among other vehicles. Additionally, the data from many vehicles may be aggregated and additional data mined from the raw measurements. In this way the sensor data from many vehicles is used to make better decisions for the drivers where it was not previously possible.

Generally speaking, data sharing is the sharing of any data, either raw or aggregated; however, typically this involves data collected from vehicle sensors in a raw format or aggregated. Typically this is local vehicles, but some papers extend the range outside of local neighbors. Additionally, some papers suggest sending vehicle data to a centralized location to aggregate data in a specialized way.

Kwak *et al.* suggest that a vehicle map application is supplemented with information from other vehicles on the road (Kwak, Liu, Kim, Nath, & Iftode, 2016). Once the best and alternate route is chosen, a digest of information is collected from vehicles along those routes to get a social report of the traffic along the routes. Using this digest, the drive can then make a more informed decision as to the route to take.

Jeong *et al.* suggest vehicles are used as sensors and data is sent to a centralized location where it is aggregated, and information is returned to the vehicle; in this case, vehicle route information (Jeong, Jeong, Lee, Oh, & Du, 2016).

Chaqfeh *et al.* suggest data from a region of interest is requested, where cooperating vehicles comply with the request and send it back to the requestor (Chaqfeh, Mohamed, Jawhar, & Wu, 2016).

Qiu *et al.* suggest sharing stereo video across a wireless link to allow for extending the visible range of the vehicles sensors (Qiu, Ahmad, Govindan, Gruteser, Bai & Kar, 2017). The authors conclude that given the current bandwidth of vehicles this is not feasible.

The amount of data shared is hindered by the amount of network bandwidth it is allowed. In much of the literature, data is set to a remote source to aggregate data, then send back to the requestor. This essentially treats the VC as nothing but a source of sensor data. Instead, data may be aggregated locally in a local VC. This has the advantage of keeping the data confined to a local area, thus limiting the amount of data being sent over the network. This mimics the convention CC wisdom of moving processing to where the data is, as opposed to moving the data to where the processing is being done.

8. CHALLENGES FOR VEHICULAR CLOUD RESEARCH

The main goal of this section is to identify a number of important challenges for VC research.

8.1. Identify Feasible Vehicular Cloud Applications

Various types of VC distinguished by the number of participating vehicles, existing communication bandwidth, and vehicle residency times are, in all likelihood, apt to process different types of user applications.

Let us recall that, according to Marinescu (Marinescu, 2013) cloud applications can be partitioned into three broad categories:

- Processing pipelines;
- Batch processing systems;
- Web applications.

The workloads specific to these application categories differ substantially and, not surprisingly, require different architectural support. Can VCs support these applications? Very little work along these lines has been reported in the literature. It is clear that a datacenter built on a large number of participating vehicles known to spend a predictable amount of time in the VC are suitable for demanding applications. By contrast, VCs set up in the parking lot of a convenience store where vehicles spend a short amount of time will not be able to support effectively the same application. They, however, can support a different, more lightweight, application.

In private communication to the authors, Lloyd Decker (Decker, 2016) suggested using VCs in support of the following types of applications:

- Transaction-oriented processing for decision support systems and business analytics;
- Mobile interactive applications which process large volumes of data from different types of sensors.

As an illustration, imagine a VC set up in the parking lot of a shopping mall. During the normal business hours, the various businesses at the mall can leverage the compute power of the cars in the shopping mall parking lot for business analytics, mostly involving transaction-oriented processing.

Likewise, the mall security personnel can harness the compute power of the cars in the parking lot to process various sensor inputs, including face recognition and other similar software programs intended to provide security for the shoppers at the mall.

8.2. Resource Management in Vehicular Clouds

Resource management is a fundamental ingredient of conventional CC (Barroso, Calderas, & Holzle, 2013; Hennessy & Patterson, 2012) and remains of great importance in VCs (Olariu, Eltoweissy, & Younis, 2011). In the next subsections, we will discuss important instances of resource management challenges in VCs.

8.2.1. Develop Techniques for Dynamic Resource Prediction, Discovery, Allocation, and Management

Unlike traditional clouds, VCs are much more dynamic. Even in a relatively stable situation like an airport parking garage, vehicles will be arriving and departing throughout the day. Work must be done to investigate techniques for discovering the resources that are available on each new vehicle, allocating tasks to those vehicles, and re-allocating tasks from vehicles that are departing the system. In such situations assessing the amount of available resources and assigning them to particular applications is very challenging and can only be done stochastically.

8.2.2. Virtual Machine Migration in Vehicular Clouds

VM migration is one instance of *resource management* in CC. It remains a fundamental performance attribute in VCs as well. One of the defining ways in which VCs differ from conventional clouds is resource volatility. Therefore, managing VC resources is an important but very challenging task. As vehicles enter the VC, new resources become available. When vehicles depart, often unexpectedly, their on-board resources depart with them creating a volatile environment where the task of resource management is hard.

One of the fundamental resources in the VC is the various VMs hosted by vehicles. When a vehicle leaves the VC while its hosted VM is still active a failure occurs and various strategies need to be implemented in order to minimize disruption and performance degradation. It is intuitively clear that the longer and more predictable the vehicle residency times are in the cloud, the easier it is to predict the optimal moment when VM migration has to be undertaken. By contrast, when vehicular residency times in the VC are short and, themselves, unpredictable, VM migration becomes very challenging.

The authors of this chapter are only aware of two studies of VM migration in VCs. Refaat *et al.* (2016) proposed four VM migration algorithms for a variant of VCs that they call VANET-based clouds. The second work dealing with VM migration is Florin *et al.* (Florin, Abolghasemi, Ghazi Zadeh, & Olariu, 2017) where the authors have looked at VM migration in the context of VCs running Big Data applications.

8.2.3. Server Consolidation in Vehicular Clouds

In conventional CC, server consolidation is one of the fundamental performance boosting and cost-reduction strategies (Buyya, Vecchiola, & Selvi, 2013). The idea behind server consolidation is to migrate VMs from lightly loaded servers so that they can be powered off (to save energy), or to migrate VMs from overloaded servers to improve processing or response times. Many strategies for VM consolidation to optimize a given objective functions have been proposed in the CC literature (Beloglazov, Abawajy, & Buyya, 2012; Cardosa, Singh, Pucha, & Chandra, 2012; Teng, Yu, Li, Deng, & Magoules, 2017).

It is known that server consolidation can be reduced to *bin-packing* (Song, Z., Chen, & Luo, 2013) which is a known NP-hard problem. Thus, general exact solutions for server consolidation are very unlikely to exist. This has motivated the search for heuristics of which some are quite efficient (Beloglazov, Abawajy, & Buyya, 2012; Tang & Pan, 2014). Server consolidation in CC is still an active area of investigation.

However, somewhat surprisingly, in spite of its fundamental importance, server consolidation has not been addressed in the context of VCs. The only work-in-progress that we know of is (Zadeh, 2018).

8.2.4. Job Assignment in Vehicular Clouds

As mentioned before, the dynamically changing availability of compute resources due to vehicles joining and leaving the VC unexpectedly leads to a volatile computing environment where the task of assigning incoming user jobs to vehicles is quite challenging.

To get a feel for the problem, assume that a job was just assigned to a vehicle currently in the VC. If the vehicle remains in the VC until the job completes, all is well. Difficulties arise when the vehicle leaves the VC before job completion. In this case, unless special precautionary measures are taken, the work is lost and the job must be restarted, taking chances on another vehicle, and so on, until eventually the job is completed. Clearly, losing the entire work of a vehicle, in case of a premature departure, must be mitigated. One possible method is to use checkpointing, a strategy originally proposed in databases. This strategy requires making regular backups of the job's state, and storing them in a separate location. In the event a vehicle leaves before job completion, the job can be restarted, on a new vehicle, using the last backup.

Alternatively, reliability and availability can be enhanced, as it is often done in conventional CCs and other distributed systems, by employing various forms of redundant job assignment strategies (Barroso, Calderas, & Holzle, 2013; Bhagavathi, Looges, Olariu, & Schwing, 1994; Deng, Huang, Han, & Deng, 2010; Egwutuoha, Chen, Levy, & Selic, 2012; Ko, Hoque, Cho, & Gupta, 2010; Malik & Huet, 2011; Olariu, Eltoweissy, & Younis, 2007; Olariu, Schwing, & Zhang, 1991; Olariu, Schwing, & Zhang,1992; Park, Yu, Chung, & Lee, 2011; Zomaya, Clements, & Olariu, 1998). Indeed, the volatility of resources in VCs suggests employing job assignment strategies wherein each user job is assigned to, and executed by, two or more vehicles in the VC (Ghazizadeh, Florin, Ghazi Zadeh, & Olariu, 2016).

It is intuitively clear that the longer and more predictable the vehicle residency times in the VC, the easier it is to assign jobs to vehicles in the VC. Surprisingly, there has not been work done on evaluating conditions under which checkpointing leads to better results than redundancy-based job assignment or vice-versa.

It is also clear that the same job assignment strategy will have to take into account the type of job, in terms of compute resources needed and the estimated job completion time. As stated in Florin *et al.* (Florin, Ghazi Zadeh, Ghazizadeh, & Olariu, 2017), a high-redundancy job assignment works well for jobs of a duration between two and eight hours even if the vehicle residency times are relatively short. However, a more comprehensive study is necessary if performance of VC is to be enhanced. This promises to be an exciting area for future work.

8.3. Vehicular Cloud Federations and Ecosystems

Recently, Marinescu *et al.* (Marinescu, Paya, Morrison, & Olariu, 2017) have made the point that given its current trajectory, the complexity of CC ecosystems will evolve to where traditional resource management strategies will struggle to remain fit for the purpose. Unlike conventional clouds where, for cost efficiency purposes, the compute resources are as uniform as possible (Barroso, Calderas, & Holzle, 2013), in VCs the compute resources of individual vehicles are heterogeneous. This, in conjunction with the possible federation of VCs, proposed for the first time by (Eltoweissy, Olariu, & Younis, 2010; Olariu, Khalil, & Abuelela, 2011), and the emergence of new services that these VC will have to offer, will make it increasingly hard to manage VC resources efficiently.

To the best of our knowledge, the important aspect of VC federation and the creation of VC ecosystems has not been addressed in the VC literature.

8.4. Promoting Reliability and Availability in Vehicular Clouds

The tremendous popularity of Internet services that people use every day, including social networks, various forms of online gaming, location-specific services, online shopping, and many others, have led architects to the design of huge datacenters.

It is also important to note that these Internet services have hundreds of millions of users worldwide, a customer base that grows exponentially every year. One of the main reasons for the emergence of CC was the need for providing this huge customer base with a high-quality experience supported by at least a "four-nine" (i.e., 99.99%) availability.

The only way of achieving the desired level of reliability and availability is through service redundancy. Indeed, a typical Google datacenter comprises between 50,000 and 100,000 commodity servers along with the required electrical, cooling, and networking infrastructure.

It has been pointed out repeatedly (Henessy & Patterson, 2012) that the phenomenal success of CC was due, to a large extent, to the system reliability and availability of datacenters. For example, Amazon and Google are striving to make customer experience more enjoyable by insisting on extremely high system availability under all imaginable conditions.

There is a huge variety of workloads in a typical datacenter. However, the most prevalent are searches and multiple services, say, in support of enhancing a customer experience. Managing the contents of a shopping cart and various queries launched by a prospective customer require the ability to efficiently store and recover customer preferences, a history of previous purchases, returns, personal data and so on. This information must be maintained by the datacenter and must be made available in a fraction of a second. The common wisdom is that unhappy customers will not be return customers.

Another type of workload is associated with queries launched by a customer and involving a composite service. For example, a customer may be interested in restaurants in New York. The search algorithm must traverse all available postings of each such item and must return the answers in a given, often customer-aware, way. A successful such service presupposes that the data is stored reliably in the datacenter and that it is accessible in a fraction of a second irrespective of how many servers are down at the moment.

Importantly, in both examples above, the total user-perceived latency needs to be a fraction of a second. Consequently, any file system adopted by the datacenter must support latency reduction. However, high throughput is also an important performance metric because a highly popular service needs to support many thousands of simultaneous queries.

The high availability requirements in such a system can only be supported through redundancy of storage (in addition to execution redundancy, where each user query may be executed by two or more servers) (Barlow & Proschan, 1967). Storage redundancy implies that virtually each data item must be stored at several locations in the network of datacenters, oft-time in datacenters at various locations around the world.

However, as mentioned before, the dynamically changing availability of compute resources due to vehicles joining and leaving the VC unexpectedly leads to a volatile computing environment where promoting system reliability and availability is quite challenging. Yet, it is clear that if VCs are to see an adoption rate and success similar to CCs, reliability and availability issues must be addressed in VCs.

To get a feel for the problem, assume that a job was just assigned to a vehicle currently in the VC. If the vehicle remains in the VC until the job completes, all is well. Difficulties arise when the vehicle leaves the VC before job completion. In this case, unless special precautionary measures are taken, the work is lost and the job must be restarted, taking chances on another vehicle, and so on, until eventually the job is completed. Clearly, losing the entire work of a vehicle, in case of a premature departure, must be mitigated. One possible method is to use *checkpointing*, a strategy originally proposed in databases. This strategy requires making regular backups of the job's state, and storing them in a separate location. In the event a vehicle leaves before job completion, the job can be restarted, on a new vehicle, using the last backup.

While this seems simple and intuitive, to the best of our knowledge, checkpointing has not been implemented in VCs. Perhaps, one of the reasons is that as shown by (Ghazizadeh, 2014) in VCs most checkpoints remain unused and introduce a very large network traffic overhead.

Alternatively, reliability and availability can be enhanced, as it is often done in conventional CCs, by employing various forms of redundant job assignment strategies (Barroso, Clidaras & Holzle, 2013; Bhagavathi, Looges, Olariu & Schwing, 1994; Malik & Huet, 2011; Olariu, Eltoweissy, & Younis, 2007; Zomaya, Clemens, & Olariu, 1998).

Indeed, the volatility of resources in VCs suggests employing job assignment strategies wherein each user job is assigned to, and executed by, two or more vehicles in the VC (Ghazizadeh, Florin, Ghazi Zadeh, & Olariu, 2016)

As already hinted to, the longer and more predictable the vehicle residency times in the VC, the easier it is to ensure system reliability and availability. Recently, (Ghazizadeh, Florin, Ghazi Zadeh, & Olariu, 2016) have studied redundancy-based job assignment strategies and have derived analytical expressions for the corresponding MTTF. Similarly, (Florin, Ghazizadeh, Ghazi Zadel, & Olariu, 2015) have looked at the problem of reasoning about reliability issues in Military Vehicular Clouds (MVC), a version of VCs that suits the needs of tactical applications.

Building on the results of (Ghazizadeh, Florin, Ghazi Zadeh, & Olariu, 2016; Florin, Ghazizadeh, Ghazi Zadeh, & Olariu, 2017) have developed an analytical model for estimating the job completion time in VCs assuming a redundant job assignment strategy.

Quite recently, (Florin, Ghazi Zadeh, Ghazizadeh, & Olariu, 2017) have studied the reliability of VCs with *short vehicular residency times*, as is the case of shopping mall parking lots or the short-term parking lot of a major airport. They showed how to enhance system reliability and availability in these types of VCs through a family of redundancy-based job assignment strategies that attempt to mitigate the effect of compute resource volatility. They offered a theoretical prediction of the corporate MTTF and availability offered by these strategies in terms of the MTTF of individual vehicles and of the MTTR.

To the best of our knowledge, theirs is the first paper in the literature that is looking at evaluating the MTTF and availability in VCs built on top of vehicles with a short residency time. Extensive simulations using data from shopping mall statistics have confirmed the accuracy of their analytical predictions.

8.5. Large-Scale Implementations of Vehicular Clouds

As already mentioned in this chapter, to the best of the authors' knowledge, no large-scale implementation of VCs has been reported in the literature. We are aware of two attempts at addressing this challenge: one is purely simulation-based and was reported in (Florin, Abolghasemi, Ghaza Zadeh, & Olariu, 2017). The other is a recent testbed that we discuss next.

The testbed of Lu *et al.* (Lu, Li, Huang, Lu, Deng, & Chowdhary, 2016) contains 19 robot vehicles of four types: seven VC-trucks, one VC-van, one VC-sedan, and 10 VC compacts. The types of robot vehicles vary in payload, mobility, sensing, computing capabilities, cost, and battery life and their mobility performance can be configured. They have used an independent WiFi network for the purpose of the robot vehicle management. Each robot vehicle is equipped with WiFi interface and also LTE modem, in case the robot is out of the range of the WiFi network. In addition, several WiFi routers are deployed as roadside units (RSU). They have also allowed interfaces to be used as WiFi access points, vehicle-to-vehicle communication, or vehicle-to-infrastructure communication.

In this testbed, the cloud resources contain robot vehicles, on-board mini cloud, and remote cloud in the data center. The remote cloud consists of Linux servers managed by OpenStack and provides a cloud management user interface, centralized control and storage. A Kernel-based Virtual Machine (KVM) was installed on the robot vehicles to provide basic virtualization environment. Users can create virtual machines (VM) or import existing VM images into mini cloud, and run their programs and applications on these virtual machines. Live migration is also allowed from a robot vehicle to another.

As an example of a vehicle cooperation application, an implementation of the longitudinal platooning algorithm was done. Vehicles driving in the same direction form a line cluster on the road with relatively small separation distance and all vehicles except the leader will drive automatically. The platooning software component registers a method and an event on its local message broker. Then other members directly connect to the header through V2V communication, and call the header join method. The join method then returns the name of the robot vehicle at platoon tail. The platooning algorithm regularly calculates desired speed based on predecessor's states, separation distance as well as the header's state to maintain stability of the whole cluster.

The testbed of (Lu, Li, Huang, Lu, Deng, & Chowdhary, 2016) is interesting work-in-progress and proof-of-concept prototype for VC-bots, an evolving platform for testing vehicular network and VC applications. To the best of our knowledge this is the first-time researchers have implemented a testbed for VCs and it is a proof that this area of research should be explored and studied further.

8.6. Supporting Big Data Applications in VCs

Science and engineering could greatly benefit from cloud computing as many applications in these areas are compute-intensive and data-intensive. One of the significant research challenges in VCs is to identify conditions under which VCs can support Big Data applications. It is apparent that Big Data applications, with stringent data processing requirements, cannot be supported by ephemeral VCs, where the residency time of vehicles in the cloud is too short for supporting VM setup and migration.

Quite recently Florin *et al.* (Florin, Abolghasemi, Ghazi Zadeh, & Olariu, 2017) have identified sufficient conditions under which Big Data applications can be effectively supported by datacenters built on top of vehicles in a parking lot. This is the first time researchers are looking at evaluating the feasibility of the VC concept and its suitability for supporting Big Data applications. The main findings of (Florin, Abolghasemi, Ghazi Zadeh, & Olariu, 2017) are that if the residency times of the vehicles are sufficiently long, and if the interconnection fabric has a sufficient amount of bandwidth, then Big Data applications can be supported effectively by VCs.

In spite of this result, a lot more work is needed to understand what it takes for VC to be able to support, in a graceful way, data- and processing-intensive applications.

8.7. Vehicular Cloud Support for Smart Cities

Visionaries depict the smart cities of the future as fully connected entities supported by various forms of *road-side infrastructure* as well as advanced in-vehicle resources such as embedded powerful computing and storage devices, cognitive radios and multi-modal programmable sensor nodes (Hatch, 2013; Lakakis & Kyriakou, 2015; Litman, 2015; Townsend, 2013). As a result, in the near future, vehicles equipped with computing, communication, and sensing capabilities will be organized into ubiquitous and pervasive networks with a significant Internet presence while on the move. This will revolutionize the driving experience making it safer, more enjoyable, and more environmentally friendly.

The rise of Internet of Things (IOT) and adoption of smart cities create opportunities for creative and efficient management and utilization of the available resources. One of the characteristics of smart cities is the interconnectivity of the city's infrastructure, which allows data to be collected from various human-generated or machine-generated sources. Future vehicles with powerful on-board computers, communication capabilities, and vast sensor arrays are perfect candidates in this hyper-connected environment to utilize into a fluid Vehicular Cloud capable of performing large-scale computations.

The way the authors see it, the main challenge for VCs in the context of Smart Cities should be aligned with the 2015-2019 strategic priorities recently spelled out by US-DOT (USDOT, 2015). In order to show relevance of VCs to Smart Cities, Marinescu *et al.* (Marinescu, Florin, & Olariu, 2017) proposed to achieve the following objectives:

- **Enhance Urban Mobility by Exploring Methods and Management Strategies That Increase System Efficiency and Improve Individual Mobility Through Information Sharing:** VCs should combine detailed knowledge of real-time traffic flow data with stochastic predictions within a given time horizon to help: (1) the formation of urban platoons containing vehicles with a similar destination and trajectory; (2) adjust traffic signal timing in order to reduce unnecessary platoon idling at traffic light; and (3) present the driving public with high-quality information that will allow them to reduce their trip time and its variability, eliminate the conditions that lead to congestion or reduce its effect.
- **Avoid Congestion of Key Transportation Corridors Through Cooperative Navigation Systems:** Congestion-avoidance techniques that become possible in Smart Cities environments will be supplemented by route guidance strategies to reduce unnecessary idling and will limit environmental impact of urban transportation.
- **Handling Non-Recurring Congestion:** VCs will explore strategies to efficiently dissipate congestion, by a combination of traffic light retiming and route guidance to avoid more traffic buildup in congested areas.

To meet this challenge, it is important to develop techniques for building and organizing a VC over a wireless network at traffic intersections. While VCs can be set up in a proactive manner, the true usefulness of the VC concept is shown by applications where the VC is set up in an ad hoc manner to help mitigate events that have occurred. Communications to the vehicles participating in the VC will be through wireless devices. It is, thus, essential to investigate the communication needs of and develop protocols in support of the wireless VC model. Such is the case, for example, in an effort to mitigate the effects of a downtown traffic jam involving hundreds of vehicles. The owners of these vehicles will gladly grant the municipality access to the on-board resources in their vehicles so that the pooled computing resources of the cars in the congestion can be used to enable a city-wide re-timing of the traffic signals.

9. CONCLUSION

Vehicular clouds represent a transformative, next-generation computing system whose potential cannot be fully achieved with currently available technology. Work in this area will enable further research into the coordination between tomorrow's vehicles and the computational, sensing, storage, and networking resources that they will possess. Enabling this interaction faces many challenges, including the requirement of wireless communications and the mobility of individual vehicles that affects the dynamics of groups of vehicles. It is important to explore how VCs can be used to improve traffic flow through traffic optimization simulations. We anticipate many other applications of this technology that could have a long-ranging impact on our society. Vehicular clouds are a compelling extension of the conventional CC paradigm. Because Amazon and Google have made CC well known to the general public and vehicles are ubiquitous, the VC concept is an ideal example of a cyber-physical system whose basic ideas can easily be explained to the public at large.

VC research is still a maturing area of research within the community. There is not yet a standard agreed upon definition for what a VC is, as is evident from the range of models mentioned in this work. As research continues, there needs to be a good effort on the side of the researchers to ensure the given proposed model is realistic in light of possible dynamic mobility, restricted bandwidth, and unforeseen issues in the expected route of the vehicles on the road.

Further work needs to be done to determine creative ways to fully utilize the additional resources a vehicle has, as well as a realistic look at what they can provide given limited bandwidth, possible short residency times, and non-homogenous resources.

REFERENCES

Abuelela, M., & Olariu, S. (2010). Taking VANET to the Clouds. *Proc. ACM MoMM.*

Aloqaily, M., Kantarci, B., & Mouftah, H. T. (2016). Multiagent/multiobjective interaction game system for service provisioning in vehicular cloud. *IEEE Access: Practical Innovations, Open Solutions, 4*, 3153–3168. doi:10.1109/ACCESS.2016.2575038

Arif, S., Olariu, S., Wang, J., Yan, G., Yang, W., & Khalil, I. (2012). Datacenter at the airport: Reasoning about time-dependent parking lot occupancy. *IEEE Transactions on Parallel and Distributed Systems, 23*(11), 2067–2080. doi:10.1109/TPDS.2012.47

Barlow, R. F., & Proschan, F. (1967). Mathematical Theory of Reliability. John Wiley and Sons.

Baron, B., Campista, M., Spathis, P., Costa, L. H., de Amonim, M. D., Duarte, O. C., ... Viniotis, Y. (2016). Virtualizing vehicular node resources: Feasibility study of virtual machine migration. *Vehicular Communications, 4*, 39–46. doi:10.1016/j.vehcom.2016.04.001

Baron, B., Spathis, P., de Amorim, M. D., & Ammar, M. (2016). Cloud storage for mobile users using pre-positioned storage facilities. *Proc. 2nd Workshop on Experiences in the Design and Implementation of Smart Objects*, 11–16. 10.1145/2980147.2980151

Barroso, L. A., Calderas, J., & Holzle, U. (2013). *The datacenter as a computer: An introduction to the design of warehouse-scale machines* (2nd ed.). San Rafael, CA: Morgan & Claypool.

Beloglazov, A., Abawajy, J., & Buyya, R. (2012). Energy-efficient resource allocation heuristics for efficient management of datacenters for cloud computing. *Future Generation Computer Systems, 28*(5), 755–768. doi:10.1016/j.future.2011.04.017

Bhagavathi, D., Looges, P. J., Olariu, S., Schwing, J. L., & Zhang, J. (1994). A fast selection algorithms on meshes with multiple broadcasting. *IEEE Transactions on Parallel and Distributed Systems, 5*(7), 772–778. doi:10.1109/71.296326

Brik, B., Lagraa, N., Ghamri-Doudane, Y., & Lakas, A. (2016). Finding the most adequate public bus in vehicular clouds. *Proc. 2016 International Conference on Wireless Networks and Mobile Communications*, 67-74. 10.1109/WINCOM.2016.7777193

Buyya, R., Vecchiola, C., & Thamarai Selvi, S. (2013). *Mastering Cloud Computing: Foundations and Applications Programming*. Elsevier.

Cardosa, M., Singh, A., Pucha, H., & Chandra, A. (2012). Exploiting spatio-temporal tradeoffs for energy-aware MapReduce in the cloud. *IEEE Transactions on Computers*, *61*(12), 1737–1751. doi:10.1109/TC.2012.166

Chaqfeh, M., Mohamed, N., Jawhar, I., & Wu, J. (2016). Vehicular cloud data collection for intelligent transportation systems. Smart cloud networks & systems (scns), 1–6. doi:10.1109/SCNS.2016.7870555

Decker, L. (2016, October). private communication.

Deng, J., Huang, S.-H., Han, Y., & Deng, J. (2010, December). Fault tolerant and reliable computation in cloud computing. In *Globecom workshops (gc wkshps), 2010* (pp. 1601–1605). IEEE. doi:10.1109/GLOCOMW.2010.5700210

Egwutuoha, I. P., Chen, S., Levy, D., & Selic, B. (2012, May). A fault tolerance framework for high performance computing in cloud, *Proc. 12th IEEE/ACM International Symposium on Cluster, Cloud and Grid Computing*, 709-710. 10.1109/CCGrid.2012.80

Eltoweissy, M., Olariu, S., & Younis, M. (2010). Towards autonomous vehicular clouds. *Proc. of Ad-HocNets'2010.*

Florin, R., Abolghasemi, S., Ghazi Zadeh, A., & Olariu, S. (2017). Big Data in the parking lot. In A. Z. Kuan-Ching Li Hai Jiang (Ed.), *Big Data Management and Processing* (pp. 425–449). Boca Raton, FL: Taylor and Francis.

Florin, R., Ghazi Zadeh, A., Ghazizadeh, P., & Olariu, S. (2017). Promoting reliability and availability in vehicular clouds. *IEEE Transactions on Cloud Computing.*

Florin, R., Ghazizadeh, P., Ghazi Zadeh, A., & Olariu, A. (2015, Oct.). Enhancing dependability through redundancy in military vehicular clouds. *Proc. IEEE MILCOM'*2015. 10.1109/MILCOM.2015.7357586

Florin, R., Ghazizadeh, P., Ghazi Zadeh, A., & Olariu, S. (2017). Reasoning about job completion time in vehicular clouds. *IEEE Transactions on Intelligent Transportation Systems*, *18*(7), 1762–1771. doi:10.1109/TITS.2016.2620434

Ghazizadeh, P., Florin, R., Ghazi Zadeh, A., & Olariu, S. (2016). Reasoning about the Mean-Time-to-Failure in vehicular clouds. *IEEE Transactions on Intelligent Transportation Systems*, *17*(3), 751–761. doi:10.1109/TITS.2015.2486523

Gong, H., Yu, L., Liu, N., & Zhang, X. (2016). Mobile content distribution with vehicular cloud in urban VANETS. *China Communications*, *13*(8), 84–96. doi:10.1109/CC.2016.7563691

Hatch, D. (2013, November). *Singapore strives to become 'The Smartest City': Singapore is using data to redefine what it means to be a 21st-century metropolis.* Academic Press.

Hennessy, J. L., & Patterson, D. A. (2012). Computer Architecture: A Quantitative Approach (5th ed.). Elsevier.

Huang, C. Y., & Xu, K. (2016, July). Reliable realtime streaming in vehicular cloud-fog computing networks. *Proc. 2016 IEEE/CIC International Conference on Communications in China*, 1-6. 10.1109/ICCChina.2016.7636838

Jeong, J., Jeong, H., Lee, E., Oh, T., & Du, D. H. C. (2016, June). Saint: Self-adaptive interactive navigation tool for cloud-based vehicular traffic optimization. *IEEE Transactions on Vehicular Technology*, *65*(6), 4053–4067. doi:10.1109/TVT.2015.2476958

Kim, R., Lim, H., & Krishnamachari, B. (2016, January). Prefetching-based data dissemination in vehicular cloud systems. *IEEE Transactions on Vehicular Technology*, *65*(1), 292–306. doi:10.1109/TVT.2015.2388851

Ko, S. Y., Hoque, I., Cho, B., & Gupta, I. (2010). Making cloud intermediate data fault-tolerant. *Proc. 1-st ACM Symposium on Cloud Computing*, 181–192. 10.1145/1807128.1807160

Kwak, D., Liu, R., Kim, D., Nath, B., & Iftode, L. (2016). Seeing is believing: Sharing real-time visual traffic information via vehicular clouds. *IEEE Access: Practical Innovations, Open Solutions*, *4*, 3617–3631. doi:10.1109/ACCESS.2016.2569585

Lakakis, K., & Kyriakou, K. (2015). Creating and intelligent transportation system for smart cities: performance evaluation of spatial-temporal algorithms for traffic prediction. *Proc. 14-th International Conference on Environmental Science and Technology*.

Litman, T. (2015, December). *Autonomous vehicle implementation predictions: Implications for transport planning*. Presented at the *2015 Transportation Research Board Annual Meeting*.

Lu, D., Li, Z., Huang, D., Lu, X., Deng, Y., Chowdhary, A., & Li, B. (2016). VC-bots: a vehicular cloud computing testbed with mobile robots. *Proc. First International Workshop on Internet of Vehicles and Vehicles of Internet*, 31–36. 10.1145/2938681.2938683

Malik, S., & Huet, F. (2011). Adaptive fault tolerance in realtime cloud computing. *Proc. IEEE World Congress on Services*, 280-287.

Marinescu, D. C. (2013). *Cloud Computing, Theory and Applications.* Elsevier.

Marinescu, D. C., Florin, R., & Olariu, S. (2017). Towards platoon-aware retiming of traffic lights in smart cities. In F. Hu (Ed.), *Vehicle-to-Vehicle and Vehicle-to-Infrastructure Communications: A Technical Approach*. New York, NY: Taylor and Francis.

Marinescu, D. C., Paya, A., Morrison, J. P., & Olariu, S. (2017). An approach for scaling cloud resource management. *Cluster Computing*, *20*(1), 909–924. doi:10.100710586-016-0700-8

Meneguette, R. I., Boukerche, A., & de Grande, R. (2016, Dec). Smart: An efficient resource search and management scheme for vehicular cloud-connected system. *Proc. IEEE Global Communications Conference*, 1-6. 10.1109/GLOCOM.2016.7842271

Nakano, K., & Olariu, S. (2000). Randomized leader election protocols in radio networks with no collision detection. *Proc. International Symposium on Algorithms and Computation*, 362-373. 10.1007/3-540-40996-3_31

Olariu, S., Eltoweissy, M., & Younis, M. (2007). ANSWER: AutoNomouS netWorked sEnsoR system. *Journal of Parallel and Distributed Computing*, *67*(1), 114–126. doi:10.1016/j.jpdc.2006.08.013

Olariu, S., Eltoweissy, M., & Younis, M. (2011). Towards autonomous vehicular clouds. *ICST Transactions on Mobile Communications and Applications*, *11*(7-9), 1–11.

Olariu, S., Hristov, T., & Yan, G. (2013). The next paradigm shift: from vehicular networks to vehicular clouds. In S. Basagni & ... (Eds.), *Mobile Ad Hoc Networking Cutting Edge Directions* (pp. 645–700). New York: Wiley and Sons. doi:10.1002/9781118511305.ch19

Olariu, S., Khalil, I., & Abuelela, M. (2011, February). Taking VANET to the clouds. *International Journal of Pervasive Computing and Communications*, *7*(1), 7–21. doi:10.1108/17427371111123577

Olariu, S., Schwing, J. L., & Zhang, J. (1991). Fundamental algorithms on reconfigurable meshes. *Proc. 29-th Allerton Conference on Communications, Control and Computing*, 811–820.

Olariu, S., Schwing, J. L., & Zhang, J. (1992). Optimal parallel algorithms for problems modeled by a family of intervals. *IEEE Transactions on Parallel and Distributed Systems*, *3*(3), 364–374. doi:10.1109/71.139209

Olariu, S., & Weigle, M. C. (Eds.). (2009). Vehicular Networks: From Theory to Practice. Taylor & Francis.

Park, J. S., Yu, H. C., Chung, K. S., & Lee, E. (2011, March). Markov chain based monitoring service for fault tolerance in mobile cloud computing. *Proc. IEEE Workshop on Advanced Information Networking and Applications*, 520-525. 10.1109/WAINA.2011.10

Qiu, H., Ahmad, F., Govindan, R., Gruteser, M., Bai, F., & Kar, G. (2017). Augmented vehicular reality: Enabling extended vision for future vehicles. *Proc. 18th ACM International Workshop on Mobile Computing Systems and Applications*, 67–72. 10.1145/3032970.3032976

Rawat, R., Yan, G., Popescu, D. C., Weigle, M. C., & Olariu, S. (2009). Dynamic Adaptation of Joint Transmission Power and Contention Window in VANET. *Proc. of the IEEE Vehicular Technology Conference*.

Refaat, T. K., Kantarci, B., & Mouftah, H. T. (2016). Virtual machine migration and management for vehicular clouds. *Vehicular Communications*, *4*, 47–56. doi:10.1109/VETECF.2009.5378793

Roess, R. P., & Prassas, E. S. (2011). *Traffic Engineering* (4th ed.). Pearson Prentice Hall.

Salahuddin, M. A., Al-Fuqaha, A., & Guizani, M. (2016). Reinforcement learning for resource provisioning in the vehicular cloud. *IEEE Wireless Communications*, *23*(4), 128–135. doi:10.1109/MWC.2016.7553036

Song, W. Z. X., Chen, Q., & Luo, H. (2013). Adaptive resource provisioning for the cloud using online bin packing. *IEEE Transactions on Computers*, *73*(11), 2647–2660.

Tang, M., & Pan, S. (2014). A hybrid genetic algorithm for the energy-efficient virtual machine placement problem in datacenters. *Neural Processing Letters*, *1*(5), 1–11.

Teng, F., Yu, L., Li, T., Deng, D., & Magoules, F. (2017). Energy efficiency of VM consolidation in IaaS clouds. *The Journal of Supercomputing*, *73*(2), 782–809. doi:10.100711227-016-1797-5

Townsend, A. M. (2013). *Smart cities: Big data, civic hackers, and the quest for a new utopia*. New York, NY: W. W. Norton.

USDOT. (2015). *2015 -2019 Strategic Plan Intelligent Transportation Systems (ITS)*. Retrieved from http://www.its.dot.gov/strategicplan.pdf

Vigneri, L., Spyropoulos, T., & Barakat, C. (2016). Storage on wheels: Offloading popular contents through a vehicular cloud. *Proc. 17-th IEEE International Symposium on a World of Wireless, Mobile and Multimedia Networks*, 1-9. 10.1109/WoWMoM.2016.7523506

Yan, G., Wen, D., Olariu, S., & Weigle, M. C. (2013). Security challenges in vehicular cloud computing. *IEEE Transactions on Intelligent Transportation Systems*, *14*(1), 284–294. doi:10.1109/TITS.2012.2211870

Zadeh, A. G. (2018). *Virtual machine managent in vehicular clouds* (PhD Thesis). Department of Computer Science, Old Dominion University.

Zheng, K., Meng, H., Chatzimisios, P., Lei, L., & Shen, X. (2015, December). An smdp-based resource allocation in vehicular cloud computing systems. *IEEE Transactions on Industrial Electronics*, *62*(12), 7920–7928. doi:10.1109/TIE.2015.2482119

Zomaya, A., Clements, M., & Olariu, S. (1998). A framework for reinforcement-based scheduling in parallel processor systems. *IEEE Transactions on Parallel and Distributed Systems*, *9*(3), 249–260. doi:10.1109/71.674317

KEY TERMS AND DEFINITIONS

Big Data: A general term for large data sets, the computation algorithms used to analyze them and the results learned from the analysis.

Cloud Computing: A model for providing a remote computing infrastructure, platform, or software as a service to a user for payment based on usage, typically over the internet.

Cyber Physical System: A system with a symbiotic cyber and physical component; the physical world is sensed and used as an input to the cyber system. The cyber system reads the inputs and makes a decision that makes physical changes.

Road-Side Infrastructure: Communications and network support for vehicular networks that is not is road-side, rather than contained in a vehicle.

Smart City: A concept for a future city that has the infrastructure for ubiquitous connectivity that enables a better experience for its citizens.

Utility Computing: A model for providing compute resources with a quality of service to a user for payment based on usage.

VANET (Vehicular Ad Hoc Network): A network of vehicles within wireless range, formed to allow for automated driving, provide additional information to the driver, and even entertainment.

Vehicular Cloud: A dynamic group of vehicles whose excess computing, communication, storage, and sensing resources are coordinated and dynamically allocated to authorized users.

Virtual Machine Migration: A process where the virtual machine of a host is moved to another host.

Chapter 2
Vehicular Cloud Computing:
Architecture and Applications

T. Kavitha
AMC Engineering College, India

ABSTRACT

Cloud computing provides a way to avail hardware and software to provide services over the networks to users, whereas in mobile cloud computing, mobile devices are a part of cloud users and service providers. Vehicular ad hoc networks (VANETs) are developed with the help of mobile ad hoc networks (MANET) with the difference that the data are exchanged among mobile vehicles to enhance the road safety and traffic efficiency. VANETs safety applications includes traffic signal violation, curve speed warning, emergency brake lights, pre-crash sensing, collision warning, left turn assist, lane change warning, and stop sign assist. But due to the requirement of smaller size, one of the significant challenges of the vehicular network is the limited resource with respect to memory, computation power, and bandwidth. It leads to slow data processing capability. However, the demanding requirement of emerging application is complex computation with high storage capacity. This necessity can be fulfilled by sharing the available resource among all the vehicles which are in near proximity. It can be achieved by integrating the vehicular network with cloud computing which is called as vehicular cloud computing (VCC). VCC is a new technological shift that can take the advantages of cloud computing to afford the services to drivers of VANET. But VCC is still in early stage and due to its unique features and applications it has become a significant emerging research area which needs to be explored further.

INTRODUCTION

Communication and information technology is the source of power for most important innovations in the society, in specific automotive industry. In the last two decades, wireless communications have changed our lifestyle by allowing the exchange of information from anywhere at any time. The use of this wireless communication systems in vehicles is expected to be a significant resource to improve vehicles and to create a safer infrastructure for road transport which is confirmed by a number of different projects initiated towards vehicular networks by national and international organization.

DOI: 10.4018/978-1-5225-3981-0.ch002

From the last two decades, the industrialists and academicians were focusing their research towards road safety and traffic management. The first outlook of researchers was that radio-equipped vehicles could increase the awareness of road condition which is then expanded with internet and its services. A classic car or truck today is expected to have at least some of the devices like an on-board computer, a GPS device, a radio transceiver, a short-range rear collision radar device, a camera, a variety of sophisticated sensing devices and Event Data Recorder (EDR). EDR collects transactional data from most of the vehicle subassemblies and able to store retrievable data (Mohamed, Stephan, & Mohamed, 2010). In 2006 the National Highway Traffic Safety Administration (NHTSA) declared its intent to standardize the various EDR devices provided by car and truck manufacturers.

As technology is moving closer and closer to packing sophisticated resources in individual vehicles, manufacturers are focusing their attention to provide the transport infrastructure which will ensure journey to be safe and enjoyable. It is assured by providing information about traffic congestion, accidents, hazardous road conditions, weather conditions, and the location of facilities (e.g., gas stations and restaurants) (Rodolfo, 2016). In this context efficiency is obtained by increasing the capacity of the road network, reducing pollution and congestion, making travel time more predictable, reducing operating costs for vehicles, providing more efficient logistics management, improving management and control of road networks, and increasing efficiency of public transport systems (Rodolfo, 2016). Journey becomes more enjoyable by providing access to the Internet, with its tourist information/advertising, the ability to download files, and chat. These applications together known as an intelligent transport system (ITS), (Salim, 2012) which aims to apply intelligent technologies of information and communication to provide innovative services of traffic management and transport to improve efficiency, safety and the fun. ITS helps to avoid traffic congestions and accidents which are increased with the population growth and caused several undesirable effects like long travel time, air pollution, and fuel consumption. Many efforts have been devoted by ITS organizations in order to find ubiquitous solutions by developing vehicular networking and traffic communications. But, the communication, storage and computing resources available in the vehicles are not utilized efficiently. These underutilized resources of vehicles on the road can be pooled and rented to customers with the help of cloud known as Vehicular Clouds which are technologically feasible and will provide benefits of societal and technological impacts. The goal of this chapter is to facilitate readers to better understand the fundamental of emerging paradigm of vehicular cloud computing mechanisms to lead their research activities.

The rest of this chapter is organized as follows: First chapter introduces the background of vehicular cloud computing such as VANET, cloud computing, and Mobile cloud computing. Then it gives about the overview of VCC and its working. Then VCC architecture, framework and its design principles are explained. Specifically in detail about the integration of the Mobile cloud computing with VANET, cloud based middle ware for VANET, and cloud-assisted information dissemination system in VANET. The sensing platforms for VCC are discussed. Then the potential applications scenarios and types of applications are also discussed. At last, this chapter ends with conclusion.

BACKGROUND

A variety of information technologies have developed which ensures the intelligent vehicles, roads, and traffic infrastructures such that all vehicles are associated. In addition to sophisticated array of sensing and actuators, computation capabilities, huge storage devices, the availability of virtually unlimited power supply and growing communication technologies which is used to interconnect vehicles and road infrastructures to provide access to Internet will make vehicles perfect candidates for housing powerful on-board computers, may act as networked computing centers on wheels.

Guerrero-Ibanez et al., (2015) has discussed about the current and future impact of emerging technologies on transportation systems with respect to four fundamental principles, sustainability, integration, safety, and responsiveness, which will play a fundamental role in achieving the main objectives (access and mobility, environmental sustainability, and economic development).

An efficient innovative ITS must use the most recent technological advances. Development in wireless communication networks has made Inter-Vehicular Communications (IVC) and Road-Vehicle Communications (RVC) possible in Mobile Ad Hoc Networks (MANETs). This leads to a new type of MANET known as the Vehicular Ad Hoc Network (VANET). VANET is a special case of MANET. However, in VANET, the nodes are moving on predefined roads, and vehicle velocities are restricted by speed limits and the situation of road traffic and traffic control mechanism. Furthermore, future vehicles can equip with rechargeable energies or solar cells, so energy restriction, which exists in MANET, will not occur in VANET. It can be concluded that VANET is a special case of MANET, which is very flexible, and relatively easy to manage. It employs a combination of Vehicle-to-Vehicle (V2V) and Vehicle-to-Infrastructure (V2I) communications (Hartenstein & Laberteaux, 2008) have been proposed to give drivers advance notification of traffic events as shown in Figure 1. In V2V systems, each vehicle is responsible for inferring the presence of an incident based on reports from other vehicles. For instance, Dedicated Short Range Communications (DSRC) or Wireless Access for Vehicular Environments (WAVE) is specifically designed for Vehicle-to-Vehicle, Vehicle-to-Infrastructure, and Vehicle-to-Device systems (V2X) communications. The IEEE 802.11p, called Wireless Access in Vehicular Environments (WAVE) (Uzcategui & Acosta-Marum 2009), is currently a popular standard for DSRC. Besides, the Long-Term Evolution (LTE), LTE-Advanced and Cognitive Radio (CR) (Yu, Zhang, Liu, Xie, Song, & Guizani, 2012; Xie, Liu, Zhang, & Yu, 2010) are all fairly competitive technologies for vehicular networking (Wang, Song, Han, & Jiao; Wang, Song, & Han, 2013).

Requirement to inform fellow drivers of actual or imminent road conditions, congestion, hazardous driving conditions and other similar traffic-related concerns was the main force for the interest in VANET. Here, Vehicles can collect, transmit, and interpret information through a wide range of sensors, cameras, computers, and communication resources, and this can assist in the acquisition of data and help the drivers to make decisions to take appropriate actions (Zingirian, & Valenti, 2012). Therefore, most VANET applications focus on the traffic status reports, collision avoidance, emergency alerts, cooperative driving, and other similar concerns. (Lee, Lee, & Gerla, 2009; Yang, & Bagrodia, 2009; Subir, Raymond, & Francois, 2006; Fatima, 2005). Therefore, it can effectively reduce road accidents, improve safety, and manage city traffic with high efficiency. In addition value added services like, it can able to share information between cars, such as weather forecast, gas station and restaurant addresses. When Connecting to Internet as terminal networks, via VANET it is possible to download music. Rapid converge of Intelligent Transportation Systems (ITS) and VANET leading to the materialization of Intelligent Vehicular Networks which creates the safe, secure, and robust ubiquitous computing environment that will ultimately spread through our highways and city streets. But due to the requirement of smaller size, one of the significant challenges of the vehicular network is the limited resource with respect to memory, computation power and bandwidth. It leads to slow data processing capability.

VANET leverages vehicles as mobile nodes moving on the road and stationary Road Side Unit (RSUs) installed on the road side and/or hot spots (for instance in an intersection of a cross-road) as shown in Figure 1. The main roles contributed in VANET are management entities, certification authorities, revocation authorities, and end users (vehicular nodes). It is expected that the advancements in vehicular technology will equip the high-end and middle-end vehicles with rich set of computational, communication, and storage resources in the near future. VANET employs architecture-less communication among the vehicular nodes and infrastructure-based communication between vehicles and RSUs using existing infrastructure. The results from these communications are fed to safety and non-safety applications (Rasheed, Rezaeifar, & Oh, 2015).

Cloud computing which is considered to be business model rather than a technology, started from the fact that businesses may find it useful to rent the infrastructure and sometimes the needed software to run their applications. This idea has been emerged, at least in part, by ubiquitous and relatively low-cost high-speed Internet, virtualization and advances in parallel and distributed computing and distributed databases. It refers to both services accessed via, and delivered through, the vast universe of internet in a seamless manner, and the hardware and system software in remote datacenters that provide those services. Due to the features of cloud computing, it becomes very attractive to the end-user for many of reasons. It has been adopted by a large number of infrastructure providers who have a large installed infrastructure that often goes under-utilized (Mohamed, Stephan, & Mohamed, 2010). Cloud computing provides dependable and secure data storage center, immense possibility for internet application, infinite

Figure 1. VANET architecture diagram
Md AliAl Khairul, Md Fakhrul Alam, & Esfar- E- Alam, 2012.

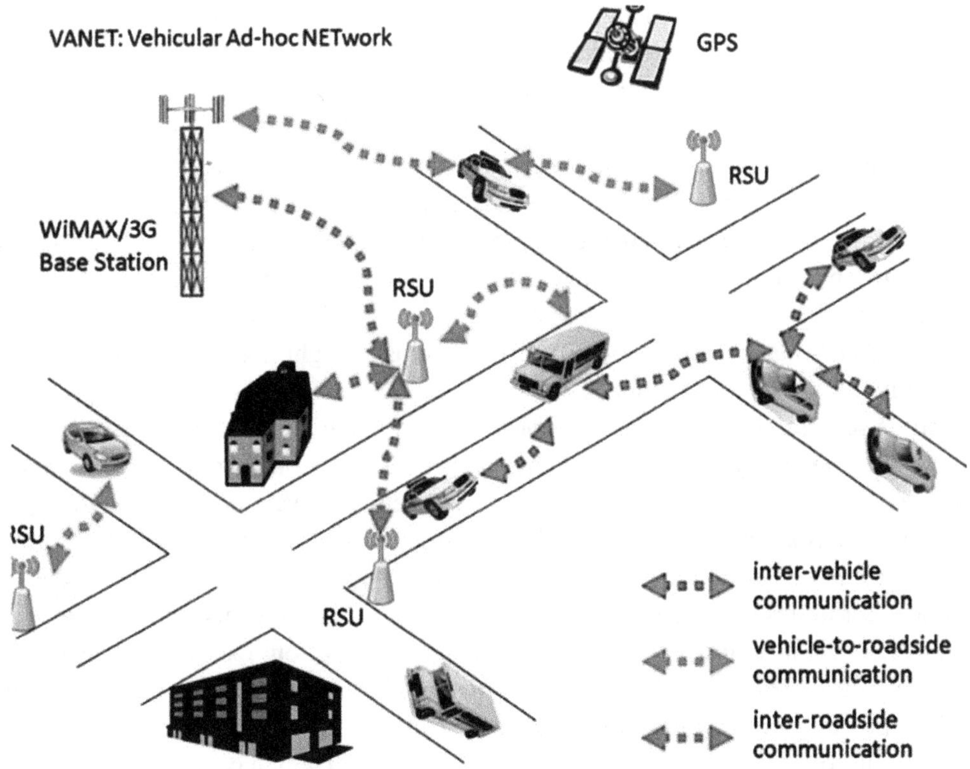

space for storing and managing data, powerful computing capacity and also information technology (IT) services for users to complete all kinds of application. It is also known as 'utility computing' which is based on 'pay-as-you-go' service (Armbrust, Fox, Griffith, Joseph, Katz, & Konwinski, 2010). Users will change their habit of using computer totally, from services centered by desktop to services centered by Web. The purpose of cloud computing is to combine all the resources, and let anyone can use it. From the most basic significance, cloud computing is to utilize software and data of the internet (Ouyang & Jing, 2008). There are three delivery models for cloud services: Software as a Service (SaaS) where customers rent software hosted by the vendor; Platform as a Service (PaaS) where customers rent infrastructure and programming tools hosted by the vendor to create their own applications; and Infrastructure as a Service (IaaS) where customers rent processing, storage, networking and other fundamental computing resources for all purposes (Hoover & Martin, 2008).

Mobile Cloud Computing is a recent emerging paradigm in the information technology field (Sanaei, Abolfazli, Gani, & Buyya, 2013). The environment where mobile devices act as a service providers and cloud users is known as mobile cloud computing (Abolfazli, Sanaei, Ahmed, Gani, & Buyya, 2013). Here mobile phones are going beyond the scope of PC, as the popular method of Web access method. Because data of interest are scattered across many mobile observers, and will require in local data aggregation and query resolution using software specialized for the local context. These mobile devices form the Mobile Clouds that offload the Internet Clouds from tasks which cannot perform in a timely or efficient manner (Fernando, Loke, & Rahayu, 2013). Due to high processing speed and storage capacity of mobile devices, local processing becomes more efficient. On the other hand, effective communication is achieved by keeping the locally relevant information on the mobiles instead of uploading it to the internet due to the scarcity of urban spectrum. Mobiles upload to the internet cloud only the contents of global, long lasting information and data that need too complex or too energy consuming to process in the mobile cloud (Shiraz, Gani, Khokhar, & Buyya, 2012). This Mobile cloud reduces the communication delay, spectrum costs and provides the opportunity to new scenarios (Niroshinie, Seng, & Wenny, 2013).

Mobile Cloud Computing leads to three main scenarios known as Mobile Vehicular Cloud, Mobile Personal Cloud and Mission Oriented Mobile Clouds (Mario, 2012). Mobile Vehicular Cloud mainly used for safe driving, urban sensing, content distribution, mobile advertising and intelligent transportation. Mobile Personal Cloud consists of smart phones and tablets that provide support for mobile social computing, mobile healthcare, urban sensing and entertainment. Mission Oriented Mobile Clouds consist of both humans and robots for goal oriented activities (e.g., terrain exploration, disaster recovery, surveillance, environment monitoring, crowd sourcing, etc.). Mobile Mission Clouds are pre-defined, while vehicular and personal clouds are opportunistic clouds brought together based on coincidence of needs or by geographical proximity criteria.

The demanding requirement of emerging application of VANET is complex computation with high storage capacity. This necessity can be fulfilled by sharing the available resource among all the vehicles which are in near proximity. The number of vehicles on streets, roadways and parking lots spend hour to days which are treated as plentiful and underutilized computational resources are simply wasted. These underutilized resources of vehicles can be used for providing public services which makes vehicles the perfect candidates for nodes in a cloud computing network. Vehicle owners who are going to park for long time or who are stuck in traffic congestion may agree to rent out excess on board resources and get benefit economically. These resources can be utilized to help city traffic authorities to run complex simulations designed to remove congestion by rescheduling the traffic lights of the city. Vehicles will help local consultants to resolve traffic incidents in a timely fashion which is very difficult with the traffic management centers alone due to the lack of adequate computational resources (Md, Mehdi, Abdullah, & Rajkumar, 2014). It can be achieved by integrating the vehicular network with cloud computing which is called as vehicular cloud computing (VCC). Mohammad et al., (2015) has compared the Cloud Computing, Mobile Cloud Computing and Vehicular Cloud Computing.

VEHICULAR CLOUD COMPUTING

The concept of a Vehicular Cloud (VC) has been introduced by Eltoweissy et al. (2010); Abuelela, and Olariu, (2010), suggested to take conventional VANETs to the cloud; Olariu et al. (2011) which leverages the on board resources in participating cars. Vehicular Cloud Computing (VCC) can be defined as:

... a group of largely autonomous vehicles whose corporate computing, sensing, communication and physical resources can be coordinated and dynamically allocated to authorized users (Eltoweissy, Olariu, & Younis, 2010).

VCC is a new technological shift which can takes the advantages of cloud computing to afford the services to drivers of VANET with existing infrastructure itself. VCC focus the convergence of ITS to MCC and a technically feasible incorporation of the ubiquitous sensing of WSN, ITS and MCC for better road safety and secured intelligent urban traffic systems (Tekbiyik & UysalBiyikoglu, 2011).

Thus, the main focus of VCC are to provide computational services at low cost using existing transportation infrastructures along with novel techniques to regulate, control and manage the traffic; collect, process and disseminate information to drivers to minimize traffic congestion, accidents, travel time and environmental pollution; and to ensure uses of low energy and real-time services of software, platforms, and infrastructure with QOS to drivers (Gerla, 2012). So, ITS need to evolve towards seamless integration with heterogeneous technological system which can collect large amount of data, process it and then take appropriate actions based on real time information. These actions include influencing a vehicle's or driver's behavior by providing information like traffic status, occurrence of accidents etc., so that reducing stress of the drivers and minimizing or preventing road accidents which can help to regulate traffic flow and reduce accidents, redirect traffic away from accidents and alert emergency services as soon as an incident occurs using all the collected information, and help reduce fuel consumption and greenhouse gas emissions. Finally, with the help of self-organized autonomous resources, vehicles will provide on demand services in real time to resolve occurrence of large, serious problems of unpredictable events at unexpected time.

Many car companies like Ford, Toyota and General Motors's research group have adopted the vehicular cloud computing technologies to enable the features like real time traffic updates, information, turn-by-turn driving directions, business, sports, weather, and news through voice commands driver health well-being, and smart electrification (Ford-Comp, 2010); access to telecommunications information, streaming music, energy management, and GPS services, while on the road (Squatriglia, 2011); predicting best time for charging when energy cost is less expensive, check and monitor battery level for calculation how far it can go (Gayomali, 2011); connectivity includes in-vehicle Wi-Fi hot spots for mobile devices, entertainment options like streaming video for passengers, real-time updates, Internet radio options and faster application downloads (Harris, 2013).

Vehicular cloud has numerous unique features compared to cloud computing (Gu, Zeng, & Guo, 2013) as follows:

1. **Instability:** In resource availability, since there is no guarantee on the vehicles behavior.
2. **Heterogeneity:** In vehicles provides resources in heterogeneous types and capabilities which makes the challenges in the interoperation and coordination of these heterogeneous resources.
3. **Resilience:** This implies that without Internet connection, some vehicular cloud services are still available.
4. **Selflessness:** A vehicular cloudlet which has a cooperative community having many selfless vehicles. Any vehicle in this community willingly contributes its resources and enjoys the services from the community meanwhile.
5. **Greenness:** This enables the realization of green computing by leveraging excessive resources on vehicles such that unnecessary hardware deployment and investment are avoided.

VEHICULAR CLOUD COMPUTING WORKING

Computing and networking models becomes two integral parts which are efficiently supported by VCC and Information-centric networking (ICN) respectively to support emerging VANET applications and services (Lee, Gerla, & Oh, 2014). Integration of VCC and ICN called as the vehicular cloud networking (VCN). It creates a common virtual platform on which efficient collaboration is done.

The main advantage of vehicular cloud is that it does not require any additional infrastructures because cloud vehicles can use the existing transportation and communication infrastructures of cloud computing and VANETs as shown in Figure 4. The motivation behind vehicular cloud has come that how the capabilities of third-party providers to support non-safety-related applications in VANETs that include location-based services, Internet access, vehicular peer-to-peer communications, on-line gaming, and other forms of mobile entertainment.

VCC operations consist of six main tasks (Lee, Gerla, & Oh, 2014):

1. Cloud resource discovery,
2. Cloud formation,
3. Task assignment and result collection,
4. Content publishing and sharing,
5. Cloud maintenance, and;
6. Cloud release.

In Cloud resource discovery phase, vehicle which runs the application becomes a cloud leader and it recruits member from its vicinity that can provide their resources to construct a vehicle cloud for the application. Having the requirement of the set of necessary resource types, the cloud leader broadcasts a resource request message, RREQ, to nodes within its search range. Nodes willing to share their resources send a resource reply message, RREP, back to the leader with information on their resource capabilities. Based on the received RREP, cloud leader selects cloud member using some criteria to form the cloud. The cloud leader splits the application into sub tasks and distributes it to cloud members by considering their resources. Cloud leader maintains this information in cloud table. Once the task has completed, each cloud member node returns the results to leader node. Now leader node aggregates the results and saves/publishes it. This published content can be shared over the entire network. If leader not having sufficient capability to process and store the results of each member node, it may create another cloud for it. Now any vehicle can send a request to find them. Cloud leader or any cloud member node can send a reply with the contents. Here, a cloud member can send a leave message to a leader on its leaving which can be replaced by another node that has sent RREP earlier in the first phase. This new node now assigned with task and updates this information in the cloud table. When the leader itself moves out of the cloud, it releases the resources of all cloud members by sending the cloud release message to all members.

VANET clouds classified into three architectures as shown in Figure 2 namely vehicular clouds (VC), vehicles using clouds(VuC) or road side cloud, and hybrid vehicular clouds (HVC) central cloud (Yu, Zhang, Gjessing, Xia, & Yang, 2013; Zeinab, Rezaeifar, & Oh, 2015). VC is further divided into two scenarios from movement standpoint as shown in Figure 3. Static clouds refer to the stationary vehicles providing cloud services (renting out storage or processing resources). In case of static VANET clouds, the infrastructure (communication, storage, and process) can be rented out to make revenue as well. IaaS and data storages services are feasible for such arrangements. On the other hand, dynamic clouds are formed on demand in ad hoc manner. VuC connects the VANET to traditional clouds where VANET users can use cloud services on the move such as infotainment, traffic information, and etc. In HVC, vehicular clouds interact with traditional cloud for services exchange. The vehicles and RSUs serve as gateways on the VANET part thereby communicating with the gateways of the traditional clouds.

Data generated from sensors available in vehicle are stored in data storage, which is shared between cloud participants upon receiving an external query. The technological advancement helps to connect, read and control the sensor directly with the external system; sensor can also able to self-actuate and detect events in the physical world. The computing resource is similar to that of the Internet cloud except that its capacity is limited because it is a collection of mobile resources. Here, Vehicles are internet-worked via purely peer-to-peer connections, so each vehicle negotiates the level of resource sharing directly with each other. To improve efficiency Lee et al., (2014) suggested that one vehicle in a cloud can be elected as a broker based on some selected metrics (e.g., connectivity to vehicles) to mediate the process of resource sharing as well as other cloud operations. Stationary cloud member RSU can do ef-

Figure 2. Taxonomy of VANET cloud

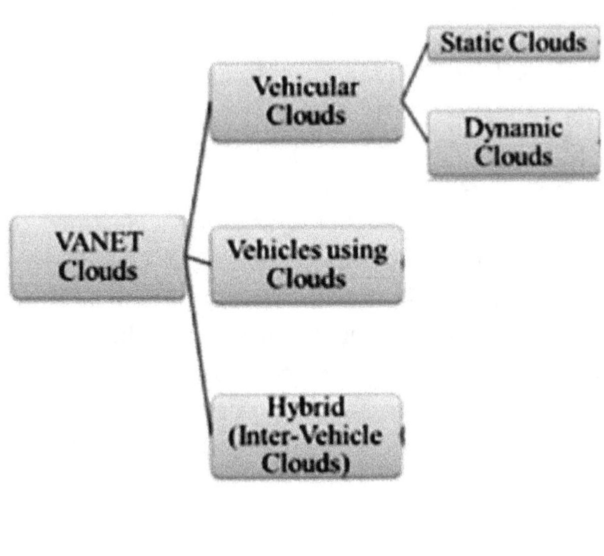

Figure 3. a. Static cloud; b. Dynamic cloud

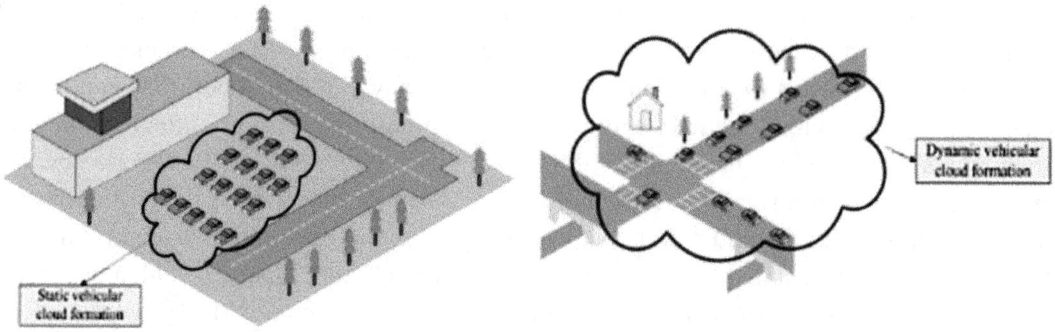

fective content discovery, which may not have enough storage and computing power, but can be a good candidate for the negotiator role due to reliable connections to vehicles. Set of Road Side Units (RSUs) which are located at the certain predefined places are attached to a central cloud (i.e. Via Internet) and provide some services for vehicles. This vehicle cloud is created temporarily by interconnecting RSUs and available resources in the vehicles commonly known as data storage, sensors and computing which is shown in Figure 7.

In case a driver indicates the abnormal behavior on the road an Emergency Warning Messages (EWMs) will be generated and sent to the cloud storage and surrounding vehicles, which contains the geographical location, speed, acceleration and moving direction of the offender. The second component of the communication layer is vehicle-to infrastructure (V2I), which is responsible for exchanging the operational data among vehicles, infrastructures and the cloud over wireless networks such as 3G, satellite or internet. The V2I component is used to augment the safety level of vehicles on highways by reducing the percentage of crashes, delays and congestion improve mobility, and provide Wireless Roadside Inspection (WRI) to automatically inspect commercial vehicles (Bordley, Cherry, Stephens, Zimmer, & Petrolino, 2012).

VEHICULAR CLOUD COMPUTING ARCHITECTURE

Yu et al., (2013) proposed a hierarchical architecture for Vehicular network that consists of three interacting layers: vehicular cloud, roadside cloud, and central cloud as shown in Figure 8. The Vehicular cloud computing architecture proposed by Md et al., (2014) relies on three layers: 1.) inside-vehicle; 2.) communication and; 3.) cloud as illustrated in Figure 9. Architecture of VCC having three layers is explained in detail in the order as shown in Figure 5.

The end-users represent the basic entity of this layer either a customer or an ITS-Cloud datacenter and are characterized by a certain level of mobility (Gu, Zeng, & Guo, 2013). The entity represented by a vehicle driver or passenger, can establish its service request through its communication and computing device like Smartphone, Laptop, Onboard computer, GPS, Wi-Fi, Internet access point etc. As a result, the end-user receives its service response through the upper layers. The end-user can play the cloud datacenter role which belongs to the cloud layer as temporary cloud. It is considered as computing entity which helps the others to perform their service requests against a service price.

Figure 4. VCC system architecture
Chul Sur, Youngho Park & Kyung Hyune Rhee, 2016.

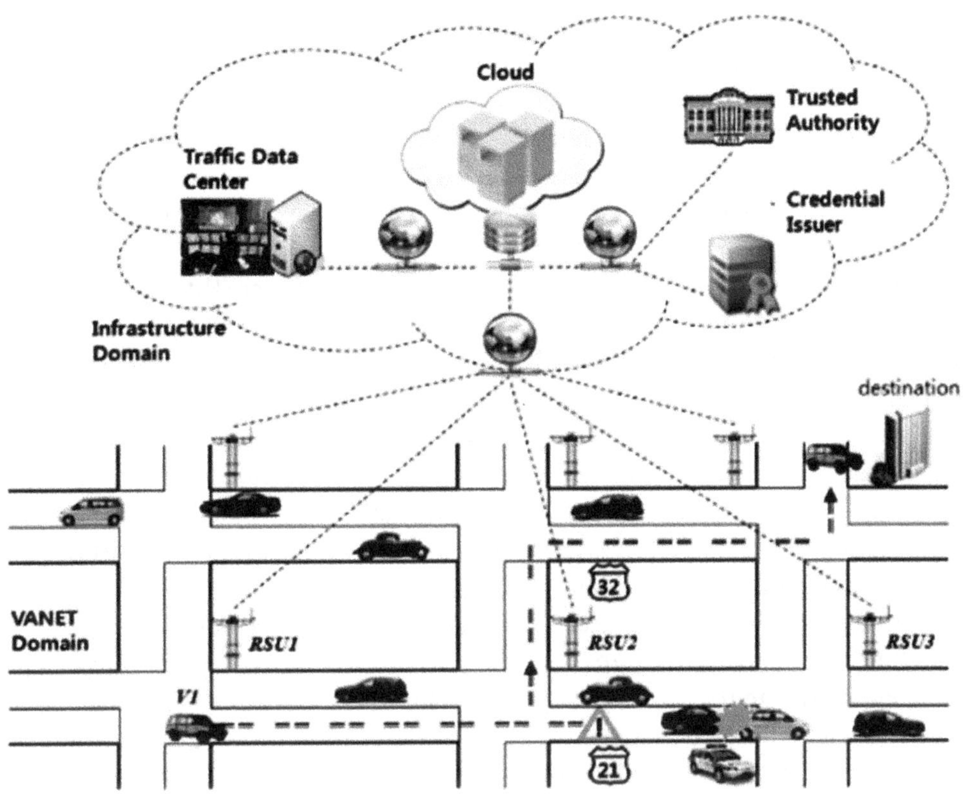

Figure 5. Layers of VCC

Layer 2	CENTRAL CLOUD application, cloud infrastructure, cloud platform
Layer 1	COMMUNICATION LEVEL Road side cloud (V2I) Vehicular cloud (V2V) Communication inside-vehicle (Device level)
Layer 0	END-USERS' LEVEL (customer or data center)

The tendency of car development is changing from improving driver's safety to increasing their comfort and convenience degree. As a result, more and more tiny and smart devices have been deployed inside the vehicles or even attached to the drivers in order to provide real time measurement and protection services. The next layer of this architecture is called communication. Sasikala and Deepti (2013) has given the classification of V-Cloud communication taxonomy as shown in Figure.6 which is based on the communication objects, communication level.

This layer consists of several communication devices and networks: Internet gateways and wireless networks. At this level, all connection technologies (the physical layer, data link layer, medium access control, routing protocols, etc.) are defined and fixed according to the communication device used and the type of client in the lower layer and the data center in the upper layer (Bitam, Mellouk, & Zeadally, 2015). V-Cloud communications are divided into in-car communication module, vehicle-to-vehicle (V2V) communication module and vehicle-to-infrastructure (V2I) communication module.

In the *communication inside-vehicle*, there are various devices in this layer ranging from sensors, actuators, GPS, smart phone and pc-based devices with more flexibility and stronger capabilities, can communicate with each other in a self-organized manner and the embedded sensors and actuators in the car can form a body area sensor network (BASN) through multi-hop manner. BASN also formed by the sensors attached to the drivers to collect physical information like EEG, ECG, heart rate etc. Sasikala and Deepti (2013) have given the flow of data processing of BASN from sensor to actuator and output as shown in Figure 6. Raw data like temperature, pressure, image and biomedical information collected from sensor are stored in a repository for future processing. In the repository, raw data are further classified into low level context (like temperature, pressure) and high-level context (like gesture, activity) based on pre-processing techniques then they are formally processed based on techniques like data mining, reasoning, k-means clustering etc. Finally, outputs are to be delivered for the required actions like stop the car, turn slow and send alarm signals to the driver.

By using GPS localization technique, drivers can reduce their travelling time on road and control room can direct the vehicles (like bus, taxi) in a more efficient way especially during peak hours. Smart phone can enhance driver's convenience based on its strong sensing, processing and communication functions via some customized application to interact with the Internet and other BASN devices like bio-sensors and video sensors etc. PC-based devices like in-car navigator provide various services such as road traffic monitoring, entertainment (audio, video, TV), Internet and resource sharing with other vehicles in a friendly and interactive manner. Other devices like Bluetooth, WiFi can also provide additional support and services to drivers based on their standard communication interface. Devices responsible for monitoring the health and mood of the driver and collecting information inside the car can also be used such as pressure and temperature by using body sensors, environmental sensors (Gerla, Weng, & Pau, 2013), smart phone sensors, the vehicle's internal sensors, inertial navigation sensors (INS), and driver behavior recognition (Oliver & Pentland, 2000) to predict the driver's reaction and intentions. Then, the information collected via sensors should be sent to the neighborhood vehicles and cloud for storage or for use as input for various software programs in the application layer.

The vehicle-to-vehicle (V2V) systems via DSRC which forms a local cloud established among a group of cooperative vehicles. An inter-vehicle network, i.e., a vehicular ad hoc network (VANET), is formed by V2V communications. Vehicles can work in an autonomous way solely relying on the V2V communication capabilities. It can be either the service providers to enrich exiting cloud services by providing various on-road-information or be the service consumers to enjoy existing centralized Internet cloud services via data centers using vehicle-to-infrastructure (V2I) communications. The vehicles in

Figure 6. Three tier architecture: a. Flow of Data processing; b. Communication taxonomy; c. cloud service taxonomy

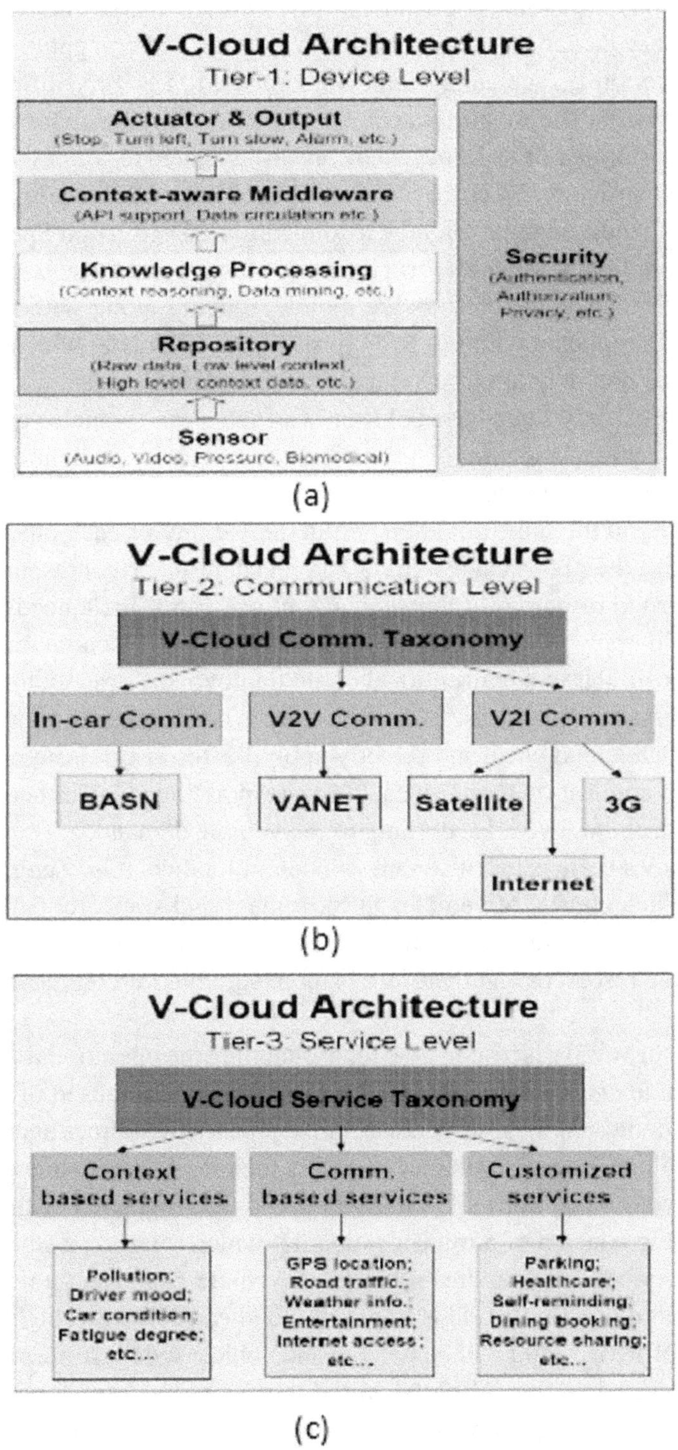

a group are viewed as mobile cloud sites and they cooperatively create a *vehicular cloud* or *cloudlet* to share resources and information with each other as given in Figure 8. Vehicular cloud consists of set of vehicles moving on the road uses some applications which involves large amounts of data to download/upload and store temporary data, need more bandwidth, storage and computation resources.

But due to vehicles mobility, vehicular cloud implementation is very different from a cloud in a traditional computer network. The mobility attribute of the VC implies that the long-term pooling of vehicles resources is almost impossible. Hence, there are no dedicated resources available, form vehicular cloud based on the cluster structure. Yu et al., (2013) proposed two customization strategies for vehicular clouds known as Generalized Vehicular Cloud Customization (GVCC) and Specified Vehicular Cloud Customization (SVCC). In GVCC vehicles form dynamic clusters and those ones that are more suitable become cluster head (CH). Dynamic clusters are mobile, moving along with the high-speed vehicles and the vehicles will communicate with the RSU through the CH node, which is also responsible for the creation, maintenance and deletion of a vehicular cloud (Arkian, Atani, & Kamali, 2014). Arkian et al., (2014) suggested the fuzzy logic based cluster head selection. Vehicles grouped into clusters according to the vehicle location, direction of travel and their velocity. Neighborhoods information are collected from HELLO messages broadcasted by each vehicle which has its current speed and position data. The vehicles moving in the same direction and in the vicinity of each other come under a cluster. All vehicles will virtualize their physical resources and register the virtual resources in the cluster head. Then, it schedules all virtual resources of the vehicular cloud. If a vehicle needs some resources of the vehicular cloud, it should apply to the cluster head. With respect to resource management, the cluster-based vehicular cloud is similar to a conventional cloud deployment strategy in which cloud resources are scheduled by the controller. But in case of SVCC, there is no cloud controller. A vehicle will specify some vehicles as candidate cloud sites, and directly apply for resources from candidate cloud sites. If the resource requirement application is approved, the corresponding vehicles become cloud sites, which will customize virtual machines (VMs) according to the vehicle demand.

The mobility of vehicles defines two different vehicular cloudlets (Gu, Zeng, & Guo, 2013), known as mobile and static, which consist of vehicles in movement and statically parked ones, respectively. The mobile vehicular cloudlet makes them suitable for data ferrying service. Static vehicular cloudlet is with comparatively stable resources and therefore is more suitable for computation or storage resource provision.

Vehicle-to infrastructure (V2I), is responsible for exchanging the operational data among vehicles and infrastructures. A *roadside cloud* is uses the V2I communication composed of dedicated local servers and roadside units. The dedicated local servers virtualize physical resources and act as a potential cloud site. RSUs provide radio interfaces for vehicles to access the cloud. A roadside cloud is accessible only by the nearby vehicles whose locations are within the radio coverage area of the cloud site's RSU. Yu et al., (2013) proposed the concept of a roadside cloudlet which refers to a small-scale roadside cloud site that offers cloud services to bypassing vehicles. A vehicle can select a nearby roadside cloudlet and customize a transient cloud for use. This transient cloud can serve a vehicle for short time, once it moves out of the range of RSU, cloud will be deleted and vehicle will customize a new cloud from next RSU cloudlet in its moving direction. Vehicles in the transient cloud are serviced by virtual machine (VM) which consists of interfacing known as VM-base in the roadside cloudlet and the VM-overlay in the vehicle. Customization of VM is done with the help of these interfaces. To have continuity of cloud service when vehicle moves, the customized VM should be transferred to next nearest roadside cloudlets known as VM migration.

Figure 7. Vehicular cloud resources
Lee, Lee, Gerla, & Oh, 2014.

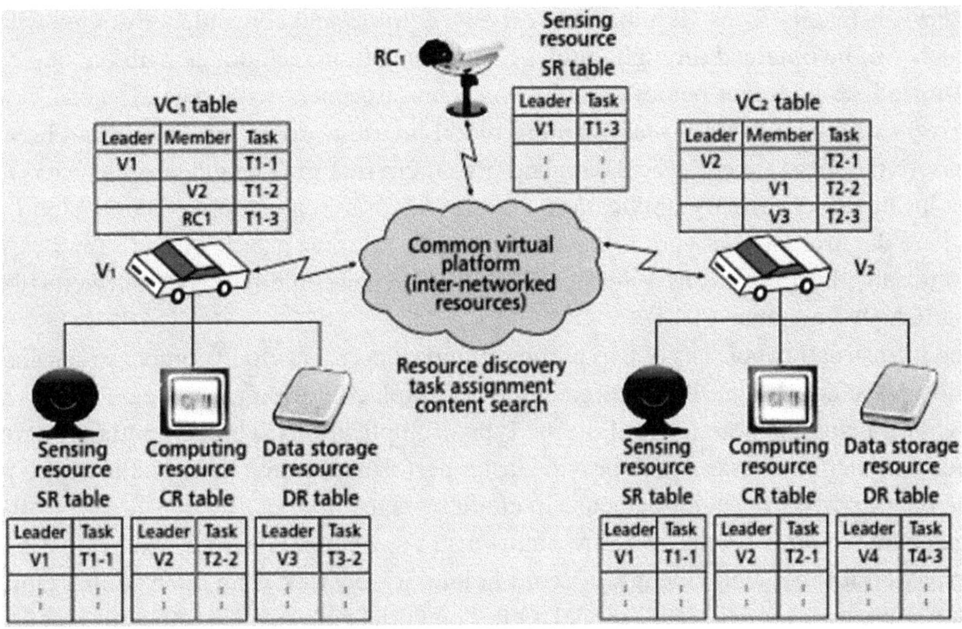

Figure 8. VCC architecture
Yu, Zhang, Gjessing, Xia, & Yang, 2013.

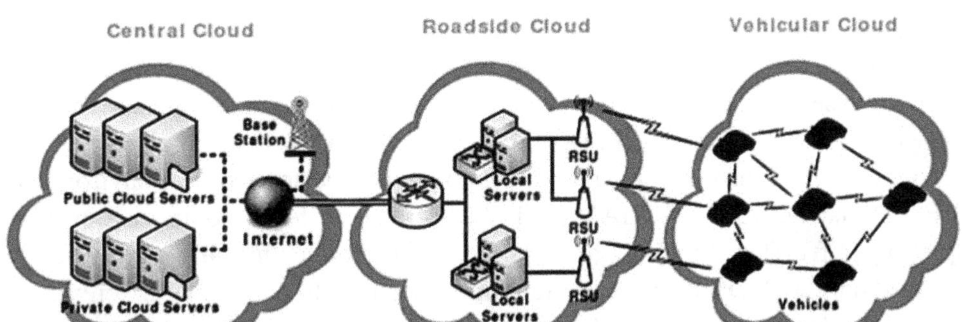

Central Cloud has much more resources compared to a vehicular cloud and a roadside cloud. It can be driven by either dedicated servers in vehicular networks data center or servers in the Internet as shown in Figure 8. A central cloud is mainly used for complicated computation in short time, massive data storage, and global decision. The cloud layer consists of three internal layers: application, cloud infrastructure, and cloud platform (Salim & Abdelhamid, 2012; Md, Mehdi, Abdullah, & Rajkumar, 2014).

In the application layer, various applications and services are considered as real-time services or cloud primary services, which are accessible remotely by drivers, such as fuel feedback, human activity recognition, health recognition and environmental recognition. V-Cloud services shown in Figure 6 (Sasikala & Deepti, 2013) are classified into context based, communication based and customized services. Context based services mainly focus on improving driver's safety and health which include driver status monitoring, road pollution monitoring, car itself performance monitoring etc. Communication based services mainly focus on improving driver's comfort and convenience degree which include services which will be obtained through wireless communication techniques with outside car counterparts. Customized services can be developed for certain customers with special needs. For example, for those drivers who have heart disease or high blood pressure, some bio-sensor can be attached on them to measure their real time physical data and give them real time healthcare services. The various components in the cloud primary application services layer are called as vehicular cloud computing services and are deployed as Network as a Service (NaaS), Storage as a Service (STaaS), Cooperation as a Service (CaaS), Information as a Service (INaaS), and Entertainment as a Service (ENaaS) (Md, Mehdi, Abdullah, & Rajkumar, 2014).

The cloud infrastructure consists of two parties: cloud storage and cloud computation. The data gathered by the inside-vehicle layer will be stored in the geographic information system (GIS), a road traffic control device or a storage system based on the type of applications. The computation part is used to calculate the computational tasks which provide faster performance, for example, the health recognition sensors send data to driver behavior database in cloud storage.

The open source or commercial software platforms (Yu, Zhang, Gjessing, Xia, & Yang, 2013) that could be employed for the deployment of a central cloud. Openstack is an open source cloud platform using Infrastructure as a Service (IaaS) model. Other potential commercial platforms are Amazon Web Services, Microsoft Azure and Google App Engine.

SENSING PLATFORMS

The success of connected vehicles will depend on sensing platforms with capabilities to access, collect, and process accurate sensor data. Sensing platforms include intra-vehicular sensor and urban sensing technologies, which are two of the most relevant sensing platforms used to collect information about traffic conditions. Some of the communication technologies used in vehicular environments includes DSRC/WAVE, cellular networks, WiMAX, and 802.11p. These technologies can enable operations aimed at improving traffic flow, highway safety, and other ITS applications in a variety of environments. Vehicles equipped only with DSRC can operate in infrastructure-free mode (V2V only), infrastructure mode (V2I), and mixed mode (V2V and V2I) (Figure 10). Vehicles equipped with other broadband wireless access technologies (i.e. LTE, WiMAX) can communicate with each other via the Internet (Figure 10). For instance, people with smart phones and Internet access can set up a peer-to-peer (P2P) overlay network via the Internet. Finally, when vehicles have both DSRC and other broadband wireless access methods, there will be a mixed access scenario (Figure 10).

APPLICATIONS OF VEHICULAR CLOUD COMPUTING

Goal of this section is to illustrate the power of the VCC through some applications. Mohamed et al., (2010) has outlined about application scenarios that are comes under two application categories. Under traffic management synchronizing traffic lights after clearing an accident, autonomous mitigation of recurring congestion, sharing on-road safety precautions and dynamic management of HOV lanes were

Figure 9. VCC architecture components
Md AliAl, Khairul, Md Fakhrul Alam, & Esfar- E- Alam, 2014.

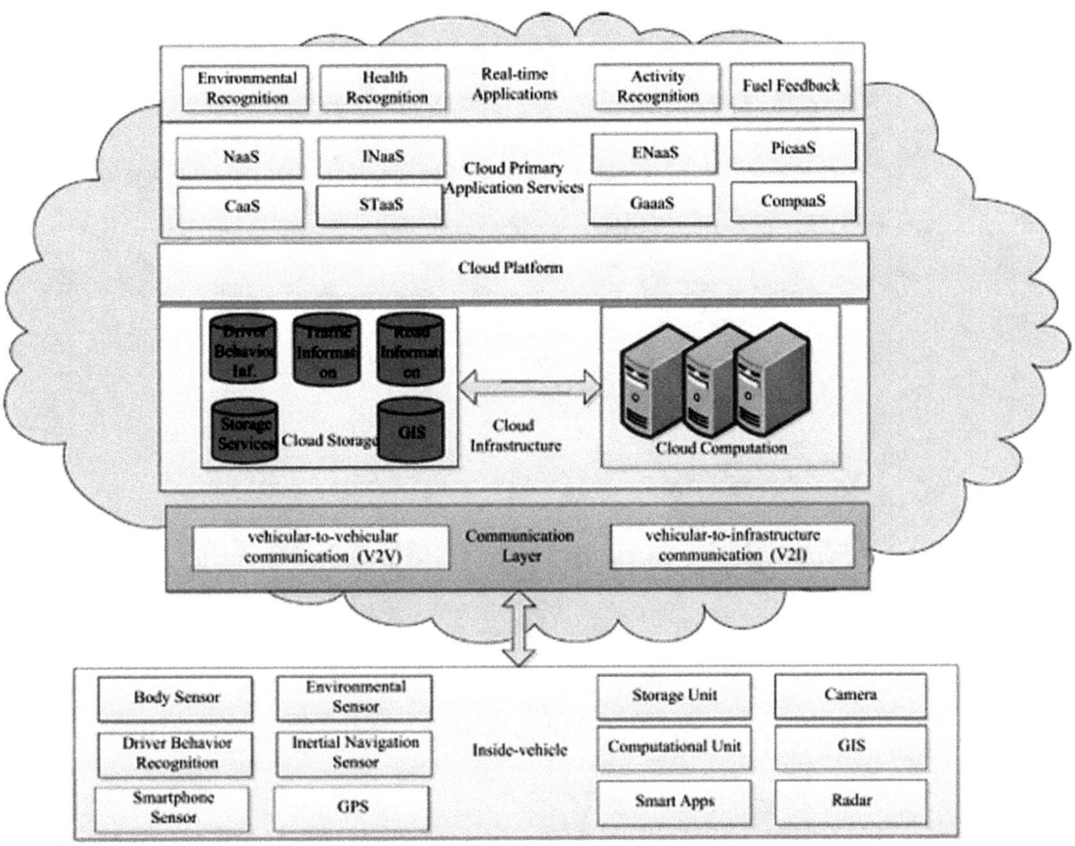

explained. And asset management covers mobile experimental and analytics laboratory in support of homeland security, augmenting the capabilities of small businesses, efficient tasking of law enforcement officers, dynamic management of parking facilities, dynamic asset management in planned evacuations and AVCs in developing counties. Sasikal and Deepti (2013) have discussed about the real-time services of road traffic monitoring and healthcare monitoring under vehicle cloud architecture. Md et al., (2014) have given several possible implementation scenarios and the application outcome of VCC. These are airport as a datacenter, Parking lot data cloud, Shopping mall data center, Dynamic traffic light management, Optimizing traffic signals, Self-organized high occupancy vehicle (HOV) lanes, Managing evacuation, Road safety message, Easing frequent congestion, Managing parking facilities and VC in developing countries perspective. VANET-Cloud can provide several services which are classified in to different types as follows (Bitam, Mellouk, & Zeadally, 2015; Salim & Abdelhamid, 2012):

- **Safety Applications in a Very Large Area:** VANET-Cloud can provide much information to other vehicle concerning collision notification and avoidance, crash prevention and safety, traffic incident management, and so on. This information can leads to take safety measures

Figure 10. Wireless vehicular networking scenarios
Guerrero-Ibanez, Zeadally, & Contreras-Castillo, 2015.

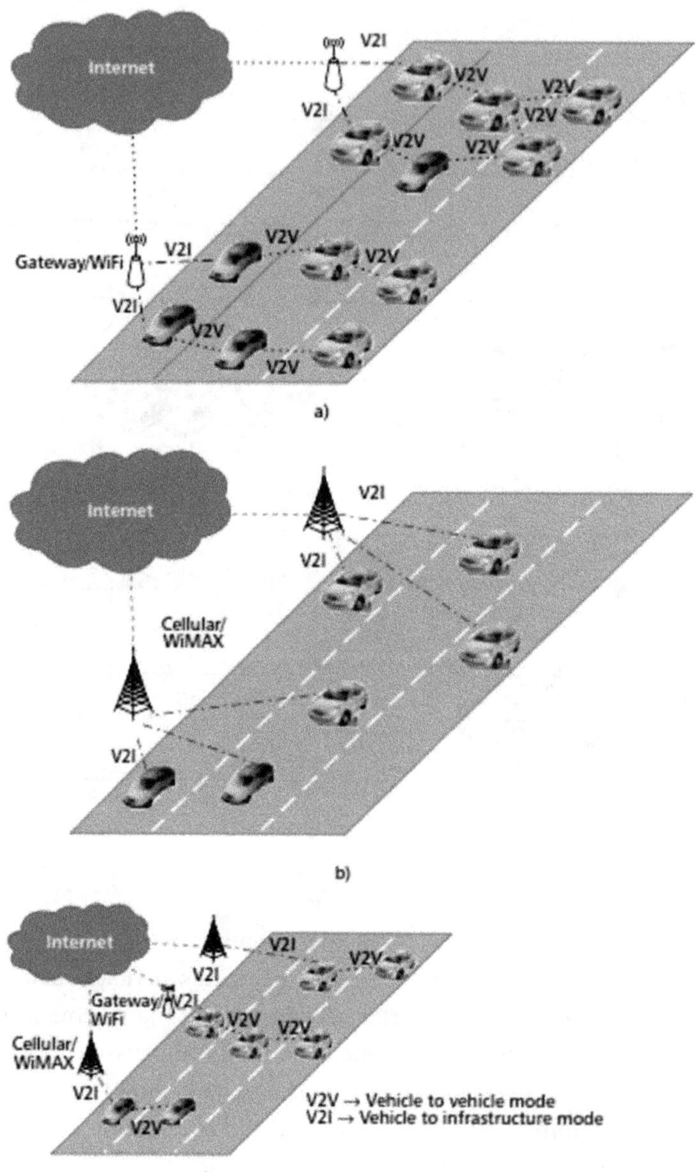

- **Vehicular Software and Applications:** VANET-Cloud components (i.e., vehicles and data centers) offer to end users about various vehicular software and applications that help to discover geographic positions and real-time information concerning car parking places and availability, full station, service station, hotels, and so on.
- **Web Services and Applications:** VANET-Cloud allows services of business's web server over the Internet for clients. Web services (i.e., email, meeting organization, appointments) are advertised over a large computing capacity provided by VANET-Cloud which is much faster and cheaper using the vehicles' onboard computers.

- **Processing and Cloud Backup:** Using data centers and vehicles' computing resources, VANET-Cloud can be used to handle compute-intensive problems encountered in vehicular areas. Therefore, computational complexity in terms of computational time can be reduced or overcome compared to the use of traditional computing resources alone. VANET-Cloud also guarantee the data storage in data centers using standard data formats.
- **Business and Research Applications:** The e-commerce is easier in using VANET-Cloud due to the cloud approach which is based on the digital buying services that are rarely available in VANETs. Vehicular providers (e.g., drivers, passenger) can provide commercial applications against a usage fee. Through the use of a large number of vehicular onboard computers, VANET-Cloud ensures an efficient computing platform that could be used to conduct academic and experimental studies.
- **Load Balancing Capability:** VANET-Cloud helps spreading a service workload between the cloud datacenters to reach several advantages like the scalability, avoiding the servers overloading, ensuring reliability of the computing in the case of starting multiple processing copies in different datacenters, minimizing the response time etc.
- **Correspondence Application:** VANET-Cloud offers the personal information and knowledge management such as email transmission and archiving, appointments and meeting organization, project management features, files sharing, chat etc.

CHALLENGES AND OPPORTUNITIES

Vehicular cloud computing concept arises as a promising model to be deployed on vehicular networks. Practical implementations of vehicular cloud computing has several technical challenges for its success. IoT-based vehicular data clouds must be efficient, scalable, secure, and reliable for the successful deployment at a large scale. Existing algorithms and mechanisms are unsatisfactory to meet all these requirements at the same time. Open challenges in R&D from both a scientific and an industrial view, the important challenging research fields identified by the experts are related to the topics heterogeneous vehicular networks, vehicular networking applications, field operational tests, and scientific foundations of inter-vehicle communication (Falko, Hannes, Onur, & Ozan, 2014).

Storing personal information in virtual resources in cloud, led to free access over such data. So, users should have more control to decide what information could be exposed to others (e.g. service providers). Hence, security and privacy aspects are of supreme importance and more difficult due to high mobility, frequent topology changes, dynamicity of the environment and the heterogeneity of vehicles. Establishing trust relationships between drivers is a vital part of trustworthy communication. In addition, limited storage on mobile devices, inadequate battery life, scalability and service availability are another set of research challenges in vehicular cloud computing. Selective network coding methods to enhance the reliability and efficiency of information dissemination is the other interesting challenge. Furthermore, service delay which is very crucial for some application and service metering which is used in deciding cloud service providers are still unaddressed/ less explored issues.

Mohamed et al., (2010) has presented some of the research issues along the three systems engineering dimensions, namely structure, function, and behavior (operation and policy).

- **Structural Challenges:**
  - **Elastic Mobile Architecture:** VC networking and associated protocol architecture must be developed to accommodate changing application demands and resource availability on the move.
  - **Resilient AVC Architecture in the Wild:** VC basic structural and composed building blocks must be designed to withstand structural stresses induced by the inherent instability in the operating environment. Research is needed on architectures enabling vehicle visualization and migration of virtual vehicles.
  - **Service-Oriented Network Architecture:** Contemporary layered network architectures have proven limited in face of evolving applications and technologies. The adoption of service-oriented component-based network architectures with intrinsic monitoring and learning capabilities will be the replacement for the layered architecture.
- **Functional Challenges:**
  - **Enabling AVC Autonomy:** Research is needed on developing a trustworthy base, negotiation and strategy formulation methodology (game theory, etc.), efficient communication protocols, data processing and decision support systems, etc.
  - **Managing Highly Dynamic Cloud Membership:** There is a critical need to efficiently manage mobility, resource heterogeneity (including sensing, computation and communication), trust, and vehicle membership (change in interest, change in location, resource denial and/or failure, etc.).
  - **Cyber-Physical Control:** AVCs can be defined by their aggregated cyber and physical resources. Their aggregation, coordination and control are non-trivial research issues.
  - **Cooperation between AVCs:** AVCs have to coordinate and to solve the traffic problem collectively for the efficient traffic management, since they cannot proceed to selfishly that benefits each of them individually.
- **Operational and Policy Challenges:**
  - **Trust and Trust Assurance:** The establishment of trust relationships between the various players is a key component of trustworthy computation and communication. The problems of assuring emergent trust and security in AVC communication and information need to be addressed. The task of establishing proactively, a basic trust relationship between vehicles is possible and may be even desirable.
  - **Contract-Driven vs. Ad-Hoc AVC:** It was anticipated that AVCs will be largely contract-driven, where the owner of the vehicle or fleet consents to renting out some form of excess computational or storage capacity. At the same time, mobility concerns dictate that in addition to the contract-based form of AVC, there should be possible to form an AVC in an ad hoc manner as necessitated by dynamically changing situations like those discussed in the scenarios above.
  - **Effective Operational Policies:** In order for the AVCs to operate and inter-operate seamlessly, issues related to authority establishment and management, decision support and control structure, the establishment of accountability metrics, assessment and intervention strategies, rules and regulations, standardization, etc. must all be addressed. Dealing with these will require a broad participation and must involve local, state or even federal decision makers.

- **AVC Utility Computing:** There is a need for economic models and metrics to determine reasonable pricing and billing for AVC services.

Falchetti et al., (2015) has grouped the technical challenges into six categories: mobility, latency, reliability, criticality, topology, and security. Nageswara Rao and Reddy (2015), has discussed about the issues with respect to: 1.) scalability and technology integration; 2.) Performance, reliability and quality of service; 3.) Security and privacy; 4.) Global standards for device service integration, security, privacy, architecture, and communications.

- **Mobility:** One of the main properties of vehicular communication is the high mobility of its agents. This is usually taken as a negative issue, as the Doppler effect is hard to compensate for.
- **Latency:** Due to the speed of the vehicles, actions must be taken promptly, as the physical mechanics of the environment are very dynamic. Any warning and coordination must happen instantly if a collision is to be averted. Also, this coordination time must be stable and predictable; otherwise, there's no way to know if another car is not listening or the network is just slow. This translates into a need for low, controllable, and predictable latency.
- **Criticality:** There are many different types of digital data flowing through the network. Some of these digital data streams are more critical than others, so they should be prioritized.
- **Topology:** Another big difference of vehicular networks compared to traditional networks is that their topology changes quickly with time, as vehicles come in and out of the reach of each other abruptly. Therefore, the network should be able to handle these topology changes gracefully and efficiently.
- **Security and Privacy:** All network systems must ensure a minimum level of security and privacy for its users. The system must be resilient against attempts to gather unauthorized private information from the network's users, as well as malicious tampering to the network architecture. VANET is an emerging paradigm that is expected to increase the public safety standards and enhance the safety level of drivers/passengers and pedestrians on roads through a variety of applications. Safety applications rely on cooperation between neighbours by exchanging localization details between vehicles. An adversary could take advantage of this and publish false information to provoke havoc in the system. Trust relationships are hard to be built in vehicular clouds because of the large and dynamically changing number of vehicles on the road. Balanced security measures are needed to enhance the security and trust of cloud services without limiting the flexibility of the system. In addition, many drivers do not want their vehicle locations to be tracked or monitored due to the worries about their privacy.
- **Scalability:** The effectiveness of a vehicular cloud depends on its scalability to handle a dynamically changing number of vehicles. In addition to handling regular traffic, vehicular clouds must be able to handle traffic spikes caused by special events or situations. More development on optimization algorithms that coordinate virtual machines, storage space, and network bandwidth to balance server workload and improve computing resource utilization on the vehicular clouds is needed.
- **Technology Integration:** As new devices and technologies are coming out each year, developing effective IoT Middleware that supports integration of these new technologies and devices with existing in-vehicle technologies from automobile manufacturers will be a challenge.

- **Performance, Reliability, and Quality of Service:** Because of high mobility in vehicles, the vehicular networking and communication is often intermittent or unreliable. More new mechanisms are needed to enhance the communication reliability with reduced traffic overhead. Cross layer data synchronization mechanisms should also be designed to minimize the traffic overhead between layers. Acceleration data compression algorithms for resource-constrained sensors, actuators, and other Internet-access devices need improvement to be able to effectively and efficiently compress a large amount of raw data generated. Multiple processing units and support in different cloud data centers are needed to minimize the service response time, improve availability and stability of the service, and increase cloud reliability and fault tolerance. Real-time evaluation mechanisms regarding the performance, reliability, and service quality on vehicular data clouds will have to be further developed.

Lack of global standards for device, service integration, security, privacy, architecture, and communications: Global standards are essential to avoid conflicts between locally developed vehicular data clouds. However, as there are a number of stakeholders involved in vehicular data clouds, and complex dependencies among these stakeholders also exist, it is challenging to establish global standards to lower the complexity and make vehicular data clouds more compatible and cost effective. Further efforts on standardization are needed to coordinate various efforts and resources for implementing vehicular data clouds.

Bitam et al.,(2015) has addressed about the: 1.) Security and privacy issues in the vehicular cloud; 2.) Sensing and aggregation of data; 3.) Green VANET-Cloud; 4.) Communication and coordination between VANET-Cloud sub-models; 5.) Interoperability and standardization; 6.) Quality of the Provided Services and; 7.) VANET-Cloud resource allocation and sharing.

- **Security and Privacy Issues in the Vehicular Cloud:** As a distributed system, the major challenges for the vehicular cloud are security and privacy. In such a system, users input their data and run their applications at data centers managed by other people. The problem becomes even more challenging when temporary VANET-Cloud onboard computers are considered as servers. Consequently, security and privacy issues should focus on ensuring data integrity, controlling data access, preventing data loss, protecting the confidential data of users, and so on. The nature of mobile nodes of vehicular cloud causes the several security challenges in the VCC. Md et al., (2012) addressed security challenges such as authentication; secure location information, message confidentiality, the safety of messages and securing vehicular communication.
- **Sensing and Aggregation Data:** In the VANET context, cloud computing faces an important challenge, which is the sensing and aggregation of data as a service. This kind of service can be provided by the temporary VANET-Cloud using sensors of vehicles and digital equipment vehicular passengers. New research solutions are required to efficiently sense and aggregate various types of sensor data, including traffic data, drivers' health information, information about the environment (disasters, fire, etc.), movements of vehicles and citizens on roads, and so on. The sensed data can be processed to yield results that can help to cope with critical or undesirable situations.
- **Green VANET-Cloud:** Another major issue in VANET-Cloud is the improvement of energy efficiency. Cloud data centers consume an important amount of energy each year (Gerla, Weng, & Pau, 2013). VANET-Cloud providers can contribute to reduce energy cost in data centers using

alternative servers such as onboard computers located in vehicles. The use of mobile computing resources can also ensure the redistribution of the energy consumed because these vehicles are self-powered and help to minimize the energy consumption of data centers of the entire cloud compared to the energy consumption when using only centralized data centers. To achieve this result, researchers need to investigate energy efficient solutions by combining permanent and temporary servers.

- **Communication and Coordination Between VANET-Cloud Sub-Models:** Continuous and preferred communication within permanent and temporary sub-models, in addition to the communication and coordination between these two sub-models, are also VANET-Cloud challenges. Vehicles have limited lifetime in the network of the cloud, and require a strategy to ensure continuous and efficient communication between VANET nodes as VANET-Cloud entities. We also need to develop new communication protocols that perform data exchange within or between VANET-Cloud components.
- **Interoperability and Standardization:** Since VANET-Cloud is based on diverse stationary and mobile computing resources such as data centers or onboard computers, many steps should be taken to address the interoperability challenge to allow these different entities to work together. Furthermore, standardization can be seen as a good solution to address the interoperability issue so that a consensus can be reached among the VANET-Cloud players (i.e., developers, vehicle manufacturers, etc). As a result, interoperability issues (heterogeneous operating system, application programming interface, etc.) among VANET-Cloud devices and software will not cause any additional cost for the client or for the provider.
- **Quality of the Provided Services:** VANET-Cloud aims to provide different services requested by the client with a certain level of quality, known as service level agreements (SLAs). Consequently, the provider is responsible for guaranteeing the availability, reliability, and performance of the cloud resources. SLAs are highly constrained in the VANET-Cloud environment because an important part of the cloud is non-permanent, especially when vehicles move and change the vehicular network topology. The definition of the SLA specification and SLA evaluation throughout the service execution should be investigated in the future.
- **VANET-Cloud Resource Allocation and Sharing:** Resource allocation and sharing are important issues that also need to be addressed, particularly for mobile cloud entities such as vehicles and passenger devices. This mobility can affect the performance of VANET-Cloud applications. An optimal dynamic combination of mobile computing resources can improve the offered service. In such cases, a reserved resource should be ready to be used with a reduced response time.

CONCLUSION

One of the significant challenges of the vehicular network is the limited resource with respect to memory, computation power and bandwidth. It leads to slow data processing capability. However, the demanding requirement of emerging application is complex computation with high storage capacity. This necessity can be fulfilled by sharing the available resource among all the vehicles which are in near proximity. It can be achieved by integrating the vehicular network with cloud computing which is called as vehicular cloud computing (VCC). VCC is a new technological shift which can takes the advantages of cloud computing to afford the services to drivers of VANET.

The chapter has started with the need for vehicular cloud network and then what was the scenario before VCC are discussed which includes about the VANET, cloud computing, and Mobile cloud computing. Working principles and different architecture proposed by researchers are explained in detail. The technologies involved in the sensing platforms for VCC are discussed. Then the potential applications scenarios and types of applications are also discussed.

REFERENCES

Abolfazli, S., Sanaei, Z., Ahmed, E., Gani, A., & Buyya, R. (2013). Cloud-based augmentation for mobile devices: motivation, taxonomies, and open challenges. IEEE Communications Surveys and Tutorials. doi:10.1109/SURV

Abuelela, M., & Olariu, S. (2010). Taking VANET to the clouds. *Proceedings of the 8th international conference on advances in mobile computing and multimedia (MoMM'10)*, 6–13. 10.1145/1971519.1971522

Arkian, H. R., Atani, R. E., & Kamali, S. (2014). FcVcA: A fuzzy clustering-based vehicular cloud architecture. *2014 7th International Workshop on Communication Technologies for Vehicles (Nets4Cars-Fall)*, 24-28. 10.1109/Nets4CarsFall.2014.7000907

Armbrust, M., Fox, A., Griffith, R., Joseph, A. D., Katz, R., & Konwinski, A. (2010). A view of cloud computing. *Communications of the ACM*, *53*(4), 50–58. doi:10.1145/1721654.1721672

Bitam, S., Mellouk, A., & Zeadally, S. (2015, February). VANET-cloud: A generic cloud computing model for vehicular Ad Hoc networks. *IEEE Wireless Communications*, *22*(1), 96–102. doi:10.1109/MWC.2015.7054724

Bordley, L., Cherry, C. R., Stephens, D., Zimmer, R., & Petrolino, J. (2012). Commercial motor vehicle wireless roadside inspection pilot test, Part B: Stakeholder perceptions. *91st annual meeting of the transportation research board.*

Eltoweissy, M., Olariu, S., & Younis, M. (2010). Towards autonomous vehicular clouds. *Ad Hoc Networks*, 1–16.

Falchetti, A., Azurdia-Meza, C., & Cespedes, S. (2015). Vehicular cloud computing in the dawn of 5G. *2015 CHILEAN Conference on Electrical, Electronics Engineering, Information and Communication Technologies (CHILECON)*, 301-305. 10.1109/Chilecon.2015.7400392

Falko, D., Hannes, H., Onur, A., & Ozan, T. (2014, June). InterVehicle Communication: Quo Vadis. *IEEE Communications Magazine*, *52*(6), 170–177. doi:10.1109/MCOM.2014.6829960

Fatima, B., & Shoad, A. K. (2005). Traffic Estimation and Real Time Predication Using Ad Hoc Networks. *Proceedings of IEEE International Conference on Emerging Technologies 2005.*

Fernando, N., Loke, S. W., & Rahayu, W. (2013). Mobile cloud computing: A survey. *Future Generation Computer Systems*, *29*(1), 84–106. doi:10.1016/j.future.2012.05.023

Ford-Comp. (2010). *University of Michigan Reveal Students' Vision for Future of In-Car Cloud Computing Apps*. Ford Online.

Gayomali, C. (2011). *Toyota and microsoft team up to bring cloud computing on the road.* Time Tech.

Gerla, M. (2012). Vehicular cloud computing. *Proceedings of the 11th annual Mediterranean ad hoc networking workshop (Med-Hoc-Net)*, 152–5. 10.1109/MedHocNet.2012.6257116

Gerla, M., Weng, J.-T., & Pau, G. (2013). Pics-on-wheels: Photo Surveillance in the Vehicular Cloud. In *Proceedings of the International Conference on Computing, Networking and Communications (ICNC).* IEEE. 10.1109/ICCNC.2013.6504250

Gu, L., Zeng, D., & Guo, S. (2013). Vehicular cloud computing: A survey. *2013 IEEE Globecom Workshops (GC Wkshps)*, 403-407. doi: 10.1109/GLOCOMW.2013.6825021

Guerrero-ibanez, J. A., Zeadally, S., & Contreras-Castillo, J. (2015). Integration challenges of intelligent transportation systems with connected vehicle, cloud computing, and internet of things technologies. *IEEE Wireless Communications*, *22*(6), 122–128. doi:10.1109/MWC.2015.7368833

Harris, C. (2013). *General motors connects cars to 4 G LTE Tech page one*. Academic Press.

Hartenstein, H., & Laberteaux, K. P. (2008). A tutorial survey on vehicular ad hoc networks. *Communications Magazine, IEEE*, *46*(6), 164–171. doi:10.1109/MCOM.2008.4539481

Hoover, J.N., & Martin, R. (2008). Demystifying the Cloud. *Information Week Research & Reports*, 30–37.

Le, L., Festag, A., Baldesari, R., & Zhang, W. (2009). CAR-2-X Communications in Europe. In Vehicular Networks: From Theory to Practice (pp. 4.1–b8.32). Taylor and Francis.

Lee, E., Lee, E. K., Gerla, M., & Oh, S. Y. (2014, February). Vehicular cloud networking: Architecture and design principles. *IEEE Communications Magazine*, *52*(2), 148–155. doi:10.1109/MCOM.2014.6736756

Lee, U., Cheung, R., & Gerla, M. (2009). Emerging Vehicular Applications. In S. Olariu & M. C. Weigle (Eds.), *Vehicular Networks: From Theory to Practice* (pp. 6.1–6.30). Boca Raton, FL: Taylor and Francis. doi:10.1201/9781420085891-c6

Mario, G. (2012). Vehicular Cloud Computing. *Vehicular Communications and Applications Workshop*, 152-155.

Md, W., Mehdi, S., Abdullah, G., & Rajkumar, B. (2014). A survey on vehicular cloud computing. *Journal of Network and Computer Applications*, *40*, 325–344. doi:10.1016/j.jnca.2013.08.004

Md AliAl, M., Khairul, A., Md Fakhrul Alam, O., & Esfar- E- Alam, A. (2012). *Proceedings of the World Congress on Engineering and Computer Science 2012,* 1.

Mohamed, E., Stephan, O., & Mohamed, Y. (2010). Towards Autonomous Vehicular Clouds - A Position Paper. ADHOCNETS 2010. *LNICST*, *49*, 1–16.

Mohammad, P., Mohammad, U. F., Mahboob, U. A., & Khaleel, U. R. (2015). Vehicular Cloud Computing: leading towards tomorrow's Internet of Vehicles. *Journal of Wireless Sensor Networks, 1*, 1-0006.

Nageswara Rao, S., & Reddy, B. R. M. (2015). Developing Data Cloud Services in Various Environments. International Journal & Magazine of Engineering, Technology, Management and Research, 2(8).

Niroshinie, F., Seng, W. L., & Wenny, R. (2013). Mobile cloud computing: A survey. *Future Generation Computer Systems*, *29*(1), 84–106. doi:10.1016/j.future.2012.05.023

Olariu, S., Khalil, I., & Abuelela, M. (2011). Taking VANET to the clouds. *International Journal of Pervasive Computing and Communications*, *7*(1), 7–21. doi:10.1108/17427371111123577

Oliver, N., & Pentland, A. P. (2000). Driver behavior recognition and prediction in a smartcar. Proceedings of the SPIE-The International Society for Optical Engineering, 280-90.

Rasheed, H. (2015). A Paradigm Shift from Vehicular Ad Hoc Networks to VANET-Based Clouds. *Wireless Personal Communications*, *83*(2), 1131–1158. doi:10.100711277-015-2442-y

Rodolfo, I. M. (2016). A Vehicular Cloud-Based Framework for the Intelligent Transport Management of Big Cities. *International Journal of Distributed Sensor Networks*, *12*(5).

Salim, B., & Abdelhamid, M. (2012). ITS-Cloud: Cloud Computing for Intelligent Transportation System. *Globecom 2012 – IEEE, Communications Software, Services and Multimedia Symposium*, 2054-2059.

Sanaei, Z., Abolfazli, S., Gani, A., & Buyya, R. (2013). Heterogeneity in mobile cloud computing: Taxonomy and open challenges. *IEEE Communications Surveys and Tutorials*. doi:10.1109/SURV.2013.050113.00090

Sasikala, G., & Deepti, A.R. (2013). Real Time Services for Cloud Computing Enabled Vehicle Networks. *IOSR Journal of Computer Engineering, 11*(1), 8-14.

Shiraz, M., Gani, A., Khokhar, R., & Buyya, R. (2012). A review on distributed application processing frameworks in smart mobile devices for mobile cloud computing. *IEEE Communications Surveys and Tutorials*, *15*, 1–20.

Squatriglia, C. (2011). *Toyota, microsoft to bring the cloud to cars*. Wired.

Subir, B., Raymond, T., & Francois, D. (2006). Vehicle-to-Vehicle Wireless Communication Protocols for Enhancing Highway Traffic Safety. *IEEE Communications Magazine*, 74–82.

Sur, C., Park, Y., & Rhee, K. H. (2016). An efficient and secure navigation protocol based on vehicular cloud. *International Journal of Computer Mathematics*, *93*(2), 325–344. doi:10.1080/00207160.2014.934685

Tekbiyik, N., & Uysal-Biyikoglu, E. (2011). Energy efficient wireless unicast routing alternatives for machine-to-machine networks. *Journal of Network and Computer Applications*, *34*(5), 1587–1614. doi:10.1016/j.jnca.2011.02.005

Uzcategui, R. A., & Acosta-Marum, G. (2009, May). WAVE: A Tutorial. *IEEE Communications Magazine*, *47*(5), 126–133. doi:10.1109/MCOM.2009.4939288

Wang, T., Song, L., & Han, Z. (2013). Coalitional Graph Games for Popular Content Distribution in Cognitive Radio VANETs. *IEEE Transactions on Vehicular Technologies, 62*(8), 4010 – 4019. DOI: 10.1109/TVT.2013.2245353

Wang, T. (n.d.). *Popular Content Distribution in CR-VANETs with Joint Spectrum Sensing and Channel Access*. IEEE.

Xie, S., Liu, Y., Zhang, Y., & Yu. (2010). A Parallel Cooperative Spectrum Sensing in Cognitive Radio Networks. *IEEE Transactions on Vehicular Technology, 59*(8), 4079-4092.

Yang, Y., & Bagrodia, R. (2009). Evaluation of VANET-based Advanced Intelligent Transportation Systems. *Proc. ACM VANET*. 10.1145/1614269.1614273

Yu, R., Zhang, Y., Gjessing, S., Xia, W., & Yang, K. (2013). Toward cloud-based vehicular networks with efficient resource management. *IEEE Network, 27*(5), 48-55. doi: 10.1109/MNET.2013.6616115

Yu, R. Y., Zhang, Y., Yi, L., Xie, S., Song, L., & Guizani, M. (2012, April). Secondary Users Cooperation in Cognitive Radio Networks: Balancing Sensing Accuracy and Efficiency. *IEEE Wireless Communications Magazine*, *19*(2), 2–9. doi:10.1109/MWC.2012.6189410

Zingirian, N., & Valenti, C. (2012). Sensor clouds for intelligent truck monitoring. *Proceedings of the IEEE Intelligent Vehicles Symposium (IV '12)*, 999–1004.

KEY TERMS AND DEFINITIONS

ITS (Intelligent Transportation Systems): A technology, application, or platform that improves the quality of transportation, or achieves other outcomes based on applications that monitor, manage or enhance transportation systems.

MANET (A Mobile Ad Hoc Network): Generally defined as a network that has many free or autonomous nodes, often composed of mobile devices or other mobile pieces that can arrange themselves in various ways and operate without strict top-down network administration.

V2I (Vehicle to-Infrastructure Communication): Vehicle-to-infrastructure (V2I) communication, collectively referred as V2X, is a wireless technology aimed at increasing road safety and improve traffic management.

V2V (Vehicle-to-Vehicle Communication): The wireless transmission of data between motor vehicles. The goal of V2V communication is to prevent accidents by allowing vehicles in transit to send position and speed data to one another over an ad hoc mesh network.

VANET (Vehicular Ad Hoc Network): A web of vehicles within wireless range, formed to allow for automated driving, provide additional information to the driver, and even entertainment.

Vehicular Cloud: A dynamic group of vehicles whose excess computing, communication, storage, and sensing resources are coordinated and dynamically allocated to authorized users.

ENDNOTES

1. CICAS, "Official website cooperative intersection collision avoidance systems (CICAS) project," http://www.dot.state.mn.us/guidestar/2006201 0/cicas.html.
2. VICS, "Official web site of vehicle information and communication system (VICS) Japan," http://www.vics.or.jp/english/vics/index.html.
3. PATH, Official Website Path, 2014, http://www.path.berkeley.edu/research.
4. National Highway Traffic Safety Administration, Traffic Safety Facts (2006), http://www-nrd.nhtsa.dot.gov.
5. Ouyang Jing, how can the Cloud Computing -IBM Tivoli general manager Alfred Zollar interview, November 2008 11.

Chapter 3
Challenges and Opportunities in Vehicular Cloud Computing

Zeinab E. Ahmed
University of Gezira, Sudan

Rashid A. Saeed
Sudan University of Science and Technology, Sudan

Amitava Mukherjee
Globsyn Business School, India

ABSTRACT

Vehicular ad-hoc networks (VANET) have become an important research area due to their ability to allow sharing resources among the users to carry out their application and provide services of transport and traffic management. VANET communication allows exchange of sensitive information among nearby vehicles such as condition of weather and road accidents in order to improve vehicle traffic efficiency through Intelligent Transportation Systems (ITS). Many technologies have been developed to enhance ITS. Recently, vehicular cloud computing (VCC) has been developed in order to overcome the drawbacks VANET. VCC technology provides low-cost services to vehicles and capable of managing road traffic efficiently by using the vehicular sources (such as internet) to make decisions and for storage. VCC is considered as the basis for improving and developing intelligent transportation systems. It plays a major role in people's lives due to its safety, security, trust, and comfort to passengers and drivers. This chapter investigates the vehicular cloud computing. The authors first concentrate on architectures. Then, they highlight applications and features provided by VCC. Additionally, they explain the challenges for VCC. Finally, the authors present opportunities and future for VCC.

DOI: 10.4018/978-1-5225-3981-0.ch003

INTRODUCTION

For the past few years, Intelligent Transportation Systems (ITS) attracted the attention of many researchers for the purpose of improving the traffic monitor, road safety, and signals utilization. Vehicular Ad-hoc Networks has been proposed as ITS environment due to its ability to managing traffic, enhance road safety using GPS information, computing power, and media. A VANET is communicating among the vehicles by using a wireless network to provide services of transport and traffic management (Whaiduzzaman, Sookhak, Gani, & Buyya, 2014). There are two components of VANET architecture: hardware and software. There are three types of communication in vehicular ad hoc network: 1.) vehicles-to-road infrastructure (V2R); 2.) vehicle-to-vehicle (V2V), and; 3.) vehicles-to-sensors (V2S) communication (Kumar, Singh, Bali, Misra, & Ullah, 2015; Eltahir, & Saeed, 2015). In the V2V, the vehicles communicate with another by using On Board Units (OBU), while the vehicles communicate in the V2R with infrastructure units like road side units (RSUs) (Eltahir, Saeed, Mukherjee, & Hasan, 2016).

Many technologies have been found to enhance Intelligent Transportation Systems (ITS). Some of the solutions to face the challenges of VANET were proposed such as Cloud computing and later than appearance Mobile Cloud Computing and Vehicular Cloud Computing (VCC) (Whaiduzzaman, Sookhak, Gani, & Buyya, 2014). In Cloud Computing, users share resources such as applications, location, and storage over the Internet. The increasing of mobile applications and mobile devices new technique appeared called mobile cloud computing to overcome shortages of Cloud Computing. Vehicular Cloud Computing (VCC) has a big effect on the ITS specially when using the resources of vehicles like computing power for instant decision making, the internet, storage, GPS, and sharing information on the cloud. VCC has many benefits such as low energy, real-time services of software, platforms, and infrastructure with QOS to passengers and drivers. And also VCC prove, better road safety, and secured intelligent urban traffic systems. The list of acronyms which appeared in this chapter is given in Table 1.

This chapter is organized as follows:

- Firstly, offers an overview of vehicular ad-hoc networks.
- Discusses the cloud computing and mobile cloud computing.
- Provides an overview of Vehicular Cloud Computing (VCC).
- Present and discuss about related work.
- Explains the architectures and organization of VCC.
- Focuses on the applications of the VCC.
- Discusses the challenges.
- Finally, talk about opportunities and the future for VCC.

VEHICULAR AD-HOC NETWORKS (VANET)

A VANET is a communications between the vehicles to exchange sensitive information like a condition of weather and road accidents to improve vehicle traffic efficiency through ITS. The main aim of VANET is to obtain the highest safety on the road (Sugumar, Rengarajan, & Jayakumar, 2016). Vehicles, Infrastructure Domain, and Road Side Unit (RSU) are the contents of VANET architecture (Bhoi, & Khilar, 2014). In a VANET, the communications between vehicles by using wireless equipment such as OBUs for

Table 1. List of acronyms

CaaS	Cooperation as a Service
CC	Cloud Computing
DaaS	Data as a service
DSRC	Dedicated Short Range Communication
INaaS	Information as a Service
ITS	Intelligent Transportation Systems
MCC	Mobile Cloud Computing
NaaS	Network as a Service
OBU	On Board Unit
PaaS	Platform as a Service
RSU	Road Side Unit
SaaS	Software as a Service
STaaS	Storage as a Service
V2I	Vehicle-to-Infrastructure
V2S	Vehicles-to-Sensors
V2V	Vehicle-to-Vehicle
VANET	Vehicular Ad-Hoc Networks
VCC	Vehicular Cloud Computing
VCN	Vehicular Cloud Network
WAVE	Wireless Access in Vehicular Environment

V2V and RSUs for V2R. This communication happens when the Dedicated Short Range Communication (DSRC), which is a standard, developed by the USA, and enabled IEEE 802.11p. A VANET classified into two types based on network topologies: infrastructure-less (ad-hoc) and infrastructure-based (Cushman, Rawat, Chen, & Yang, 2016). The V2V communication is transmitting data by using broadcasting or multicasting. The broadcasting is classified into naive and Intelligent, naive broadcasting which generate the collision, but the collision is decreased when using intelligent broadcasting. In a VANET to avoid accidents or congested, the vehicle broadcasts messages called beacons (Rajput, Abbas, & Oh, 2016) contain very important information like location, direction and a speed of it. This beacon makes a VANET is attractive areas for engineers and researcher, many papers focus on a VANET application and characteristics. The VANET had many important characteristics like high mobility, a dynamic environment, and relatively low antenna heights on the vehicles and RSU (Viriyasitavatm, Boban, Tsai, & Vasilakos, 2015). These characteristics make the VANET facing more issues and challenges. However, a VANET has some shortcomings, therefore new technology appeared called a cloud computing.

CLOUD COMPUTING (CC)

Nowadays, cloud computing (CC) is an active area because delivering services to users over the internet through sharing resources such as applications and storage. In the CC, when tasks sharing by users, it can carry out their tasks fast and efficiently as well as decrease the computation time (Bajpai & Singh, 2016). Addition, the providers of cloud service manage a network infrastructure for collection the computing power enormous from a number of servers. Cloud computing has some main characteristics like (Verma, & Kaushal, 2011; Habib, Hauke, Ries, & Muhlhauser, 2012):

- Multi-tenancy or sharing of resources
- On-demand self-service
- Ubiquitous network access
- Rapid elasticity
- Measured service

CC has several features which provided advantages for a business domain such as No up-front investment, highly scalable, lowering operating cost, easy access and reducing business risks and maintenance expenses (Zhang, Cheng, & Boutaba, 2010). Moreover, it has many advantages such as connecting services, shared architecture, flexible nature, and metering architecture (Hindia, Reza, Dakkak, Awang Nor, & Noordin, 2014). The cloud computing architecture separates into 4 layers: the hardware layer, the infrastructure layer, the platform layer and the application layer, demonstrated in Figure 1. CC is categorized into four models: public, private, hybrid and community. Each of these models is divided into three service models: Software as a Service (SaaS), Platform as a Service (PaaS), and Infrastructure as a Service (IaaS) (Cardoso, Moreira, & Somoes, 2014).

MOBILE CLOUD COMPUTING (MCC)

In mid-2007, a new technique was appeared which called mobile cloud computing. and it become attractive area for business domain because of a spread use of a mobile applications with low cost. When the processing and storage for data was occurring outside of the mobile device, this infrastructure called mobile cloud computing. The applications moved from mobile phones to the cloud, to share applications, data storage and mobile computing among the mobile devices (Singh, Kaur, & Sandhu, 2015). So, the Mobile cloud computing using in a wide range of application like crowd computing, natural language processing, sharing GPS, image processing, sensor data applications, querying, sharing Internet access, and multimedia search (Fernando, Loke, & Rahayu, 2013).

Mobile cloud computing has several advantages: a) all users anywhere and anytime; can be access to services; b) the services provide information to users such as location and context; c) Improves processing power and capacity of data storage; d) extending the battery lifetime; e) Mobile computing has found many solutions to overcome cloud computing problems (Singh, Kaur, & Sandhu, 2015; Shravanthi, & Guruprasad, 2014; Dinh, Lee, Niyatio, & Wang, 2013). Most of the mobile applications, such as mobile commerce, mobile learning, and mobile healthcare, benefited from the MNC advantages. The architecture of mobile cloud computing classified based on data centers layer and the cloud service standard model that includes Platform as a Service (PaaS), Infrastructure as a Service (IaaS), and Software as a Service (SaaS) (Singh, Kaur, & Sandhu, 2015; Fernando, Loke, & Rahayu, 2013). Although it has several advantages and services provided to users, it faces many challenges like computing offloads, low bandwidth, quality of services, and security for mobile users.

Figure 1. Cloud computing architecture

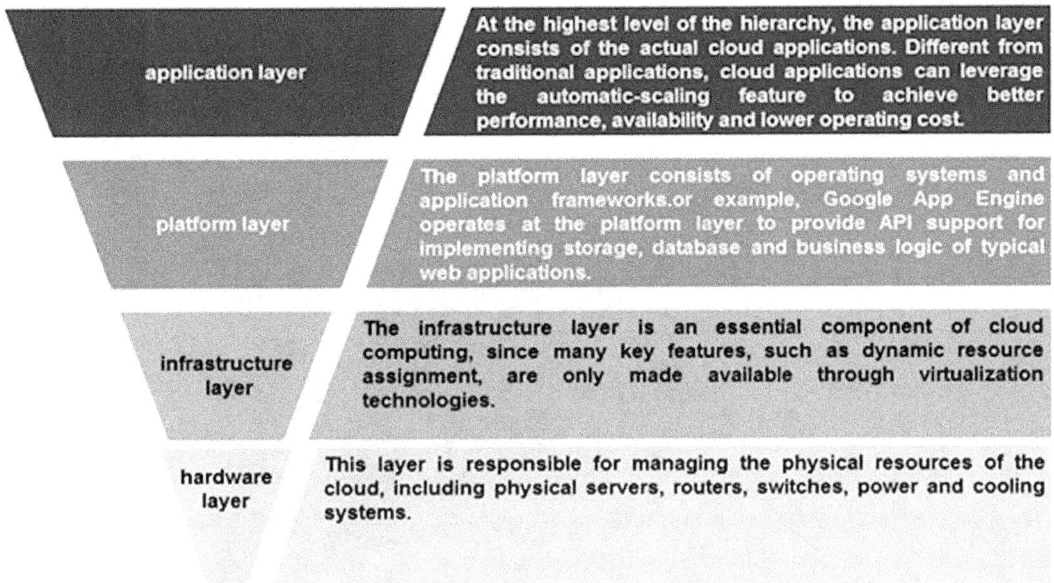

VEHICULAR CLOUD COMPUTING (VCC)

Vehicular cloud computing (VCC) is an extension of mobile cloud computing (MCC). In VCC vehicles, the resources and services could arrive in real time for information from anywhere. This makes passengers and drivers access numbers of new applications to provide various services (Othman, Madani, & Khan, 2014; Ahmad, Noor, Ali, & Qureshi, 2016). So, VCC considers the basis for improving and developing intelligent transportation systems and rich environment for researchers. Additionally, it plays a major role in people's lives due to providing safety, security, trust, and comfort to passengers and drivers. The basic objective of VCC is to provide low-cost services to drivers, and to reduce accidents, congestion of traffic. The traditional cloud computing provided services such as software, storage, and computing resources, but in the vehicular cloud appeared other services. Figure 2 shows the vehicular cloud computing, which exchanges information between the vehicles or between cloud data center, and vehicles for the vehicular through RSUs in order to computing or storing.

The VCC services classified into three types: 1.) Network-based services; 2.) Sensing-based services, and; 3.) Cooperation-based services (Mekki, Jabri, Rachedi, & ben Jemaa, 2016).

- **Network-Based Services or (NaaS):** Where some vehicles on the road have Internet, and it can be used by others vehicles which that haven't Internet.
- **Cooperation as a Service (CaaS):** When the driver and passengers access to the applications and use the services like a road condition, and traffic information of this type (Sharma, & Kaur, 2015).
- **Sensing-Based Services:** Where vehicles can share sensing information between them to improve the knowledge of drivers (Mekki, Jabri, Rachedi, & ben Jemaa, 2016).

Figure 2. Vehicular Cloud Computing

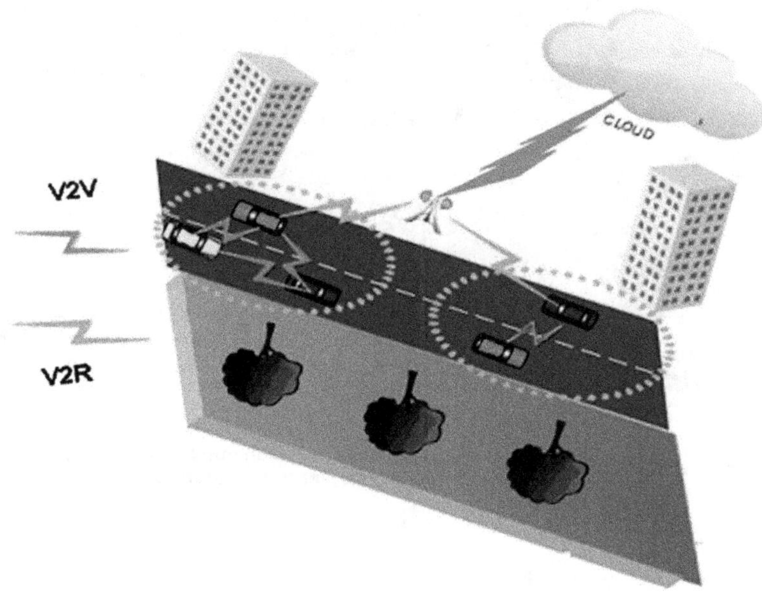

The vehicular cloud infrastructure provides services to several domains such as scientific application, social networking, education, and business etc. Vehicular Clouds provide services, like Infrastructure as a Service (IaaS), Platform as a Service (PaaS), Application as a Service (AaaS), Software as a Service (SaaS), and Storage as a Service (STaaS), to vehicle when connected via OBUs to get unlimited storage and computing power (Ahmad, Noor, Ali, & Qureshi, 2016). The VCC improves the road traffic management and or by reducing the risk of life, cost and time.

RELATED WORK

Recently, Baby et al. proposed a new method known as Vehicular Cloud for Roadside scenarios (VCR). VCR architecture has separated into two types: the public cloud and private cloud communication. The main aim is to provide safety and non-safety services in vehicular applications through the public and private vehicular cloud to fulfill max benefit (Baby, Sabareesh, Saravanaguru, & Thangavelu, 2013). Chang, Yao-Chung, et al. used Software-Defined Networking (SDN) architecture to assist a vehicular cloud serving system and or by achieving highly efficient services. The design based on SDNBroker system which is an SDN application to schedules resources using the linear programming algorithm and improves network routing using Dijkstra's routing (Chang, Chen, Ma, & Chiu, 2015).

An intelligent transport system (ITS) had a high impact on people's lives because it provide several applications, safety, and other information to the passenger and drivers as well as manage traffic of vehicles. So, the author has (Meneguette, 2016) proposed VehIcular Cloud Transport Management (VIC-TiM) to a management a big city as well as to provide mechanisms for storage of information and allow carrying out heterogeneous communication among the several vehicles. Azizian et al (2016) proposed a new solution to solve the service delivery problem more efficiently by using VC based on transmission scheduling methods. Distributed clustering algorithm (DHCV) was used to fulfill this proposed through creating a cloud of vehicles have fixed size and regulate vehicles inside cloud according to mobility (Azizan, Cherkaoui, & Hafid, 2016). After that applied transmission scheduling methods to improve delay and throughput for the service delivery.

In (Garai, Rekhis, & Boudriga, 2015) the authors proposed a cloud Communication-as-a-Service (CaaS) have three aims in their: firstly, provide a continuous communication to vehicles when the move outside area uncovered through Roadside units, secondly overcome from resource limit, and lastly supply high Quality of Service (QoS). The architecture was separated into three layers based on a tree topology: Vehicular Cloudlet (V-Cloudlet), Roadside Cloudlet (R-Cloudlet), and central cloud (CC).

Merging vehicular social networks and mobile wireless communications became future of networks, so, Li, Ting, et al. proposed a new method to support marketers known as On Selecting Vehicles as Recommenders for Vehicular Social Networks (SV-VSNs) (Li, Zhao, Liu, & Huang, 2017). Three algorithms to improve were present efficiencies for marketers based on benefit factor, coverage factor, and geographical track to select vehicles in vehicular social networks. The obtained result was effective when applied in real traffic in some cities.

Recently, location-based services (LBS) use become widespread to assist people and make life very easy, but facing some issues such as privacy. To solve this problem, Zhu et al. (2016) suggests a new scheme called efficient and privacy-preserving LBS query scheme in the outsourced cloud (EPQ) (Zhu, Lu, Huang, Chen, & Li, 2016). The EPQ depends on the encryption mechanism especially spatial range query algorithm over ciphertext (SRQC). EPQ proved excellent for facing the most security threats, and adding an effective performance in the cloud server.

Road traffic complexity leads to one of the important issues which are a collision. Many research puts approaches to solving this issue, in (Riaz, & Niazi, 2016) offers a comprehensive study for a survey of vehicular cyber-physical systems (VCPS) from the collision-avoidance perspective. In addition to aid researchers through giving solutions to avoid a collision and understanding communication of vehicular cyber-physical systems.

VEHICULAR CLOUD COMPUTING ARCHITECTURES

The architecture of Vehicular cloud computing was based on three layers, which are: inside-vehicle layer, cloud computing layer and the communication layer, as shown in Figure 3.

- **Inside-Vehicle Layer:** Also called an on-board layer. In an Inside- vehicle layer, before establishing communication among the vehicles, firstly the vehicle sense the environment, road condition and collecting information inside the car or other parameters by using a number of sensors such as vehicles' internal sensors and smartphone sensors (Ghafoor, Mohammad, & Lloret, 2016). The collated information either store on the cloud or considered as input for other software programs in the application layer.
- **Communication Layer:** The communication layer is the next layer used to enable communication between vehicles and vehicular clouds through cellular communication devices such as 3G or 4G, Wi-Fi, WiMAX, Wireless Access in Vehicular Environment (WAVE), IEEE 802.11p or Dedicated Short Range Communication (DSRC) (Mekki, Jabri, Rachedi, & ben Jemaa, 2016). The communication on this layer is classified into two: vehicle-to-vehicle (V2V) and vehicle-to-

infrastructure (V2I). The connection among the vehicles needs an equipment such as On-Board Units (OBUs) and GPSs to make the internet available on moving devices. In V2V, communicate between vehicles is directly established to provide traffic safety and related applications for the passengers and drivers. Unlike the V2I where the communication is indirect among vehicles, infrastructures and the cloud to exchange Information over wireless networks like satellite, 3G (Whaiduzzaman, Sookhak, Gani, & Buyya, 2014).

- **Cloud Computing Layer:** The last layer called cloud computing layer. It is split into three sub-layers:
    - Cloud services.
    - Cloud applications.
    - Cloud infrastructure.

Cloud infrastructure consists of computation cloud and cloud storage. Cloud storage, which is responsible for storing all information that is collected in the on-board layer (Sharma, & Kaur, 2015). Cloud computation plays an important role in improving network performance by using computational tasks such as the health and behavior of the driver in cloud storage. Cloud applications consist of many applications and services used by passenger and driver these applications like health, human activity, and environmental recognition. Figure 4 illustrates most details for vehicular cloud computing architecture.

Several services such as Infrastructure as a Service (IaaS), Platform as a Service (PaaS), Application as a Service (AaaS), Software as a Service (SaaS), Network as a Service (NaaS), Information as a Service (INaaS), Data as a service (DaaS), and Storage as a Service (STaaS), consider basic parts of cloud services. Several vehicles on the road, one of them have internet and another doesn't have the internet. The vehicles have internet can provide access the Internet for other vehicles that need Internet this process know as Network as a Service (NaaS) (Sharma & Kaur, 2015).

The services have permission to exchange information among the vehicles in the same cloud. Often the drivers want the important kind of information like news of large events, advance warnings, and road conditions. The Information as a Service (INaaS) response for supplying these information to a driver.

Figure 3. Layers of the architecture of vehicular cloud computing

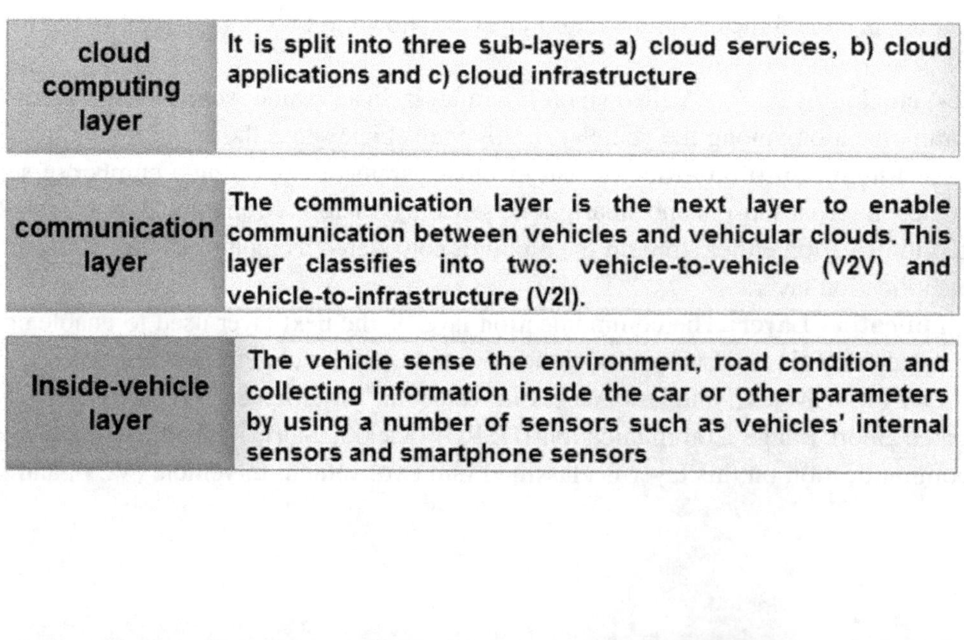

When the passenger and drivers on the road used all hardware to run the applications at the same moment, this leads to need an additional storage. So, some vehicles have a higher capacity for storage and it can provide the repository for storage to other vehicles that have low storage capability. It knows virtual network hard-disk and also called Storage as a Service (STaaS) (Mershad, & Artail, 2013).

ORGANIZATION OF VCC INFRASTRUCTURE

Vehicular cloud computing can classify into two types: static VC and dynamic VC (Ghafoor, Mohammed, & Lloret, 2016; Hussain, Son, Eun, Kim, & Oh, 2012; Olariu, Khalil, & Abuelela, 2011).

1. **Static VC:** Earlier, people spend a lot of time in shopping, airport, work, and hospital and when parking their vehicles. Most companies are thinking to take benefit from advantages of parking because those vehicles consider idle computing resources. Static VC creates center data storage when combining between a computer cluster, storage resources, and computational power of the participating vehicles. Figure 5 shown static VCC.
2. **Dynamic VC:** Dynamic formation of VC because of the high mobility of the vehicles and the speedily change among the networks. One of the vehicles on the cloud, known as cloud head, is responsible to invite all near vehicular to join for formation dynamic VC, as shown Figure 6.

APPLICATIONS OF VEHICULAR CLOUD COMPUTING

Because of sharing resources among the vehicular and RUS with clouds, the VCC provides a vast range of applications like in an airports as a data center, traffic management, road safety message, managing

Figure 4. Vehicular cloud computing architecture

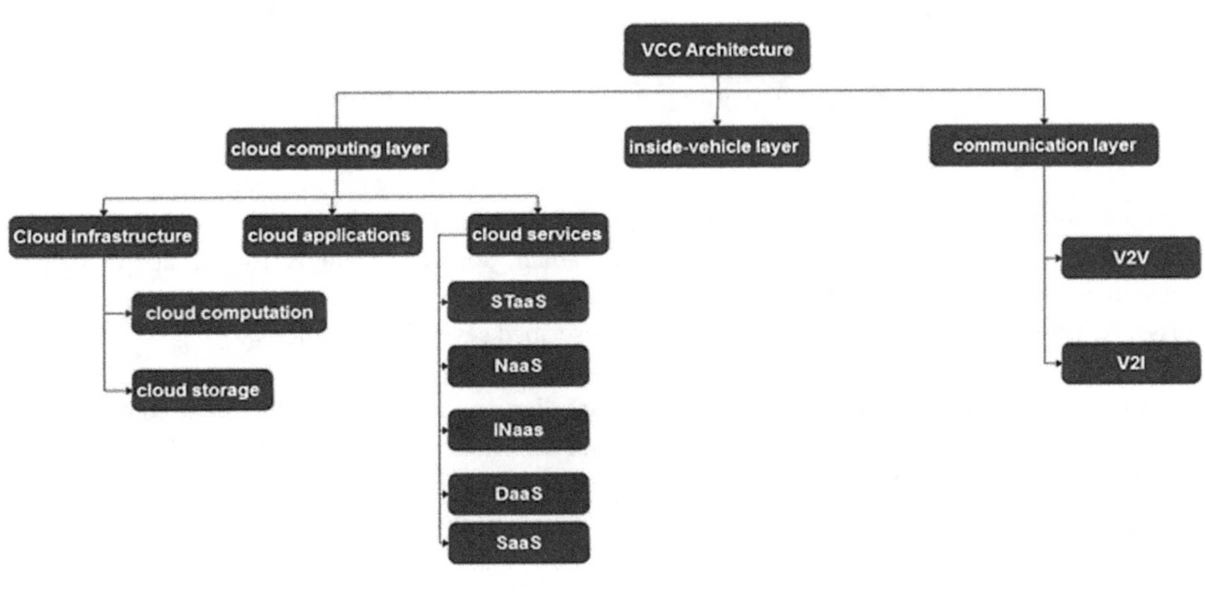

Figure 5. Static VCC

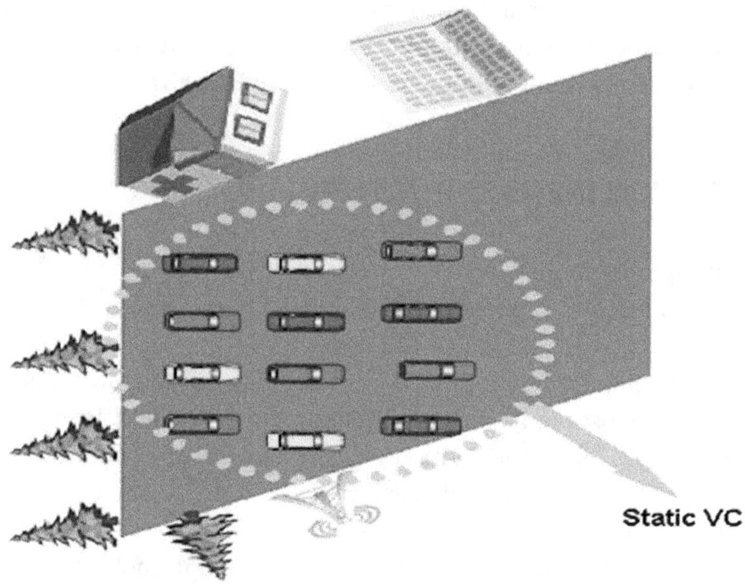

Figure 6. Dynamic VCC

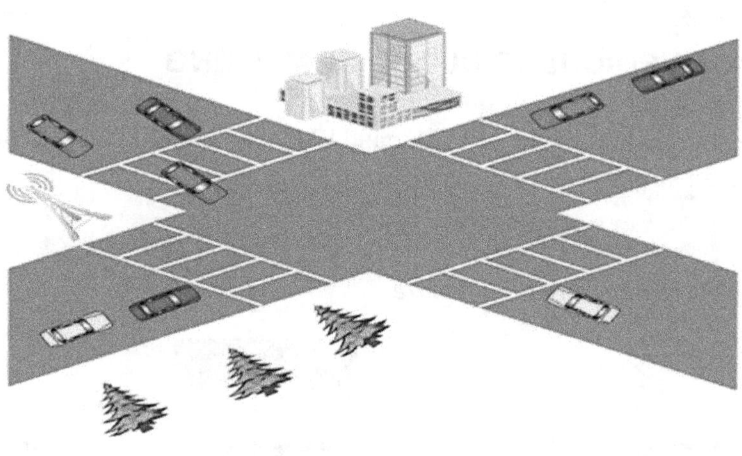

parking facilities, real-time navigation, vehicle maintenance, accident alerts at intersections, parking lot data cloud, and managing evacuation etc. (Whaiduzzaman, Sookhak, Gani, & Buyya, 2014;Ghafoor, Mohammed, & Lloret, 2016; Yan, Wen, Olariu, & Weigle, 2013; Gu, Zeng, & Guo, 2013; Olariu, Eltoweissy, & Younis, 2011). The following paragraph described several applications:

1. **Road Safety Message:** It is one of the important alerts on road because the VCC contributes to road safety. Modern vehicles have a sensor, to monitor road and provide road safety for drivers and passengers, to query the state of roads such as risk, flooding areas, overcrowding, temperature, and speed, to evaluate the situation of other vehicles and alert drivers.

2. **Traffic Management:** The result of the increasing number of vehicles, the traffic management plays an important role on the road to find the best solution for several problems which face drivers and passengers such as wastes the valuable time and energy of human, threatens the health of citizens and needs the huge computational effort to be solved. One of the solutions which provide to manage the road, drivers receives reports about status for traffic (e.g., congestion) from VCs. Reduce frequent congestion is one of the important application types for traffic management. In the congestion, some drivers want to transform the road (Yan, Wen, Olariu, & Weigle, 2013). This decision depends on the last information on the vehicular and itcloud provide an efficient solution to the drivers. This information being available after calculating the impact of the local road and the cause of the congestion in the traffic flow.
3. **Real-Time Navigation:** Navigation is a static geographic map for vehicles in traditional networks, but the VCC has a virtual-reality application which the drivers can move freely without hurdles. To provide a real-time navigation for vehicles. the 3D space for interaction with the virtual environment was used.
4. **Management Parking:** In the populated areas, it is difficult for drivers to find parking space. The VCC allows drivers to book parking easily supplying near and suitable space because all information related locations parking found in the cloud.
5. **Managing Evacuation:** The disaster authority uses the model the road to monitor traffic to execute an evacuation. Therefore, evacuation events can be divided into cases where prior notice of an imminent event is provided. In the VCC, vehicles and vehicular cloud are taking part of the evacuation process to make decisions on the current situation.
6. **Data Center Configuration:** In mall, hospital, work, and the airport, the people are parking their vehicles for several hours or days. During this period, the vehicles are considered inefficient resources. Can benefit from parking to build a data center. The vehicular takes part of the cloud through connected to the Internet by cable. But one of the important challenges of parking data center structure is dynamic, because of time for arrival and spends per vehicles in parking unknown.
7. **Improving Traffic Signals:** Traffic signals have high significant on the road to assign a length of signal cycle and green phase lengths. Thus, VCs can maximize the signal system performance by making dynamic use of a vehicular network.

VEHICULAR CLOUD COMPUTING CHALLENGES

In the vehicular cloud computing, the mobility of nodes is very high and topology changes in the network frequently so appeared issues and challenges facing the VCC such as message confidentiality, trust management, securing vehicular communication, authentication, secure location information, the safety of messages, and the interoperability of different clouds. In this section, all these challenges can be discussed and concentrate on the issues correlated with ITS applications like security, traffic management, navigation systems, and congestion detection (Whaiduzzaman, Sookhak, Gani, & Buyya, 2014;Gu, Zeng, & Guo, 2013; Chaqfeh, Mohamed, Jawhar, & Wu, 2016; Aloqaily, Kantarci, & Mouftah, 2015; Ahmad, Kazim, Adnane, & Awad, 2015; Karagiannis, Altintas, Ekici, Heijenk, Jarupan, Lin, & Weil, 2011).

1. **Threats for Vehicular Cloud Computing:** All VCC components: vehicles, passengers, messages, wireless communication, and architectures need security against attackers. In vehicles, attackers attempting to access application unit and enter malware in it, while wireless communication considered between the vehicles or RSUs via V2V and V2R communication the best environment for the attackers, because it contains some vulnerabilities. The wireless communication threats classified into:
    a. **Denial of Service (DoS):** It blocks communication by using reject any members to forward important messages to other vehicles or the cloud.
    b. **Data Tempering (Modification):** In this type, the attacker objects the messages when the routes to other vehicles or RSUs and after that can modify and alter it.
    c. **Jamming:** DoS attack to make jamming in the channel for wireless communication which carries the messages.

There are also several threats like repudiation, information disclosure, and sybil attack. To provide secure vehicular cloud computing, communication takes considering requirements for security like authentication, integrity, and confidentiality.

2. **Authentication, Privacy, and Liability:** In VCC, the communication depends on the integrity of messages and authentication of the users. To increased authenticate between RSUs and vehicles in the vehicular networks, the vehicles used aliases to deception the attackers for not track their communication and activities. Also, exchange information between the vehicles requires being privacy and trust. The privacy is one of the most important challenges in VCC. Many researchers are working to find methods and solutions used to protect the privacy of the vehicles in a vehicular network. The passengers and drivers have controlled exchange information and determine, what the information open on networks, and what the information keeps private and can't broadcast on the networks.
3. **Service Delay and Location:** Because to the topology in the network changes quickly, the time and location play an important role on data packets sent by applications for vehicular networking. In VCC, the most applications depend on the information on the current location of a vehicle such as collision, congestion avoidance, and traffic status. The vehicular networks use devices like GPS, filtering, inter-cell position, and radar to check the location information. Many researchers proposed several methods to check the location information. These methods can classify into three types: active location, active location and general location (Whaiduzzaman, Sookhak, Gani, & Buyya, 2014).

In vehicular cloud computing, the delay can impact on price for service. So, the passengers and drivers are interested about service delay, because the increase of service delay leads up to increase the price for service.

4. **Confidentiality, Integrity, and Trust for Messages:** At VCC, messages split into four types: the first one using to send alerts or warning messages called short message, second known as media message, utilize when the vehicle needs services from other cloud or vehicles, Priority message is third one employed to end the warning messages, the last one acknowledge message after deliv-

ery of messages send these to confirm access. Most applications for a vehicular network focused verification and trust to provide a secure communication. Confidentiality is an important issue, all messages in the network can't read by an unauthorized user, but after encrypting the messages and allowing only access to authorized users. Therefore, the RSU registers all vehicles and users that enter the domain to increase confidentiality by using secure and encrypted communication. Furthermore, to safeguard the networks from security threats such as denial of service, forgery and jamming attack, must verify the integrity for receiving messages.

5. **Securing Vehicular Communication:** In the environment of a vehicular cloud, from the important providing secure communications to avoid attackers and reduce security holes like traffic tampering, privacy violations, forging messages, and preventing communication by utilizing cryptographic algorithms (Ahmad, Kazim, & Adnane, 2015; Falchetti, Azurdia-Meza, & Cespedes, 2015). The big problem facing the cryptographic algorithm is key management, so found a method for the solution this problem. The Vehicular Public Key Infrastructure (VPKI) is one of the methods key management to guarantee confidentiality and integrity of the message in vehicular networks.

OPPORTUNITIES AND FUTURE FOR VCC

The vehicular cloud computing plays a major role in people's lives due to traffic management and providing safety, security, trust, and comfort of passengers and drivers, so considered very rich environment for researchers. It considers the basis for improving and developing intelligent transportation systems. However, the development and improvement process still constrained by several factors such as challenges and issues for VCC. Several applications and services appeared in the last days result from exploiting the resources by the vehicular cloud. Future of VCC in next days benefits from underutilized resource and exploits it to the maximum extent such as to benefits, capacities of storage and computation which owned by modern vehicles, and time which wasted in the parking. So, some companies are thinking to rent resource from the parking to carry out some application or task.

In (Dressler, Handle, & Sommer, 2014), the authors present a new method, known Virtual Cord Protocol (VCP), depends on the benefit from parked vehicles to create a cloud for storing information and to provide network communication between the moving vehicles and the vehicle storing the requested data. The inter-domain routing techniques used to enable dynamic cloud services and to achieve the communication between the moving vehicles and cloud. W He, et al proposed a new architecture for IoT-based vehicular data clouds, by integrating several devices, cloud computing, and IOT to Share resources and exchange information among the passengers, vehicles, and roadside infrastructure (He, Yan, & Da Xu, 2014). The (He, Yan, & Da Xu, 2014) study and face two vehicular data cloud services: the first one is intelligent parking cloud service and the second is the mining vehicular maintenance data service. Although finding several challenges this method provides enormous opportunities for technology in the vehicle industry.

Wan et al (2014) proposes a context-aware architecture with mobile cloud support and two crucial service components, namely cloud-assisted Context-aware Vehicular Cyer-physical systems (CVCs). This paper studies context-aware dynamic parking service as a case study to improve QoS and the performance of CVCs. Addition, discuss the challenges and find potential solutions (Wan, Zhang, Zhao, Yang, & Lloret, 2014). It will consider this attractive field for researchers in the near future.

To benefit from sensors such as map, radar, and lidar the authors proposed a new method to detect the vehicles on the way or parked on the roadside to provide the best motion, the easiest method to pass the parked vehicle, and assuring safety for drivers and passengers (Mei, Nagasaka, Okumura, & Prokhorov, 2015). Additionally, assist drivers to make the decision to choose the smooth and safe path to avoid harassing from other vehicles.

The traffic accidents and traffic flow issues are major challenges for VCC, maybe occur when the vehicles parked on the way illegally. So, the authors presented the new system to monitor traffic and detect traffic flow to avoid accidents by using cumulative dual foreground differences (Wahyono, & Jo, 2017). The results appeared that the method is efficient and can be considered as part of the intelligent traffic monitoring system, but possibly generate false results when producing a noisy foreground image.

The nature VCC has open communication medium, fast topology changes, and dynamic. Thus, one of the key challenges is exchange information due to nature VCC can be fake and lead to sharing false information between drivers and passengers. Sharma et al (2015) presented a new method by using Elliptic Curve Cryptography (ECC) to provide security for communication to guarantee authentication, integrity, confidentiality, and privacy between sender and receiver (Sharma, Bali, & Kaur, 2015). The ECC has several advantages to against various attacks like spoofing attack, replay attack.

Huang, Cheng, et al presented new method, Privacy-Preserving Trust-Based Verifiable Vehicular Cloud Computing scheme (PTVC), to improve safety for traffic and offering services to passengers and drivers (Huang, Lu, Zhu, Hu, & Lin, 2016). The mechanism PTVC system merges between the best advantages of VCC, techniques of verifying, and requirements of privacy, to select the trustworthy vehicle among the vehicles to create a vehicular cloud. The PTVC scheme proves is effective, secure and robust against several attacks.

CONCLUSION

In this chapter, we first talked about vehicular ad-hoc, cloud computing, mobile cloud computing, and we listed all the related to vehicular cloud computing such as architecture, organization, application, challenge, and future. Firstly, we discussed architecture for VCC and then a detailed explanation of the tasks of each layer and clarification of types of services. After that, we talked about the type of organization for VCC infrastructure. Because of sharing resources among the vehicular and RUS with clouds, we highlight for applications providing by VCC like an airport as a data center, traffic management, road safety message, managing parking facilities, real-time navigation, vehicle maintenance, and accident alerts at intersections, etc. Addition, we will survey the issues and challenges facing the vehicular cloud computing resulted from high mobility and the fast topology changes in the network. Finally, we discussed opportunities and the future for VCC that can be the basis to open research directions.

REFERENCES

Ahmad, F., Kazim, M., & Adnane, A. (2015). Vehicular Cloud Networks: Architecture and Security. In Guide to Security Assurance for Cloud Computing (pp. 211-226). Springer International Publishing.

Ahmad, F., Kazim, M., Adnane, A., & Awad, A. (2015, December). Vehicular cloud networks: architecture, applications, and security issues. In *Utility and Cloud Computing (UCC), 2015 IEEE/ACM 8th International Conference on* (pp. 571-576). IEEE.

Ahmad, I., Noor, R. M., Ali, I., & Qureshi, M. A. (2016, September). The Role of Vehicular Cloud Computing in Road Traffic Management: A Survey. In *International Conference on Future Intelligent Vehicular Technologies* (pp. 123-131). Springer.

Aloqaily, M., Kantarci, B., & Mouftah, H. T. (2015, November). Vehicular Clouds: State of the art, challenges and future directions. In *Applied Electrical Engineering and Computing Technologies (AEECT), 2015 IEEE Jordan Conference on* (pp. 1-6). IEEE.

Azizian, M., Cherkaoui, S., & Hafid, A. (2016). An Optimized Flow Allocation in Vehicular Cloud. *IEEE Access: Practical Innovations, Open Solutions*, *4*, 6766–6779. doi:10.1109/ACCESS.2016.2615323

Baby, D., Sabareesh, R. D., Saravanaguru, R. A. K., & Thangavelu, A. (2013). VCR: vehicular cloud for road side scenarios. In Advances in Computing and Information Technology (pp. 541-552). Springer Berlin Heidelberg. doi:10.1007/978-3-642-31600-5_53

Bajpai, A., & Singh, S. (2016). A survey on Security Analysis in Cloud computing. In *Proceedings of the International Conference on Recent Cognizance in Wireless Communication & Image Processing* (pp. 249-262). Springer India.

Bhoi, S. K., & Khilar, P. M. (2014). Vehicular communication: A survey. *IET Networks*, *3*(3), 204–217. doi:10.1049/iet-net.2013.0065

Cardoso, A., Moreira, F., & Simões, P. (2014). A Survey of Cloud Computing Migration Issues and Frameworks. In *New Perspectives in Information Systems and Technologies* (Vol. 1, pp. 161–170). Springer International Publishing. doi:10.1007/978-3-319-05951-8_16

Chang, Y. C., Chen, J. L., Ma, Y. W., & Chiu, P. S. (2015, December). Vehicular Cloud Serving Systems with Software-Defined Networking. In *International Conference on Internet of Vehicles* (pp. 58-67). Springer International Publishing 10.1007/978-3-319-27293-1_6

Chaqfeh, M., Mohamed, N., Jawhar, I., & Wu, J. (2016, December). Vehicular Cloud data collection for Intelligent Transportation Systems. In Smart Cloud Networks & Systems (SCNS) (pp. 1-6). IEEE. doi:10.1109/SCNS.2016.7870555

Cushman, I. J., Rawat, D. B., Chen, L., & Yang, Q. (2016, August). Performance Evaluation of Vehicular Ad Hoc Networks for Rapid Response Traffic Information Delivery. In *International Conference on Wireless Algorithms, Systems, and Applications* (pp. 571-579). Springer International Publishing 10.1007/978-3-319-42836-9_50

Dinh, H. T., Lee, C., Niyato, D., & Wang, P. (2013). A survey of mobile cloud computing: Architecture, applications, and approaches. *Wireless Communications and Mobile Computing*, *13*(18), 1587–1611. doi:10.1002/wcm.1203

Dressler, F., Handle, P., & Sommer, C. (2014, August). Towards a vehicular cloud-using parked vehicles as a temporary network and storage infrastructure. In *Proceedings of the 2014 ACM international workshop on Wireless and mobile technologies for smart cities* (pp. 11-18). ACM. 10.1145/2633661.2633671

Eltahir, A. A., & Saeed, R. A. (2015, September). Performance Evaluation of an Enhanced Hybrid Wireless Mesh Protocol (E-HWMP) Protocol for VANET. In *Computing, Control, Networking, Electronics and Embedded Systems Engineering (ICCNEEE), 2015 International Conference on* (pp. 95-100). IEEE.

Eltahir, A. A., Saeed, R. A., Mukherjee, A., & Hasan, M. K. (2016). Evaluation and analysis of an enhanced hybrid wireless mesh protocol for vehicular ad hoc network. *EURASIP Journal on Wireless Communications and Networking*, (1): 1–11.

Falchetti, A., Azurdia-Meza, C., & Cespedes, S. (2015, October). Vehicular cloud computing in the dawn of 5G. In *Electrical, Electronics Engineering, Information and Communication Technologies (CHILECON), 2015 CHILEAN Conference on* (pp. 301-305). IEEE.

Fernando, N., Loke, S. W., & Rahayu, W. (2013). Mobile cloud computing: A survey. *Future Generation Computer Systems*, *29*(1), 84–106. doi:10.1016/j.future.2012.05.023

Garai, M., Rekhis, S., & Boudriga, N. (2015, July). Communication as a service for cloud vanets. In *Computers and Communication (ISCC), 2015 IEEE Symposium on* (pp. 371-377). IEEE 10.1109/ISCC.2015.7405543

Ghafoor, K. Z., Mohammed, M. A., & Lloret, J. L. (2016). Vehicular Cloud Computing: Trends and Challenges. *ZANCO Journal of Pure and Applied Sciences*, *28*(3), 67–77.

Gu, L., Zeng, D., & Guo, S. (2013, December). Vehicular cloud computing: A survey. In Globecom Workshops (GC Wkshps), 2013 IEEE (pp. 403-407). IEEE.

Habib, S. M., Hauke, S., Ries, S., & Mühlhäuser, M. (2012). Trust as a facilitator in cloud computing: A survey. *Journal of Cloud Computing: Advances, Systems, and Applications*, *1*(1), 19. doi:10.1186/2192-113X-1-19

He, W., Yan, G., & Da Xu, L. (2014). Developing vehicular data cloud services in the IoT environment. *IEEE Transactions on Industrial Informatics*, *10*(2), 1587–1595. doi:10.1109/TII.2014.2299233

Hindia, M. N., Reza, A. W., Dakkak, O., Awang Nor, S., & Noordin, K. A. (2014). Cloud computing applications and platforms: a survey (pp. 515–524). Academic Press.

Huang, C., Lu, R., Zhu, H., Hu, H., & Lin, X. (2016, December). PTVC: *Achieving Privacy-Preserving Trust-Based Verifiable Vehicular Cloud Computing. In Global Communications Conference (GLOBECOM)*, 2016 *IEEE* (pp. 1-6). IEEE.

Hussain, R., Son, J., Eun, H., Kim, S., & Oh, H. (2012, December). Rethinking vehicular communications: Merging VANET with cloud computing. In *Cloud Computing Technology and Science (CloudCom), 2012 IEEE 4th International Conference on* (pp. 606-609). IEEE.

Karagiannis, G., Altintas, O., Ekici, E., Heijenk, G., Jarupan, B., Lin, K., & Weil, T. (2011). Vehicular networking: A survey and tutorial on requirements, architectures, challenges, standards, and solutions. *IEEE Communications Surveys and Tutorials*, *13*(4), 584–616. doi:10.1109/SURV.2011.061411.00019

Kumar, N., Singh, J. P., Bali, R. S., Misra, S., & Ullah, S. (2015). An intelligent clustering scheme for distributed intrusion detection in vehicular cloud computing. *Cluster Computing*, 1263-1283.

Li, T., Zhao, M., Liu, A., & Huang, C. (2017). On Selecting Vehicles as Recommenders for Vehicular Social Networks. *IEEE Access: Practical Innovations, Open Solutions*.

Mei, X., Nagasaka, N., Okumura, B., & Prokhorov, D. (2015, June). Detection and motion planning for roadside parked vehicles at long distance. In *Intelligent Vehicles Symposium (IV)*, 2015 *IEEE* (pp. 412-418). IEEE. 10.1109/IVS.2015.7225720

Mekki, T., Jabri, I., Rachedi, A., & Ben Jemaa, M. (2016). Vehicular cloud networks: Challenges, architectures, and future directions. *Vehicular Communications*.

Meneguette, R. I. (2016). A Vehicular Cloud-Based Framework for the Intelligent Transport Management of Big Cities. *International Journal of Distributed Sensor Networks*, *12*(5), 8198597. doi:10.1155/2016/8198597

Mershad, K., & Artail, H. (2013). Finding a STAR in a Vehicular Cloud. *IEEE Intelligent Transportation Systems Magazine*, *5*(2), 55-68.

Olariu, S., Eltoweissy, M., & Younis, M. (2011). Towards autonomous vehicular clouds. EAI Endorsed Trans. *Mobile Communications Applications*, *1*(1), e2. doi:10.4108/icst.trans.mca.2011.e2

Olariu, S., Khalil, I., & Abuelela, M. (2011). Taking VANET to the clouds. *International Journal of Pervasive Computing and Communications*, *7*(1), 7–21. doi:10.1108/17427371111123577

Othman, M., Madani, S. A., & Khan, S. U. (2014). A survey of mobile cloud computing application models. *IEEE Communications Surveys and Tutorials*, *16*(1), 393–413. doi:10.1109/SURV.2013.062613.00160

Rajput, U., Abbas, F., & Oh, H. (2016). A Hierarchical Privacy Preserving Pseudonymous Authentication Protocol for VANET. *IEEE Access: Practical Innovations, Open Solutions*, *4*, 7770–7784. doi:10.1109/ACCESS.2016.2620999

Riaz, F., & Niazi, M. A. (2016). Road collisions avoidance using vehicular cyber-physical systems: A taxonomy and review. *Complex Adaptive Systems Modeling*, *4*(1), 15. doi:10.118640294-016-0025-8

Sharma, M. K., Bali, R. S., & Kaur, A. (2015, October). Dyanimc key based authentication scheme for Vehicular Cloud Computing. In *Green Computing and Internet of Things (ICGCIoT), 2015 International Conference on* (pp. 1059-1064). IEEE. 10.1109/ICGCIoT.2015.7380620

Sharma, M. K., & Kaur, A. (2015, September). A survey on vehicular cloud computing and its security. In *Next Generation Computing Technologies (NGCT), 2015 1st International Conference on* (pp. 67-71). IEEE

Shravanthi, C., & Guruprasad, H. S. (2014, May). *Mobile Cloud Computing as future for mobile applications. International Journal of Research in Engineering and Technology.*

Singh, P., Kaur, S., & Sandhu, S. S. (2015). An Overview of Mobile Cloud Computing.*International Journal of Computer and Communication System Engineering*, 2(2), 255–259.

Sugumar, R., Rengarajan, A., & Jayakumar, C. (2016). Trust based authentication technique for cluster based vehicular ad hoc networks (VANET). *Wireless Networks*, 1–10.

Verma, A., & Kaushal, S. (2011). Cloud computing security issues and challenges: a survey. *Advances in Computing and Communications*, 445-454.

Viriyasitavat, W., Boban, M., Tsai, H. M., & Vasilakos, A. (2015). Vehicular communications: Survey and challenges of channel and propagation models. *IEEE Vehicular Technology Magazine*, *10*(2), 55–66. doi:10.1109/MVT.2015.2410341

Wahyono, W., & Jo, K. H. (2017). Cumulative Dual Foreground Differences For Illegally Parked Vehicles Detection. *IEEE Transactions on Industrial Informatics*, *13*(5), 2464–2473. doi:10.1109/TII.2017.2665584

Wan, J., Zhang, D., Zhao, S., Yang, L., & Lloret, J. (2014). Context-aware vehicular cyber-physical systems with cloud support: Architecture, challenges, and solutions. *IEEE Communications Magazine*, *52*(8), 106–113. doi:10.1109/MCOM.2014.6871677

Whaiduzzaman, M., Sookhak, M., Gani, A., & Buyya, R. (2014). A survey on vehicular cloud computing. *Journal of Network and Computer Applications*, *40*, 325–344. doi:10.1016/j.jnca.2013.08.004

Yan, G., Wen, D., Olariu, S., & Weigle, M. C. (2013). Security challenges in vehicular cloud computing. *IEEE Transactions on Intelligent Transportation Systems*, *14*(1), 284–294. doi:10.1109/TITS.2012.2211870

Zhang, Q., Cheng, L., & Boutaba, R. (2010). Cloud computing: State-of-the-art and research challenges. *Journal of Internet Services and Applications*, *1*(1), 7–18. doi:10.100713174-010-0007-6

Zhu, H., Lu, R., Huang, C., Chen, L., & Li, H. (2016). An Efficient Privacy-Preserving Location-Based Services Query Scheme in Outsourced Cloud. *IEEE Transactions on Vehicular Technology*, *65*(9), 7729–7739. doi:10.1109/TVT.2015.2499791

Chapter 4
Resource Management in Vehicular Cloud Computing

Sawsan Elfatih Ibrahim
Gezira University, Sudan

Rashid A. Abd Elhaleem Saeed
Sudan University, Sudan

Amitava Mukherjee
IBM India Private Limited, India

ABSTRACT

Vehicular ad hoc networks are relied upon to essentially enhance movement security and transportation productivity. Accessible correspondence, stockpiling, and calculation assets of the associated vehicles are not well used to meet the administration necessities of canny transportation frameworks. Vehicular cloud computing (VCC) is a promising methodology that makes utilization of the preferences of distributed computing and applies them to vehicular systems. Vehicles have turned out to be prepared with different sensors and assets, making them competent to convey, to share assets, and to carry on helpfully. VC can be depicted from the point of view of uses and administrations offered by vehicles that have a place with a vehicular ad hoc network (VANET) due to the impediment of calculation, stockpiling, and data transfer capacity assets for complex in-vehicle sight and sound amusement applications.

OVERVIEW

The distributed computing worldview has empowered the misuse of overabundance registering power. The endless number of vehicles on avenues, roadways and parking areas will be dealt with as abundant and underutilized computational assets, which can be utilized for giving open administrations. Consistently, numerous vehicles, invest hours in a stopping carport, garage or parking area. The stopped vehicles are an unlimited unexploited asset, which is squandered basically. These elements make vehicles the ideal contender for hubs in a distributed computing system. A few vehicle proprietors may consent to lease overabundance on board assets, like colossal processing holders and storerooms who lease their abun-

DOI: 10.4018/978-1-5225-3981-0.ch004

dance limit and advantage financially. The voyagers regularly stop their autos in air terminal parking spots while they are travelling. The airplane terminal expert determination the vehicles' processing assets and take into account on request access to this parking structure server farm. So also, the drivers stuck in movement blockage will concur give their on board figuring assets to help city activity specialists run complex recreations intended to expel clog by rescheduling the movement lights of the city.

Chapter cares for resources and how to communicate with each other and managed properly to exploit them to meet the service requirements and get best results.

This chapter organized as follows. Firstly, provide a brief review of vehicular cloud networks and cloud computing in the automotive domain. Section 3, talked about types of Vehicular Networks Resources. Section 4 presents how to manage vehicular cloud resources to gain best utilization and how to manage Vehicular Traffic. Cooperative Resources Management for Vehicular cloud computing given in Section 5. Section 6 presents Comparison and Discussions for VCC Resources Management and finally concluded all topics in Section 7.

INTRODUCTION

In the recent decades, vehicular ad hoc networks VANET has been a center systems administration innovation to give security and solace to drivers in vehicular situations with mechanical headway, VANETs give drivers and travelers well-being and comfort, and besides present new applications for amusement also, condition observing. The vehicular correspondence is going to advance with developing standards. Vehicles turned out with different sensors and assets, skilled them to impart, to share assets and to act helpfully. This offered ascend to the vehicular cloud's idea (VC).

Vehicular cloud portrayed from the perspective of utilizations and administrations offered by vehicles that have a place with a Vehicular Ad hoc Network (VANET). These systems have specified consideration by the exploration group in the field of systems on the grounds advantages. They incorporate applications that emphasis on street wellbeing and how to make open streets more effective for vehicles, while offering solace and excitement to travelers all through their adventure (Sibai, Atechian, Abdo, Demerjian, &

Figure 1. Toward vehicular cloud computing

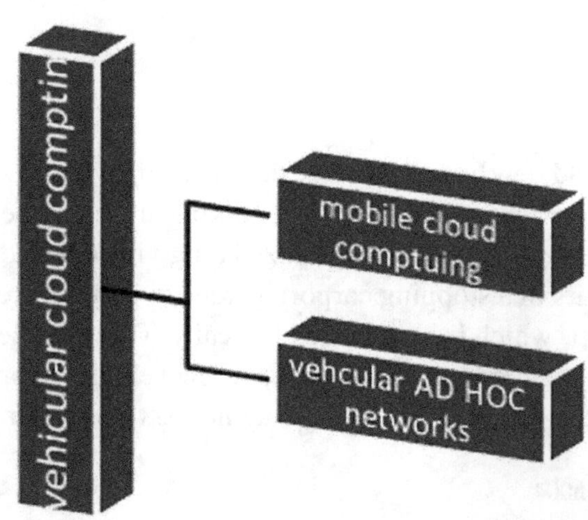

Tawil, 2015). The characteristic of VANET substance properties and the way that VANET applications devour the substance are driving the development. All the more particularly:

Vehicles create and expend an awesome measure of substance having neighborhood property importance (time, space, and purchaser). Vehicles just look for substance paying little mind to their suppliers. Vehicles team up utilizing their assets to make esteem included administrations with least assistance from the Internet framework. Cloud computing is a system getting to model that intends to straightforwardly and pervasively share an expansive number of figuring assets. These rented by a specialist co-op to computerized clients, for the most part through the Internet (Meneguette, Bittencourt, & Madeira, 2013).

Vehicular Cloud Computing

Vehicular Cloud Computing is a technology uses the advantage of cloud computing to help drivers by provide several services at low cost and no time, in other words we can say its group of cooperative vehicles. In vehicular cloud there are much resources such as computing capabilities, sensors, storage, and communication resources. Resources management means coordinate and share all resources information to introduce best services.

Vehicular Cloud Computing attracted researchers to work and study it. Vehicular Cloud Computing system divided into three frameworks; Vehicular Clouds, vehicles using clouds, and hybrid clouds (Hussain, Son, Eun, Kim, & Oh, 2012). Furthermore, a distributed and adaptive resource management proposed optimal exploitation of radio cognitive and soft-input/soft output data fusion in vehicular access networks (Papadimitratos, Fortelle, Evenssen, Brignolo, & Cosenza, 2009). Both the vehicular cloud and vehicular systems give a way to the plan of arrangements in Intelligent Transportation Systems. Vehicular networks can be seen as an expansion of Mobile Ad-hoc Networks, which have increase popularity with

Figure 2. Vehicular Cloud Computing

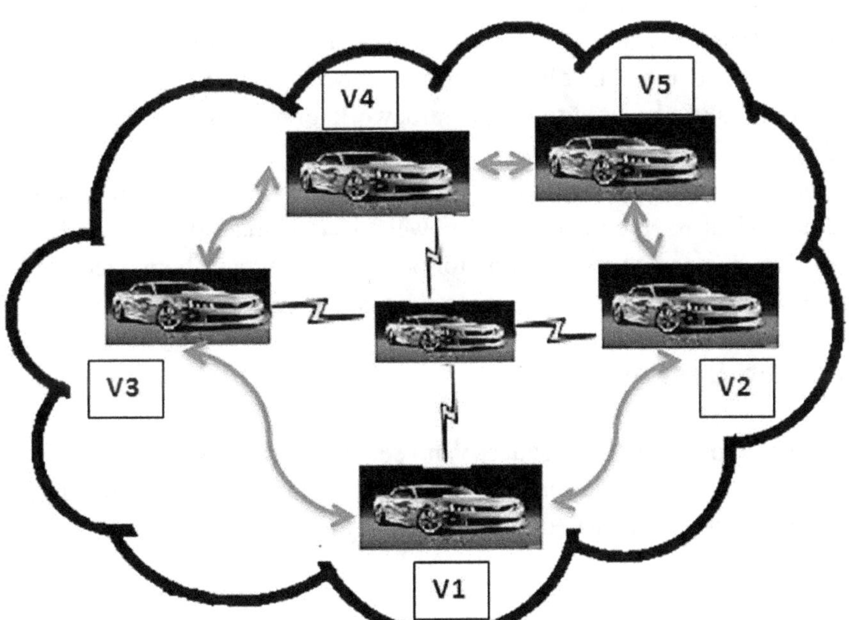

advancements solutions, technologies, and an extensive range of applications it can support. VANET can divide into two: vehicles thought stuck entities called Road Side Units (RSUs), deployed at important locations like slippery roads, dangerous intersections, or places acknowledge for ventures atmospheric.

Vehicular networks basically present two major architectures: Vehicle to Infrastructure communication (V2I) and Vehicle to Vehicle communication (V2V) (Blum, & Eskandarian, 2006).

On the off chance that the street condition or a driver's conduct is unusual, a crisis cautioning message will produced by the vehicle that watched this activity variation from the norm. Furthermore, sent to VC stockpiling pool, henceforth, all vehicles that enlisted in VC will get the crisis message, which contains land area where anomalous circumstance happens (Wan, Liu, Shao, Vasilakos, Imran, & Zhou, 2016).

Autonomous Vehicular Cloud (AVC)

The immense armada of vitality adequate vehicles that mismatch our roadways, aviation routes, and conduits, the vast majority of them with a lasting Internet nearness, including significant locally available computational, stockpiling, and detecting abilities can be thought of as a gigantic homestead of PCs moving. These traits make vehicles perfect possibility for hubs in a cloud as depicted 6previously. Without a doubt, the proprietor of a vehicle may choose to lease their in-vehicle capacities on request, or a for each example, or an everyday, per week or per-month premise, similarly as proprietors of expansive figuring offices discover it monetarily speaking to lease abundance ability to look for monetary focal points. All the more fundamentally, vehicles will self-sufficiently self-organize into mists using their corporate resources on-request and to a great extent progressively in settling basic issues that may happen out of the blue. The new vehicular mists will likewise add to disentangling some specialized difficulties of the undeniably complex transportation frameworks with their rising conduct and vulnerability. It is just a short time before the gigantic vehicular armadas on our roadways, boulevards and parking garages will be perceived as a plenteous and underutilized computational asset that can be taken advantage of with the end goal of giving outsider or group administrations. In any case, what recognizes vehicles from standard hubs in a traditional cloud is autonomy and portability. In reality, extensive quantities of vehicles invest significant measures of energy in the street and might be engaged with powerfully evolving circumstances; we contend that in such circumstances, the vehicles can possibly helpfully take care of issues that would take a brought together framework an over the top measure of time, rendering the arrangement futile.

To allude to: a gathering of to a great extent self-ruling vehicles whose corporate processing, detecting, correspondence and physical assets composed and powerfully designated to approved clients. An essential objective of the AVC to give on-request answers for occasions that have happened however can't be met sensibly with pre-allocated resources or in a proactive manner (Eltoweissy, Olariu, & Younis, 2010).

Scalability of Cloud Systems

Adaptability empowers applications to discharge and secure assets on request. Auto scaling just robotizes this procedure since human mediation is illogical as the framework develops in estimate. In cloud frameworks, adaptability can be delegated even and vertical.

1. **Flat Scaling:** This method for scaling works by repeating server machines. It adds more machine servers to the framework as indicated by the prerequisite and request of the clients or applications.
2. **Vertical Scaling:** It accomplishes versatility by expanding the limit of a server machine. It doesn't include new machines rather it builds the limit of the machines being used in the framework. In some utilization cases, the trifling adaptability benefit gave by some cloud benefit suppliers isn't satisfactory. Cloud benefit inhabitants need to expend the perfect sum of assets in light of the need of their end clients or the request of their applications. In this manner, to accomplish very effective utilization of assets, there is a need to apply dynamic versatility calculations to applications and their back-end systems (He, Yan, & Da Xu, 2014).

Resources of Vehicular Cloud Networks

Resources in the vehicle cloud recognized from the ones in the regular cloud. Every vehicle has three classifications of Resources - information stockpiling, sensors, and registering. The information stockpiling stores vehicle substance created from applications and sensors, additionally conventional sight and sound records. It bolsters information sharing between cloud members by tolerating an outside hunt inquiry and by answering with coordinated substance. The sensor can self-incite and additionally to recognize occasions in a physical world. With mechanical progression, every sensor associated specifically outside frameworks can read the sensor information and additionally control the sensor. The processing asset is like that of the Internet cloud aside from that its ability constrained in light of the fact that it is a gathering of versatile assets.

Through the participation in the gathering, the physical vehicles assets are progressively planned on request. The general asset use fundamentally upgraded.

Vehicular cloud computing introduced a completely unique, additional timely and interactive means of assembling traffic information. VCC uses vehicles and/or smartphones as mobile nodes in a very mobile unplanned network to make a mobile network. During this manner, Roadside Equipment (RSE) deployed at strategic locations will exchange info with smartphones carried by the drivers. RSE and proximate smartphones interconnected and share traffic info (e.g., traffic jam levels). Vehicles outside the range of any RSE should be still connected to the vehicle's remainder and infrastructure network via neighboring vehicles. This network will generate correct period traffic info in nice detail, based on that some elementary traffic issues relating to potency may be addressed from a complete new perspective (Meneguette, 2016).

VC used underutilized vehicle resources to form a cloud by aggregating vehicular computing resources.

The vehicular versatile cloud comprises an arrangement of portable figuring assets and vehicles, which not utilized by customary distributed computing. These assets interconnected through the vehicular system. The vehicular versatile cloud can be either settled or versatile (Abdelhamid, Hassanein, & Takahara, 2015).

Vehicular cloud networks resources are an important issue because all the cloud depends on the data comes from it. There are many types of resources as follow:

Sensors used to detect vehicle's status speedometer to measure the speed; GPS to specify the location and other sensors to obtain information about the vehicle, every sensor hub gathers perception information from the encompassing condition, does some handling locally if necessary.

The sensors ordered as application space indicated to:

1. Safety,
2. Diagnostics,
3. Convenience, and
4. Environment monitoring.

Safety sensors are the most vital being focused at diminishing mishaps and driving fatalities.

Diagnostics sensors give on-board identification of part glitch to maintain a strategic distance from further breakdowns or, on the other hand harm. Not with standing on-board benefits, the framework can incorporate an announcing capacity to bolster remote diagnostics. Convenience sensors bolster solace and comfort applications for drivers and travelers. A large portion of these sensors sent inside the vehicle compartment to give coordinate administrations for its tenants, while others are sent for giving driving help. At long last, condition monitoring sensors give cautions/notices about street dangers or reports about movement, street and climate conditions.

Cameras used to monitor the state of road clearly. Movement cameras check vehicles quantity and furthermore recognize plate numbers.

The information stockpiling stores vehicle substance produced from applications and sensors and additionally conventional mixed media records. It underpins information sharing between cloud members by tolerating an outer hunt inquiry and answering with coordinated substance. Processing asset is like that of the Internet cloud with exception of its ability restricted in light of the fact that it is a gathering of portable assets.

A vehicle can give others data acquired from its own sensors, different vehicles, RSUs, guide access to the Internet. This data may incorporate movement and ecological conditions, navigational and wellbeing data, stopping accessibility, or business notices.

Vehicles utilized to find neighboring vehicles for assessing the separation to these vehicles or educating them about their positions for precision purposes, other versatile operators can recover the perceived plate numbers to find and track lost or stolen vehicles

Vehicles can likewise utilize self-restriction systems to decide their own particular position (Gu, Zeng, & Guo, 2013).

Vehicular Cloud Networking (VCN) Operation

To make and start a cloud in vehicular systems, it requests a cloud pioneer which can be either vehicle or adjoining RSU. On the off chance that the pioneer is a vehicle to start the cloud and no adjoining RSU takes an interest in cloud development, at that point the coming about cloud is VC. Be that as it may, if the demand for cloud is started by RSU as a pioneer and neighboring vehicles reacts to its demand brings about the arrangement of IC. The cloud pioneer welcomes cloud individuals i.e. vehicles and adjacent RSUs in its region by transmitting asset ask messages (REQs) to shape a cloud. Any vehicle wishes to join the cloud reacts back to the cloud pioneer with asset answer messages (REPs) (Whaiduzzaman, Sookhak, Gani, & Buyya, 2014). At the point when cloud pioneer gets the affirmation by means of REP messages, it keeps its individuals' ID and allot diverse assignments and applications with them in like manner. The individuals discuss always with its cloud pioneer. In view of the authorization from cloud pioneer, the individuals can distribute and share the substance got from pioneer with other vehicles.

Figure 3. Resources of vehicular cloud networks

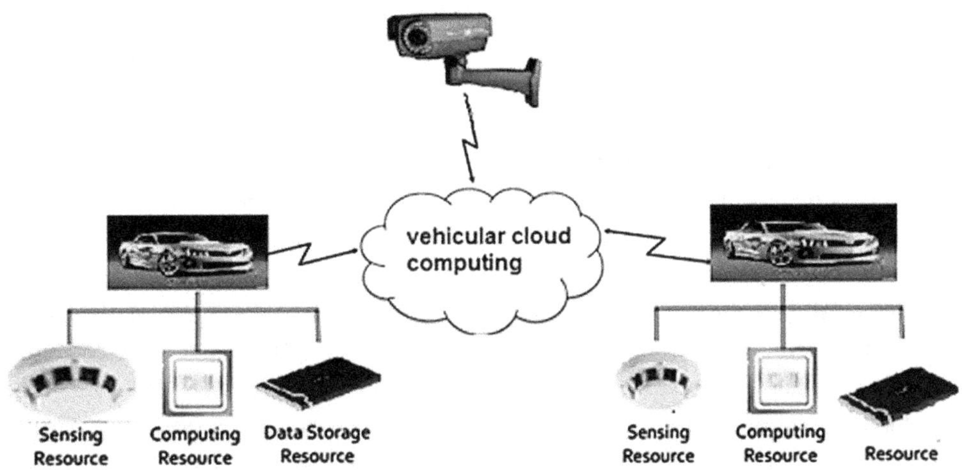

Cloud pioneer is in charge of the support of the cloud it made. Be that as it may, if any cloud part, who wishes to leave the cloud asks for the asset leaving message to cloud pioneer. All things considered, the cloud pioneer affirms the discharge of its part and enrolls new individuals by communicating REQ messages. Be that as it may, in the event that if the cloud pioneer itself not any more need to keep the cloud, it communicates the cloud discharge message also, leaves the cloud (Ahmad, Kazim, Adnane, & Awad, 2015).

Related Works

There are many studies within the literature that involve conveyance cloud computing like the study by Olariu et al (Zingirian, & Valenti, 2012) who planned a replacement construct known as conveyance cloud (VC).

VC used underutilized vehicle resources to create a cloud by aggregating conveyance computing resources. It is price noting that the planned system doesn't provide any benefits of a standard cloud capabilities as a result of this proposal is only supported vehicle resources.

An immaculate cloud framed of vehicles proposed by Zingirian and Valenti (2012) in which the creators created another sensor call benefit as an administration (SenaaS) to serve correspondence stages that supply parts of the sensor. This incorporates vehicle sensors and gadgets for vehicle following applications, for example, distributed computing assets called sensor cloud benefit. This proposition can't to make conventional cloud utilization to enhance the figuring control normally required for vehicles.

Maet al.[29] proposed the client situated Cloud Transportation System (C_TS) utilizing a driving way control. Crowdsourcing techniques and distributed computing design are used to set up a movement display that predicts by gathering, separating, and demonstrating client information. Client information (area and speed) are gathered to decide client designs in view of a crowdsourcing strategy. The information are then ascertained once more, and, to give ongoing direction, administrations are given utilizing an outer information source. Gerlaet al (2014) dispersed substance effectively to consumers through Named Data Networking (NDN), utilizing names rather than IP addresses. They proposed a model comprising of the purchaser asking for substance and suppliers giving the data to the buyers accordingly. The cloud administrations that are created and kept up by the cloud data suppliers were characterized into existing cloud and vehicle cloud administrations. The cloud assets in the vehicle can't just share information through capacity, detecting, and figuring, however can also detect and control occasion events inside the physical points of confinement of the sensor abilities. Dressleret al. (Dressler, Handle, & Sommer, 2014) utilized the system or capacity of stationary vehicles in VCC touse sit out of gear assets. These frameworks utilize a Virtual Cord Protocol (VCP) to enroll another vehicle or prohibit a current vehicle from the system. Driving and Parking modes are characterized, and when a vehicle is in Driving mode, it gets to the VCP area utilizing a between space steering for information trade. The vehicle is selected utilizing the current code in stopping mode. Between area steering chooses a vehicle to play the part of entryway for correspondence among different VCP lines, or decides which convention to utilize when information is spared to a neighborhood VCP rope. Hussain and Oh (2014) tended to the issues of security, anonymous withdrawal, and course following through VANET utilizing Clouds (VuC). The proposed strategy ensures protection through different secrecies ensuring contingent obscurity. Guide messages are spared in the cloud to use for course following and mysterious withdrawal. Foundations with expert can cooperate to follow a course chosen by a client for a specific time of time. In expansion, the organization can recognize the client. Sur et al.(2016) proposed another VCC-based secure and protection saving route convention. The proposed convention utilizes a hash-sign-switch paradigm with a trapdoor hash work. To give safe route benefit, a solitary utilize mysterious endorsement and hash key are utilized for the marking, and the course to the goal is given from a Road Side Unit (RSU) to the vehicle. What's more, for security assurance, zero-information evidence is utilized.

El Sibai et al. (Sibai, Atchian, Abdo, Demerjian, & Tawil, 2015; Eltoweissy, Olariu, & Younis, 2010) proposed an availability mindful benefit of system arrangement, where the specialist organization vehicle chosen on the premise of a few parameters, for example, the accessibility of the asked for administration and vehicles versatility. This component permits an administration requester vehicle to frame its own cloud. The administrations are traded between the administration requester vehicle and the specialist organization vehicles. The requester completes this by sending a demand message by means of communicate to its neighbors. Although the creators make are lay system, this instrument can't investigate the communicate tempest and therefore causes an overhead on the system.

The rising Internet of Things (IoT) and distributed computing innovations have appeared promising open door in settling the difficulties in the present transportation frameworks. The work by He et al. (2014) proposes a novel multi-layered vehicular information cloud stage utilizing IoT and distributed computing advancements. In this work, the creators created two vehicular cloud benefits: a vehicular information mining cloud benefit and an insightful stopping cloud benefit. It additionally displays vehicular guarantee examination in the IoT condition. Information digging models for vehicular information mining cloud administrations for example, the Naive Bayes demonstrate and a Logistic Regression display are additionally examined. Present day vehicles are getting digitalized and outfitted with savvy PC frameworks. The present vehicles have the capacity of detecting, organizing with different gadgets, correspondence with vehicles or different gadgets and information handling. The mix of these capacities with the new rising innovations caused a fast progress in the car industry. It is recommended that vehicular cloud improvement is attainable with the present innovation and bigly affect the car industry (He, Yan, & Da Xu, 2014). Programming design for a shrewd stopping cloud benefit is additionally portrayed in detail in the paper. It utilizes handsets and sensors for opening location in a parking space. Paper by Akihito et al. (Iwai, & Aoyama, 2011) manages how to assemble car cloud administrations based on different designs. The subtle elements of the both (Service Oriented Architecture) also, AUTOSAR (Automotive

Open System Architecture) are depicted in the paper. It prescribes designers to utilize these structures as an advancement stage when they create car programming frameworks.

Cloud Computing in the Automotive Domain

Cloud computing proposed to reshape vehicular programming and administrations in the car space. As progressively and more autos furnished with gadgets that can get to the web, Olariu et al. (Sibai, Atechian, Abdo, Tawil, & Demerjian, 2015) propose to incorporate existing vehicular systems, different sensors, on-board gadgets in vehicles, and cloud registering to make vehicular mists. They recommend that vehicular mists are mechanically doable and will have a huge effect on the general public once they constructed. Accordingly, both existing car programming and an assortment of data assets are virtualized and bundled as administrations to construct vehicular mists. Distinctive vehicular administrations frequently consolidated and utilized to actualize the mapping, embodiment, conglomeration, and structure and enable vehicles to collaborate with different facilitated benefits outside the vehicles. As of now, utilizing the measured approach, multilayer and SOAs to incorporate different vehicular assets and administrations has all the earmarks of being the most encouraging model furthermore, structure for building vehicular cloud benefit stages. By utilizing the particular way to deal with deteriorate a mind boggling framework into littler subsystems as indicated by their capacities, we can partition a vehicular cloud benefit stage into various utilitarian administrations and subsystems, for example, movement organization, benefit directing, data handling, vehicle guarantee investigation and mining, etc (Whaiduzzaman, Gani, & Naveed, 2014).

Dispersed registering has proposed to reshape vehicular programming and organizations in the auto space. As logically and more automobiles outfitted with devices that can get to the web, Olariu et al. (Sibai, Atechian, Abdo, Tawil, & Demerjian, 2015) propose to consolidate existing vehicular frameworks, distinctive sensors, on-board contraptions in vehicles, and cloud enrolling to make vehicular fogs. They suggest that vehicular fogs are mechanically possible and have an immense influence on overall population once they built. As needs be, both existing auto programming and a collection of resources information are virtualized and packaged as organizations to develop vehicular fogs. Unmistakable vehicular organizations are often times united and used to complete the mapping, encapsulation, mixture, and structure and empower vehicles to work together with various encouraged benefits outside the vehicles. Starting at now, using the deliberate approach, multilayer and SOAs to fuse diverse vehicular resources and organizations has every one of reserves most reassuring model. Moreover, structure for building vehicular cloud advantage stages. By using the specific approach to manage weaken a brain boggling structure into tinier subsystems as shown by their abilities, we can parcel a vehicular cloud advantage organized into different utilitarian organizations and subsystems, for instance, development association, advantage coordinating, information dealing with, vehicle ensure examination and mining, etc (Whaiduzzaman, Gani, & Naveed, 2014).

Resources Management for VCC

Resources Management means combining resources together and empowering them to give cloud administrations to general society advantage both the cloud and proprietor of vehicles.

Here the outline of how the VCC works to build up a virtual figuring stage and to empower cloud sort coordinated effort in it.

1. **Cloud Resource Discovery:** At the point when a vehicle runs an application, it turns into a cloud pioneer and selects individuals (i.e., vehicles and RSUs in the region) that can give their assets to develop a vehicle cloud for the application. The sorts and hunt scope of important assets rely on upon applications. For example, the inquiry range can be a predefined separate, a street segment, or a convergence. Once deciding the vital asset arrangement sorts, the cloud pioneer communicates a resource request for message, RREQ, to hubs inside the hunt run. Hubs, willing to share their assets, send an asset answer message, RREP, back to the pioneer with their asset data abilities.
2. **Cloud Development:** After getting RREPs, the cloud pioneer chooses hubs (cloud individuals) which it sorts out a cloud. The choice system may seek after limiting aggregate asset use in the cloud, while guaranteeing the right operation of the application.
3. **Result Collection:** The cloud pioneer parts the application into a few undertakings and appropriates them to cloud individuals, considering the accessibility and assets openness. The pioneer too keeps up a cloud table with individuals' ID and their errand task data. The cloud individuals return the outcomes back to the pioneer in the wake of finishing their assignments.
4. **Content Distributing and Sharing:** When gathering the entire errand comes about because of the individuals, the pioneer forms them to acquire yield substance and recoveries/distributes it. For the situation that the undertaking comes about handled in a solitary hub or the distributed substance can't be put away in a solitary hub, the pioneer arranges another vehicle cloud for the information preparing or content stockpiling. The distributed substances are shared over the whole systems. That is, any vehicle can send a substance demand to hunt them. After getting the demand and having the coordinated substance in information stockpiling, the cloud individuals including the pioneer can answer with the substance.
5. **Cloud Maintenance:** In a vehicle cloud, a part sends a cloud leave message to the pioneer on its taking off. At that point, the pioneer chooses a substitution among hubs that sent RREPs in the asset disclosure stage and have assets enough to finish the relegated leaving part assignment. The pioneer disseminates the assignment to the new cloud part and updates the cloud table.
6. **Cloud Release:** At the point when the pioneer no longer uses the cloud or moves out of the cloud, it discharges the assets so that different mists can reuse them. To this end, the pioneer sends a cloud discharge message to every one of the individuals what's more, evacuates the cloud part list (Whaiduzzaman, Sookhak, Gani, & Buyya, 2014).

In the Vehicular Cloud, two categories of Vehicles defined:

- A Requestor vehicle that demands one or different accessible administrations in the cloud. In this paper, requestor vehicle is initiator of vehicular cloud; in this way, it named the pioneer vehicle. The pioneer vehicle is mindful to:
    - Seek competitor vehicles willing to give administrations.
    - Start the cloud.
    - Keep up the cloud.
    - Destruct the cloud.
- A Service Provider vehicle that gives of its administrations to requester vehicle. A specialist co-op might be a stationary or, then again a versatile vehicle in VANET (ee, Lee, Gerla, & Oh, 2014).
    - Application layer contain all applications and services in the cloud.

- ◦ Resources controller provide control to resource virtualization agent, computing task scheduler and dynamic resource manager.
- ◦ Physical resources contain all resources like base stations, road side unit RSU and vehicles.

Vehicular Traffic Management

Observation and course following is a case of use completely contained in the Vehicular Cloud (the Cloud stores the information and processes the follow). The following application we look at is vehicular activity administration. Chip away at vehicular activity administration and course advancement started back in the 60's, utilizing street activity models approved by estimations and numerical programming strategies. Be that as it may, the outcomes stayed in the scholastic space and discovered close to nothing open door for execution, predominantly on the grounds that as of not long ago it was to a great degree hard to *measure the vehicular movement* continuously, and to *illuminate vehicles of new courses.* To defeat this issue, the approach taken by Department of Transportation in most recent decade to quantify the movement by instrumenting the interstates with sensors under the asphalts and camcorders. This is an exorbitant arrangement, which also identifies the street fragment activity stacks yet can't decide the movement design (source and goal of the movement). Also, the data about the "best course" passed on to drivers with bulletins, radio declaration and, all the more as of late, the Internet. Sadly, sending the same guideline to every one of the vehicles had the impact of making "course fluttering" issues and course insecurities (Wan, Zhang, Zhao, Yang, & Lloret, 2014).

Efficient Management of Vehicular Mobility Is Composed of Topology Control, Location Management, and Handoff Management

- **Topology Management:** Topology administration can be proactive and receptive. Proactive plans occasionally send flagging messages to investigate the topology data. Then again, responsive plans get the topology data just when it is required e.g., when there is another versatile hub to join the system. Since VANETs can be vast, absolutely have based topology control does not scale well in such systems. The group based topology control can fathom this impediment. In this group based topology control, vehicles are gathered into different bunches. Head of each group is in charge of intra-topology administration. These group heads facilitate among each other to deal with the whole specially appointed arrange topology. Be that as it may, because of the high speed

Figure 4. Resource management layers

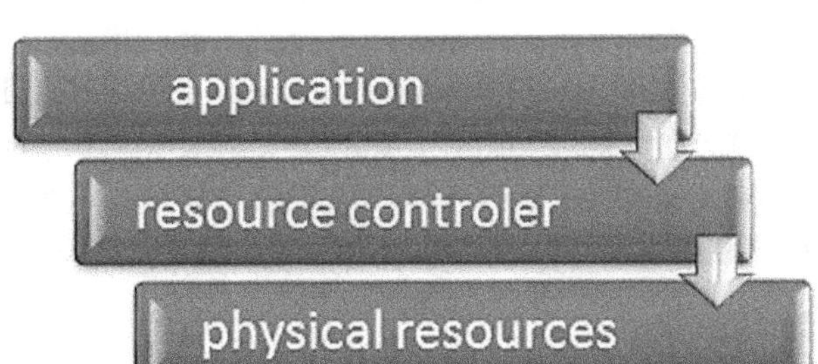

and obliged versatility e.g., moving along a straight street of vehicles, current bunching plans produced for MANETs can't accomplish the ideal execution in VANETs and the bunches could be insecure.

- **Location Management:** With interesting versatility qualities of VANETs, fundamental impromptu steering conventions can't be straightforwardly connected to VANETs because of the vast inactivity and overhead. Be that as it may, geographic directing was appeared to be productive and viable for VANETs. Utilizing geographic directing e.g., avaricious border stateless directing (GPSR), topographical directing calculation (GRA), conveying hubs are required to have the area data of each other. Accordingly, area administration conspire, which bargains with the capacity, support, and recovery of versatile hub area data, is required in VANETs. It is significant that, the area here alludes to geological area which isn't the same as the tending to area in Internet.
- **Handoff Management:** Handoff administration in vehicular impromptu systems is performed by rerouting to develop another way to the goal. At the point when a versatile hub moves, a gathering of neighbors changes and subsequently the new course of information exchange should be built up rapidly for better handoff execution. Handoff administration can be proactive and receptive which utilizes an indistinguishable ideas from those in portable specially appointed system steering. A review of directing conspires in VANETs can be found in (Zhu, Niyato, Wang, Hossain, & In Kim, 2011).

Policy and Operational Management

Motivations foundation, the accessibility measurements, the principles and the directions of the VC must deliver to work flawlessly in a choice emotionally supportive network, a control structure and an administration framework. Confronting these issues requires a more extensive group, which must connect with nearby and worldwide leaders. The measurements expected to decide the monetary models for sensible evaluating and charging of various VC administrations and elements. Therefore, we ought to offer accentuation to the accompanying components:

- **Trust Administration Assurance:** In a few circumstances, the VC may require to have expert to take nearby activity rather than a focal specialist. For instance, when the vehicles includes a car influx, for rescheduling the movement lights, a participation between the cloud development and civil or region specialist needs to advance the quick dispersal of blockage. Consequently, trust administration presence in VC can be helpful for robotized check of activities.
- **Essential Working Approaches:** Effective operational strategies required for consistent buries operation, choice support, building up responsibility measurements, institutionalization, directions, and even neighborhood and national approach making.
- **Various Mists Federation:** We are expecting sooner rather than later, a few sorts of mists will develop such as the VCC-auto cloud, sensor cloud, Smartphones cloud, all mists will communicate with each other and can interface on request continuous situation to genuine distributed computing cloud. Subsequently interoperability of distinctive sorts of mists, association, synchronization, and unwavering quality and proficiency ought to be addressed (Olariu, Hristov, & Yan, 2013).

Cooperative Resources Management for VCC

The cloud layer comprises of three inner layers: application, cloud framework, and cloud stage. In the application layer, different applications and administrations are considered as ongoing administrations or cloud essential administrations, which are open remotely by drivers, for example, fuel criticism, human movement acknowledgment, wellbeing acknowledgment and natural acknowledgment.

- **Resource Modeling:** For vehicular cloudlet without assets accessibility ensures, it is basic to display the progression of the assets. Vehicular systems basically concentrate on the versatility models of vehicles. This gives a strong establishment to demonstrate the flow of the assets in portable vehicular cloudlet. However, review that vehicular cloud additionally incorporates static vehicular cloudlet comprising of vehicles in a parking area. The practices (e.g., entry and flight) in the parking area profoundly impact the asset accessibility and it is basic to unmistakably depict such elements
- **Resource Management Strategy:** Heterogeneity and flimsiness of vehicular assets, For heterogeneity, there is a need to construct an asset depiction configuration to adequately speak to assorted assets consistently. To adapt to the self-governing administration, the depiction display should be sufficiently basic for rading between vehicles. For the flimsiness issue, a basic inquiry is the means by which to precisely and auspicious catch the elements of the assets on a correspondence channel which may likewise be temperamental.
- **Resource Allocation:** In static vehicular cloudlet, Multiple existing together vehicular cloudlet giving similar administrations. The asset allotment might be made by both the necessity attributes and asset accessibility elements to give fine-grained tuning of vehicular assets. Furthermore, such asset portion will be considered from two angles, intra-parking area tuning and between parking area tuning. Intra-parking area primarily manages the confound between the nearby asset accessibility also, the necessity. Inter-parking area tuning is to deliberately consider the asset portion over distinctive parking areas.

Cooperative Data Exchange

Some vehicular applications may need to download (substantial) information documents from the Internet. For example, an excitement application giving on-request movies/arrangement regularly requires downloading extensive documents. Despite the fact could be acknowledged by utilizing cell systems, it is not alluring since the cell system can be over-burden if each and every vehicle uses it for downloading. There is a requirement for an answer which all the more viably uses the VANET itself rather than blocking cell systems. Fortunately, the VC makes it conceivable to helpfully download and transfer information. For instance, if a specific vehicle needs to download a huge record from the Internet, it could communicate in VC. In the best case, it can happen that another vehicle in the VC happens to have asked the document before and still has it put away. Presently, this vehicle can cut the information in numerous lumps and offer it by means of different vehicles as some kind of P2P application. Along these lines, just V2V is used for spreading the information. Whenever the vehicle that asked for the information got every one of the pieces, it can reassemble it to manufacture the entire document. In any case, it is likewise likely that the asked for document is not yet exhibit in the VC. At that point it should be downloaded from the Internet first. However, with the nearness of RSUs, it is desirable over gain the information through RSUs rather than utilizing cell correspondence. Contingent upon the span of the information, a RSU or various RSUs could download it from the Internet. Once more, by part the information into different pieces, for efficiency, a RSU can contact various vehicles in the region to download the lumps. Presently the first information scattered over different vehicles in a VC. Through V2V correspondence, each lump can be conveyed to the vehicle asked for the information after which the vehicle can reassemble it once more (He, Yan, & Xu, 2014).

Human action acknowledgment utilized for a robotized investigation (or elucidation) of progressing occasions and setting video information. In the essential administrations, a few administrations are sent, for example, Network as a Service (NaaS), Storage as a Service (STaaS), Cooperation as a Service (CaaS), Information as a Service (INaaS), and Entertainment as a Service (ENaaS),

- **Network as a Service (NaaS):** Out and about while driving, numerous drivers will have web associations through cell phone systems or other fixed get to focuses. Today, PCs and tablets are interfacing with net however are underutilized, and similarly, this street organize asset may remain underutilized. These fundamental assets can be shared out and giving net access to individuals who are keen on leasing it from the drivers. The driver who is consenting to share this asset needs to publicize such data among every other vehicle around them on the roadway. Data can go among the current vehicles in the nearby nearness who can go about as a get to direct jump toward the web. Given the generally little speed contrast between autos going in the same direction.
- **Information as a Service (INaaS):** It certain that the capacity as an administration in vehicular systems not same as offering system get to or registering assets. In spite of the fact that the likely clients advantage in a split second from registering and system get to, they can utilize the capacity for reinforcement purposes or for p2p applications over an augmented time frame. This impediment might be a genuine snag for leasing auto stockpiling as are enforcement or p2p applications.
- **Storage as a Service (STaaS):** Since as of now stockpiling is effectively accessible, which means the cheapness and minimization capacity; it is normal that vehicles will stockroom different terabytes of information. In any case, in a VC it can even now happen that applications on certain vehicles surpass their own stockpiling abilities and require extra storage room to work legitimately. Consequently, it is attractive in a VC, vehicles with super uous stockpiling can represent vehicles requiring additional space, SaaS is deferent from oaring system get to or processing assets. Constancy of information in neighborhood VCs genuine deterrent for the usage of SaaS.
- **Entertainment as a Service (ENaaS):** With vehicles having the capacity to drive self-rulingly, it winds up noticeably conceivable to give the travelers (particularly the driver) a more agreeable and charming trip by giving installed stimulation. While considering ENaaS, one can consider live TV, lms/arrangement on-request, plugs, web perusing etcetera, all available in the vehicle itself (Sibai, Atechian, Abdo, Demerjian, & Tawil, 2015).
- **Cooperation as a Service (CaaS):** It enables drivers to acquire administrations utilizing exceptionally negligible framework, if the foundation exists and by V2V correspondences if no foundation exists. CaaS utilizes machine where supporter expresses their inclinations for an administration, and autos subscribed to a similar administration will give the supporter vital data about the

> administration or by reporting data to the system. Vehicles may coordinate to run applications through V2V correspondence without extra framework. Different versatile applications have diverse asset prerequisites. At times, when running certain applications, an SP may miss the mark concerning one sort of asset while it has adequate different assets. In this manner, asset sharing ought to support cloud-empowered vehicular systems. Comparative helpful thoughts can be found in vehicular postponement tolerant systems. It is vital for SPs to plan effective asset administration what's more, sharing plans to fulfill the administration quality necessities of the applications.

Some SPs coordinate to make an asset pool to share their data transfer capacity and processing assets. All the SPs can utilize the asset from the asset pool when required, subsequently expanding asset use. Participation administration enhanced and new administrations for the street clients, which, thusly, will prompt more noteworthy transport effectiveness, by improving utilization of the limit of the accessible foundation and by overseeing fluctuating requests, and expanded security, by enhancing the quality and unwavering quality of data and permitting the execution of cutting edge security applications.

Cloud-empowered vehicular systems have three layers. In the principal layer, a vehicle speaks with different vehicles or its closest RSU, which gives access to a neighborhood cloud keeping in mind the end goal to send demands. In the second layer, the cloud SP will allot assets to vehicles. In the last layer, the focal cloud is the server farm that runs portable applications on the application server. The server farm oversees and dispenses every one of the assets to the SPs. The SPs give assets to vehicles through RSUs. With the assets from the cloud enabled vehicular systems, the cell phones in the vehicles can run applications.

Cooperation as a Service (CaaS) will allow:

- A decline in the cost identified with Intelligent Transportation Systems.
- A diminishment in the seriousness of street blockages and CO2 contamination, in the meantime, time and fuel sparing. Vehicles act like portable sensors checking parameters, for example, street and climate conditions.
- Enhancing driver's wellbeing and avoiding back closure mishaps.
- VANET-Cloud asset portion and sharing: Resource distribution and sharing are imperative issues that likewise should be tended to, especially for versatile cloud substances, for example, vehicles and traveler gadgets. This versatility can influence the execution of VANET-Cloud applications. An ideal dynamic mix of portable processing assets can enhance the offered benefit. In such cases, a saved asset ought to be prepared to be utilized with a diminished reaction time.

Figure 5. Vehicular cloud computing services

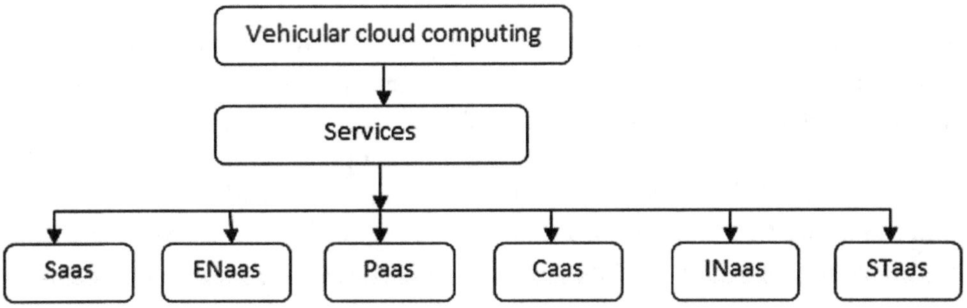

Table 1. Summary of services based vehicular data clouds

Services	Description
Network and Data Processing as a Service, i.e. Infrastructure As A Service (IAAS)	Vehicles give their systems administration and information handling capacities to different vehicles through the cloud.
Storage as a Service (SAAS)	A few vehicles may require particular applications that require huge measure of storage room. Accordingly, vehicles that have unused storage room can share their capacity space as a cloud-based administration.
Platform as a Service (PAAS)	A few vehicles may require particular applications that require huge measure of storage room. Accordingly, vehicles that have unused storage room can share their capacity space as a cloud-based administration.

Cooperation Between AVCs

To rouse the requirement for AVCs to participate, envision that in adjoining regions of a district there is a wearing occasion and in addition a stone show downtown. Both these occasions are probably going to draw a colossal group. Presently, expecting that because of terrible arranging by the region, the two occasions end at a similar time, making two unmistakable zones of gigantic clog. In such a situation, every clog occasion will trigger the arrangement of a specially appointed AVC. These two AVCs should organize and to take care of the clog issue altogether, since they can't continue to egotistically reschedule activity lights in a way that advantages each of them separately.

Decision Making of Traffic Authorities

The basic leadership handle relies on upon many elements, for example, recorded movement stream limit, street conditions, climate conditions, and movement stream anticipating. All together to make a powerful expectation, scientists on vehicular informal communities complete activity information mining to find valuable data and information from gathered enormous information. The forecast prepare relies on upon arranging the impact components and planning a choice tree. Choice references from the scholarly world can be given likewise to activity specialists. By the strategy for likelihood examination, the activity specialists powerfully orchestrate whether the street can approved to give administrations. In a few specific cases, a deadly element may straightforwardly influence the basic leadership. For instance, when a tropical storm is drawing nearer, movement specialists may quickly end administrations (Yan, Wen, Olariu, & Weigle, 2013).

Cooperative Vehicular Safety Example

Wellbeing is the most essential worry for drivers. The non-viewable pathway preferred standpoint of VANETs makes conceivable avoiding back completion mishaps. In the Electronic Brake Cautioning (EBW) application, a vehicle communicates an occasion based message when it brakes forcefully, cautioning the encompassing autos of the braking moves. The application is especially helpful when the perspective of brake lights is hindered by other vehicles. helpful vehicular security application is the Vehicle Stability Warning (VSW) application. In VSW, going before remote vehicles ready after host vehicles of up and coming potential perilous driving conditions, e.g. frigid or sleek fix of street.

Vehicle Stability Warning messages are like Electronic Brake Warning messages such that they are both occasion based, hence, requiring dependable interchanges between vehicles amid all circumstances. On the off chance that the sent crisis messages were not gotten, these applications would be ineffectual and the activity circumstance would return to the default conduct of drivers depending just on the perspective of going before vehicles. Those who peruse may discover other intriguing helpful vehicular security applications for example, On-coming Traffic Warning (OTW) and Lane Change Warning (LCW) in (Hartenstein, 2010; Mousannif, Khalil, & Al Moatassime, 2011).

Comparison and Discussions for VCC Resources Management

Contrasted and conventional mobile cloud computing (MCC) systems, helpful resource management issues are more testing in cloud-empowered vehicular systems. On one hand, not at all like conventional gadgets or terminals, vehicles have a kind of highlight in their potential high portability. Contrasted and cloud server farms, a vehicular cloud server farm is a more dynamic condition due to the portability of the associated vehicles.

Cloud computing (CC) versus vehicular cloud computing (VCC), VC in a static disposition has an indistinguishable conduct from regular cloud computing (CC). Our autos stay out and about occupied with different progressively changing circumstances or occasions each day. The event of vehicles in close region to an episode is regularly an impromptu procedure which can be called portability. Once more, assets of those vehicles that shape a VC to lighten the occasion must happen actually, which can called self-governance. Traditional mists don't have these components which viewed as a huge real normal for VCs. by breaking down these attributes.

Many mechanisms introduced to reach optimal management; here there is a survey to keep in mind these efforts:

Vehicular Cloud vs. Internet Cloud

Both offer a critical pool of assets:

- Computing, capacity, interchanges

Differences:

- **Main Vehicle Vloud Resource (and Breaking Point):** Portability
- Vehicle cloud administrations are area applicable
- **Data Sources:** From drivers or condition
- **Services:** To drivers or to group
- Vehicle cloud can be meager, discontinuous
- Vehicle cloud connects with:
    - Internet cloud
    - Pedestrian/bike (cell phones) cloud
    - Very distinctive plan of action than Internet Cloud

Table 2. Summary of Vehicular Cloud vs. Internet Cloud

	Vehicular Cloud	Internet Cloud
Offer a Critical Pool of Assets	offer	offer
Administrations	area applicable	area
Data Sources	from drivers or condition	From internet
Connections	connects with Internet cloud cell phones cloud	connects with Internet
Plan of Action	Very distinctive	distinctive

Table 3. Comparative study of CC and VCC

Characteristics	CC	VCC
Network as a Service	Yes	Yes
Storage as a Service	Yes	Yes
Corporation as a Service	Possible	Yes
Large traffic Event management	Possible	Yes
Moving Network	No	Yes
Automatic Cloud	No	Yes
Mobility of clouds	No	Yes

Authentication vs. Privacy

In a vehicular system, it's critical to tie every driver to a solitary character to keep any caricaturing assaults. For example, in the congestion avoidance scheme, to keep one vehicle from asserting to be hundreds with a specific end goal to make the hallucination of a congested street. Solid verification additionally gives important legal proof and permits utilize outside components, for example, conventional law requirement, to dissuade or anticipate assaults on vehicular systems. Be that as it may, drivers esteem their protection and are probably not going to embrace frameworks that oblige them to surrender their namelessness. For instance, in the event that we attempt to anticipate parodying in a way that uncovers each vehicle's changeless character, then we may damage drivers' protection desires. Adjusting protection worries with security needs will require systematizing legitimate, societal and handy contemplations. Most nations have generally disparate laws concerning their natives' privilege to security. Since most vehicle producers work in multinational markets, they will require security arrangements that fulfill the most stringent security laws, or that can be modified to meet their lawful commitments in each market. Verification plans should likewise weigh societal desires of security against down to earth contemplations. Most drivers would loathe a framework that enables others to track their developments, yet from a handy point of view, vehicles today are just mostly unknown. Every vehicle has a freely shown tag that extraordinarily distinguishes it (and recognizes the proprietor of the auto, offered access to the suitable records). Accordingly, singular drivers have as of now surrendered a segment of their security while driving. In a perfect world, a protected vehicular system would assemble on these current bargains as opposed to infringing any further upon a driver's entitlement to privacy (He, Yan, & Xu, 2014).

Destruction of the Cloud

The cloud destroyed in the accompanying cases:

1. The pioneer got the administration and in the end utilized it.
2. The pioneer leaves the VANET arrange.
3. The administration is intruded on: The vehicle giving the benefit can't keep on providing it, and there are no different vehicles that can give this support of the pioneer.

Challenges and Direction for Future Work

Existing mechanisms are unsatisfactory to meet all these requirements at the same time. Below is a description of some of these challenges.

1. **Adaptability and Innovation Integration:** The adequacy of a vehicular cloud relies on upon its adaptability to deal with a powerfully changing number of vehicles. Notwithstanding taking care of standard movement, vehicular mists must have the capacity to handle movement spikes or sudden requests created by extraordinary occasions or circumstances, for example, don recreations or crises. Greater advancement on streamlining calculations that facilitate virtual machines, storage room, and system data transfer capacity to adjust server workload and enhance figuring asset use on the vehicular mists is required (He, Yan, & Xu, 2014).
2. **Execution, Unwavering Quality, and Nature of Service:** As vehicles are frequently moving, the vehicular systems administration what's more, correspondence is frequently irregular or untrustworthy. All the more new instruments are expected to improve the correspondence unwavering quality with diminished movement overhead. Cross layer information synchronization systems ought to likewise be intended to limit the activity overhead between layers. Quickening information pressure calculations for asset obliged sensors, actuators, and other Internet-get to gadgets require change to be capable to viably and effectively pack a lot of crude information produced. Various preparing units and support in various cloud server farms are expected to limit the benefit reaction time, enhance accessibility and strength of the administration, and increment cloud dependability and adaptation to internal failure. Continuous assessment systems with respect to the execution, dependability, and administration quality on vehicular information mists should be additionally created (Qin, Long, Zhang, & Huang, 2013).
3. **Absence of Worldwide Models for Gadget and Administration Incorporation, Security, Protection, Design, and Communications:** Worldwide principles are fundamental to keep away from clashes between privately created vehicular information mists (Chen, Long, Zheng, Xing, & Wang, 2015). Be that as it may, as there are various partners required in vehicular information mists, and complex conditions among these partners likewise exist, it is trying to set up worldwide gauges to bring down the intricacy and make vehicular information mists more perfect and savvy. Assist endeavors on institutionalization are expected to facilitate different endeavors what's more, assets for actualizing vehicular information mists.

Notwithstanding, the primary test of vehicular systems is the means by which to keep up the network between the vehicles amid administration arrangement prepare.

CONCLUSION

- Cloud computing is improvement of the engineering's and has declared not to substitute any living innovation. Every innovation has its have position and different advances request to blend untidily in the IT situations. Sensor-Cloud empowers the sensor data to be sorted, laid in and treated in such a style, to the point that it turns cost effective, time usable and effortlessly receptive. Sensor-cloud means to supply summed components and results to the issues that fixate these advancements.
- Here investigated and highlighted a current new idea, Vehicular Cloud Computing, whose time has arrived. VCC rises up out of the meeting of capable embedded vehicle assets, propels in system portability, omnipresent detecting and distributed computing. The blend of an enormous measure of unutilized assets on board vehicles, for example, web network, stockpiling and registering force, can be leased or shared with different clients over the web, like the standard cloud assets. A few of these assets can progressively give us support to reducing movement occurrences. We likewise advocate that, when completely acknowledged and conveyed, VCCs can prompt a critical upgrade regarding wellbeing, security and monetary feasibility of society. Consequently, VCs could build up an expansive specially appointed alliance to help moderate many sorts of crises. In an arranged or impromptu departure, there is conceivable harm to the portable correspondence foundation, and unified VCs could help a choice emotionally supportive network and offer an impermanent swap for the infrastructure (Gerla, 2012).

It is basic to note that, to totally give an organization without interruption, ensure that the system term between the requestor and the authority association vehicles is sufficient for an aggregate end-to-end transmission.

A current new idea was checked on and highlighted, Vehicular Cloud Computing, whose time has arrived. VCC rises up out of the meeting of intense embedded vehicle assets, progresses in system versatility, omnipresent detecting and distributed computing. The mix of a gigantic measure of unutilized assets on board vehicles, for example, web network, stockpiling and figuring force, can be leased or shared with different clients over the web, like the standard cloud assets. A few of these assets can powerfully give us support to mitigating activity episodes. We additionally advocate that, when completely acknowledged and conveyed, VCCs can prompt a noteworthy upgrade as far as wellbeing, security and monetary practicality of the society. Hence, VCs could build up a huge specially appointed organization to help alleviate many sorts of crises. In an arranged or impromptu departure, there is conceivable harm to the portable correspondence framework, and united VCs could help a choice emotionally supportive network and offer an impermanent trade for the framework.

REFERENCES

Abdelhamid, S., Hassanein, H., & Takahara, G. (2015). Vehicle as a resource (VaaR). *IEEE Network, 29*(1), 12–17. doi:10.1109/MNET.2015.7018198

Ahmad, F., Kazim, M., Adnane, A., & Awad, A. (2015, December). Vehicular cloud networks: Architecture, applications and security issues. In *Utility and Cloud Computing (UCC), 2015 IEEE/ACM 8th International Conference on* (pp. 571-576). IEEE.

Blum, J., & Eskandarian, A. (2006). Fast, robust message forwarding for inter-vehicle communication netw. *2006 IEEE Intelligent Transportation Systems Conference.* 10.1109/ITSC.2006.1707422

Chen, J., Long, H., Zheng, Q., Xing, M., & Wang, W. (2015). An SMDP-Based Resource Management Scheme for Distributed Cloud Systems. *2015 IEEE 81st Vehicular Technology Conference (VTC Spring).* doi:10.1109/vtcspring.2015.7145692

Cordeschi, N., Amendola, D., & Baccarelli, E. (2015). Reliable Adaptive Resource Management for Cognitive Cloud Vehicular Networks. *IEEE Transactions on Vehicular Technology, 64*(6), 2528–2537. doi:10.1109/tvt.2014.2345767

Dressler, F., Handle, P., & Sommer, C. (2014, August). Towards a vehicular cloud-using parked vehicles as a temporary network and storage infrastructure. In *Proceedings of the 2014 ACM international workshop on Wireless and mobile technologies for smart cities* (pp. 11-18). ACM. 10.1145/2633661.2633671

Eltoweissy, M., Olariu, S., & Younis, M. (2010). Towards autonomous vehicular clouds. *Ad Hoc Networks*, 1–16.

Gerla, M. (2012). Vehicular Cloud Computing. *2012 The 11th Annual Mediterranean Ad Hoc Networking Workshop (Med-Hoc-Net).* doi:10.1109/medhocnet.2012.6257116

Gerla, M., Lee, E. K., Pau, G., & Lee, U. (2014, March). Internet of vehicles: From intelligent grid to autonomous cars and vehicular clouds. In *Internet of Things (WF-IoT), 2014 IEEE World Forum on* (pp. 241-246). IEEE.

Gu, L., Zeng, D., &Guo, S. (2013). Vehicular cloud computing: A survey. *2013 IEEE Globecom Workshops (GC Wkshps).* doi:10.1109/glocomw.2013.6825021

He, W., Yan, G., & Da Xu, L. (2014). Developing vehicular data cloud services in the IoT environment. *IEEE Transactions on Industrial Informatics, 10*(2), 1587–1595. doi:10.1109/TII.2014.2299233

He, W., Yan, G., & Xu, L. D. (2014). Developing Vehicular Data Cloud Services in the IoT Environment. *IEEE Transactions on Industrial Informatics, 10*(2), 1587–1595. doi:10.1109/TII.2014.2299233

He, W., Yan, G., & Xu, L. D. (2014). Developing Vehicular Data Cloud Services in the IoT Environment. *IEEE Transactions on Industrial Informatics, 10*(2), 1587–1595. doi:10.1109/TII.2014.2299233

Hussain, R., & Oh, H. (2014). A secure and privacy-aware route tracing and revocation mechanism in vanet-based clouds. 정보보호학회논문지, *24*(5), 795-807.

Hussain, R., Son, J., Eun, H., Kim, S., & Oh, H. (2012). Rethinking Vehicular Communications: Merging VANET with cloud computing. *4th IEEE International Conference on Cloud Computing Technology and Science Proceedings*. doi:10.1109/CloudCom.2012.6427481

Iwai, A., & Aoyama, M. (2011, July). Automotive cloud service systems based on service-oriented architecture and its evaluation. In *Cloud Computing (CLOUD), 2011 IEEE International Conference on* (pp. 638-645). IEEE. 10.1109/CLOUD.2011.119

Lee, E., Lee, E., Gerla, M., & Oh, S. (2014). Vehicular cloud networking: Architecture and design principles. *IEEE Communications Magazine*, *52*(2), 148–155. doi:10.1109/MCOM.2014.6736756

Lee, E., Gerla, M., & Oh, S. (2014). Vehicular cloud networking: architecture and design principles. *IEEE Communications Magazine, 52*(2), 148-155. doi:10.1109/mcom.2014.6736756

Lorido-Botran, T., Miguel-Alonso, J., & Lozano, J. A. (2014). A review of auto-scaling techniques for elastic applications in cloud environments. *Journal of Grid Computing*, *12*(4), 559–592. doi:10.100710723-014-9314-7

Ma, M., Huang, Y., Chu, C. H., & Wang, P. (2012, December). User-driven cloud transportation system for smart driving. In *Cloud Computing Technology and Science (CloudCom), 2012 IEEE 4th International Conference on* (pp. 658-665). IEEE. 10.1109/CloudCom.2012.6427600

Meneguette, R. I. (2016). A Vehicular Cloud-Based Framework for the Intelligent Transport Management of Big Cities. *International Journal of Distributed Sensor Networks*, *12*(5), 8198597. doi:10.1155/2016/8198597

Meneguette, R. I., Bittencourt, L. F., & Madeira, E. R. (2013). A seamless flow mobility management architecture for vehicular communication networks. *Journal of Communications and Networks (Seoul)*, *15*(2), 207–216. doi:10.1109/JCN.2013.000034

Mousannif, H., Khalil, I., & Al Moatassime, H. (2011). Cooperation as a Service in VANETs. *J. UCS*, *17*(8), 1202–1218.

Olariu, S., Hristov, T., & Yan, G. (2013). The Next Paradigm Shift: From Vehicular Networks to Vehicular Clouds. *Mobile Ad Hoc Networking*, 645-700. doi:.10.1002/9781118511305.ch19

Papadimitratos, P., Fortelle, A. L., Evenssen, K., Brignolo, R., & Cosenza, S. (2009). Vehicular communication systems: Enabling technologies, applications, and future outlook on intelligent transportation. *IEEE Communications Magazine*, *47*(11), 84–95. doi:10.1109/MCOM.2009.5307471

Qin, E., Long, Y., Zhang, C., & Huang, L. (2013). Cloud Computing and the Internet of Things: Technology Innovation in Automobile Service. Human Interface and the Management of Information. Information and Interaction for Health, Safety, Mobility and Complex Environments Lecture Notes in Computer Science, 173-180. doi:10.1007/978-3-642-39215-3_21

Sibai, R. E., Atechian, T., Abdo, J. B., Demerjian, J., & Tawil, R. (2015). A new software-based service provision approach for vehicular cloud. *2015 Global Summit on Computer & Information Technology (GSCIT)*. doi:.10.1109/gscit.2015.7353333

Sibai, R. E., Atechian, T., Abdo, J. B., Tawil, R., & Demerjian, J. (2015). Connectivity-aware service provision in vehicular cloud. *2015 International Conference on Cloud Technologies and Applications (CloudTech)*. 10.1109/CloudTech.2015.7337017

Sibai, R. E., Atechian, T., Abdo, J. B., Tawil, R., & Demerjian, J. (2015). Connectivity-aware service provision in vehicular cloud. *2015 International Conference on Cloud Technologies and Applications (CloudTech)*. 10.1109/CloudTech.2015.7337017

Sur, C., Park, Y., & Rhee, K. H. (2016). An efficient and secure navigation protocol based on vehicular cloud. *International Journal of Computer Mathematics*, *93*(2), 325–344. doi:10.1080/00207160.2014.934685

Towards Cloud-based Vehicular Networks - TSConIT. (n.d.). Retrieved May 17, 2017, from http://www.bing.com/cr?IG=917999F337DC48CF92519EF2CE98A5A6&CID=129E6989BD476F831B5B630CBCD76E04&rd=1&h=8vvJF-7_Y73ixMatQK8KK37JFmgw5bHLc3alqlNW1yE&v=1&r=http%3a%2f%2freferaat.cs.utwente.nl%2fconference%2f25%2fpaper%2f7577%2ftowards-cloud-based-vehicular-networks.pdf&p=DevEx,5061.1

Wan, J., Liu, J., Shao, Z., Vasilakos, A., Imran, M., & Zhou, K. (2016). Mobile Crowd Sensing for Traffic Prediction in Internet of Vehicles. *Sensors (Basel)*, *16*(1), 88. doi:10.339016010088 PMID:26761013

Wan, J., Zhang, D., Zhao, S., Yang, L., & Lloret, J. (2014). Context-aware vehicular cyber-physical systems with cloud support: Architecture, challenges, and solutions. *IEEE Communications Magazine*, *52*(8), 106–113. doi:10.1109/MCOM.2014.6871677

Whaiduzzaman, M., Gani, A., & Naveed, A. (2014). PEFC: Performance Enhancement Framework for Cloudlet in mobile cloud computing. *2014 IEEE International Symposium on Robotics and Manufacturing Automation (ROMA)*. 10.1109/ROMA.2014.7295892

Whaiduzzaman, M., Sookhak, M., Gani, A., & Buyya, R. (2014). A survey on vehicular cloud computing. *Journal of Network and Computer Applications*, *40*, 325–344. doi:10.1016/j.jnca.2013.08.004

Yan, G., Wen, D., Olariu, S., & Weigle, M. C. (2013). Security challenges in vehicular cloud computing. *IEEE Transactions on Intelligent Transportation Systems*, *14*(1), 284–294. doi:10.1109/TITS.2012.2211870

Zhu, K., Niyato, D., Wang, P., Hossain, E., & In Kim, D. (2011). Mobility and handoff management in vehicular networks: A survey. *Wireless Communications and Mobile Computing*, *11*(4), 459–476. doi:10.1002/wcm.853

Zingirian, N., & Valenti, C. (2012). Sensor clouds for Intelligent Truck Monitoring. *2012 IEEE Intelligent Vehicles Symposium*. 10.1109/IVS.2012.6232192

Chapter 5
A Secured Dynamic Privacy Preserving Scheme for Vehicular Cloud Computing

Y. Bevish Jinila
Sathyabama Institute of Science and Technology, India

K. Komathy
AMET University, India

ABSTRACT

There are rapid technological advancements in the field of transportation. Vehicles cooperate among themselves and establish an intelligent transportation system to fulfill the needs on traffic management. To make this possible, the vehicles are equipped with on board unit which shares information regarding the condition of roads, traffic, and management. Such information is stored in cloud server to be accessed by the vehicles at any time. During the message communication, the origin of the message is checked for its authenticity by the receiver. During signature verification, to preserve the privacy of the user, it is required that the ID of the vehicle should not be revealed. Security and privacy of the information stored in the cloud is a challenging task. The privacy of the vehicle user and the information should be secured from the threats. The solutions in the existing literature do not fulfill the requirements of security and privacy. In this chapter, a secured and dynamic privacy preserving scheme is proposed for vehicular cloud computing. Analysis on the proposed privacy scheme is done based on anonymity and un-linkability. The analysis shows that the proposed dynamic scheme provides better privacy compared to static schemes.

1. INTRODUCTION

A great development in vehicular industry has enhanced the vehicle capacity. The recent vehicles are equipped with varied resources namely the computing power and storage. In certain cases, the vehicles resources are not utilized properly and many vehicle resources are kept idle for a long time without use. To incorporate such issues Vehicular Networking is introduced. It is an emerging field due to its

DOI: 10.4018/978-1-5225-3981-0.ch005

unique features and applications like traffic management, road safety and infotainment. In a vehicular network, each vehicle communicates with other vehicles and the Road Side Unit (RSU). The vehicles carry on board computing facilities, storage and sensors. Several methods have evolved to make an efficient vehicular communication and effective utilization of resources. One such solution is Vehicular Cloud Computing (VCC) where all the information can be deployed in cloud and can be accessed when required. It is the incorporation of VANET and cloud computing. Each vehicle in VCC can communicate to the other vehicles or the network infrastructures by using the vehicle to vehicle or the vehicle to infrastructure network communication. The computing, sensing, communication and the storage resources of the vehicles are coordinated by VCC to balance the limitation of hardware and service requirements.

The National Institute of Standards and Technology (NIST) has provided a formal definition of cloud computing as:

Cloud computing is a model for enabling convenient and on demand network access to a shared pool of configurable computing resources that can be rapidly utilized and released with minimal management effort or service provider interaction (Mell & Grace, 2011).

The VCC architecture relies on three layers namely in-vehicle, communication and cloud. The first layer in-vehicle is responsible for collecting the information by the usage of sensors namely environment sensors, smartphone sensors, navigation sensors, temperature sensors and pressure sensors which is equipped in the On Board Unit (OBU) of the vehicle. The second layer is the communication layer which comprises of two different parts of communication namely Vehicle to Infrastructure (V-I) and Vehicle to Vehicle (V-V) communication.

The potential value of an intelligent transport system is to take as an example the study by Cintra (conducted at Harvard University in 2008), which showed that the traffic congestion cost in the city of Sao Paulo was approximately 33.5 billion reais per annum. Eighty-five percent of this cost is due to time lost in traffic, 13% is caused by fuel consumption, and 2% is due to increased air pollution. These traffic jams can be aggravated when the city hosts a major event like World Youth Day or the FIFA World Cup. The time lost in congestion is greater during city events and causes particular harm to the vehicular traffic required for emergency services such as ambulances, police cars, and fire engines as it means they take longer to arrive at these places. In addition, it can discourage visitors from attending these events. The cost of traffic congestion can be reduced by means of VANETs which provide updated and dynamic information about the conditions of road traffic. In addition, VANETs can reduce the number of road accidents while providing drivers and passengers with applications for comfortable driving, such as location services, streaming, multimedia, local news, tourist information, and warning messages about conditions on the highway and in the streets of the city.

Zingirian (2012) suggested that the vehicles can collect, transmit, and interpret information through a wide range of sensors, cameras, computers, and communication resources, and this can assist in the acquisition of data and help the drivers to make decisions to take appropriate actions. With enough sensing capacity and the ability to act in an environment, vehicles can become an important tool for smart cities, not only in vehicle traffic management but also for picking up useful real-time information that can be used in resource management.

According to Olariu et al. (2013) VCC is a new hybrid technology that consists of combination of various networks like wireless sensor networks, mobile ad hoc networks, vehicular ad hoc networks and cloud computing to provide services like vehicle navigation, traffic management and safety. It consists of three layers namely the communication network, communication channel and the cloud server.

In urban areas, environment monitoring and surveillance using video cameras and sensors are increasingly used for supporting smart traffic analysis. The fixed video cameras in light poles, roof tops and traffic lights for surveillance can be replaced by vehicles that could sense the environment. The surveillance using vehicles can be used to prevent possible attacks. For example if there is a suspicious activity reported regarding terrorist attack in a city, the vehicles in the cloud with defined driver profiles can be used for surveillance and can detect the attackers with suspected activities. After the occurrence of an incident for forensic investigation video surveillance can be used. For example, after the attack the path of a truck bomber can be reconstructed and a traffic accident can be investigated.

Surveillance and route tracking is an example of application totally contained in the Vehicular Cloud. The vehicular traffic management application is also an emerging application. Work on vehicular traffic management and route optimization was begun back in the 60's, using road traffic models validated by measurements and mathematical programming techniques. However, the results remained in the academic domain and found little opportunity for implementation, mainly because until recently it was extremely difficult to "measure the vehicular traffic" in real time, and to "inform vehicles of the new routes" Mario (2012). To overcome this problem, the approach taken by the Department of Transportation in the last decade was to measure the traffic by instrumenting the highways with sensors under the pavements and video cameras. This is a costly solution, which moreover detects the road segment traffic loads but cannot determine the traffic pattern (source and destination of the traffic). Secondly, the information about the "best route" was conveyed to drivers with billboards, radio announcement and, more recently, the Internet. Unfortunately, sending the same instruction to all the vehicles had the effect of creating "route flapping" problems and route instabilities. Everybody rushes to the newly announced route.

Day to day travelers park their vehicles in parking lots. The computational power of such vehicles is idle and gets wasted. Also, the computing power of the vehicles waiting near a traffic signal is also idle. Such On Board Unit (OBU) resources are shared by the users on a rental basis to form a vehicular cloud. But, there is a situation where an attacker can modify the data available in the cloud. For example, if an accident has occurred due to a vehicle, the information will be recorded in the cloud for accident occurrence. The driver of the concerned vehicle can enter into cloud and modify the original information. On the other hand, the information about each vehicle can be accessed by any other vehicle in the network from the cloud. This evades the privacy of the user in the network. Hence, security and privacy should be addressed before setting up a vehicular cloud.

The schemes in the literatures provide a static solution for obtaining a secured and privacy preserved cloud. To overcome the limitations addressed in the static schemes, a secured and dynamic privacy preserving scheme is adopted for the vehicular cloud. This scheme satisfies the following objectives:

- To enhance the privacy of the vehicle, dynamic pseudo ID is created to protect the origin of message based on location information of the vehicle.
- To secure the message communication from threats and trace back the ID of the vehicle user on certain conditions.

The remainder of this paper is structured as follows: Section 2 discusses the challenges and threats in a vehicular cloud. Section 3 discusses some related solutions on privacy from literatures. Section 4 explains the proposed scheme for privacy preserving vehicular cloud computing. Section 5 shows the experimental analysis of the proposed scheme. Section 6 analyzes the performance of the proposed dynamic scheme with respect to the static scheme. Finally, section 7 concludes the work and provides few future directions for further work in the field.

2. APPLICATIONS, CHALLENGES, AND THREATS

2.1 Applications

Figure 1 shows the taxonomy of vehicular cloud. The vehicular clouds are categorized into two major classes namely Vehicle to Vehicle (V-V) and Vehicle to Infrastructure (V-I) cloud. The V-V clouds are further categorized into static and dynamic cloud. Vehicles in parking areas form a cloud to share the resources known as static cloud. Moving vehicles and vehicular sensor clouds share resources and information known as dynamic cloud. The V-I cloud is further categorized into V-R and V-Cellular. V-R uses RSU for control information and V-Cellular uses 3G/LTE for communication.

The services of vehicular clouds include communication, processing, sensing and storage which are provided by the vehicles underlying hardware and software. All other services depend on these basic services. The services offered to users include Software as a service (SaaS), Platform as a Service (PaaS) and Storage as a Service (STaaS). All these require cooperation between cloud leader and cloud members.

Some applications of vehicular cloud include road traffic management, parking management. The traffic management system is classified into traffic flow management and traffic safety management. The traffic flow management include optimization of traffic like speed management, emergency vehicle management, predictive light control and green light management. The traffic safety management includes environment safety and passenger safety. The VCC based traffic management system includes smart traffic cloud, cloud transportation system, airport as a data center and shopping mall as a data center. The other applications include vehicle maintenance and planned evacuations. Whenever the vehicle manufacturers introduce a new version of application it is informed to the concerned end users for up gradation. In situations like Tsunami or hurricane vehicular cloud helps for planned evacuation of the people. This system helps for the safer rescue of the public. The VCC helps to implement the applications such that it is useful by the public.

2.2 Challenges

Due to the high mobility of nodes in a vehicular network it is subject to challenges namely authentication, secure localization, data security and access control. When the vehicles are automatically grouped to act as a cloud its computing power is shared as a resource to authorized users. It supports applications such as Intelligent Transportation Systems (ITS), smart cities and smart power grid. As addressed, vehicles share its computing power, internet access and storage in cloud. So, it is important to secure the resources available in cloud. The security and privacy challenges should be considered before setting up a vehicular cloud. This section gives an insight on the various challenges in the vehicular cloud.

- **Authentication:** The user ID and the integrity of the messages have to be verified in a vehicular cloud to ensure the authenticity. The authentication can be performed by metrics like ownership, knowledge and biometrics. However, it is challenging to authenticate vehicles due to its high mobility. To overcome this challenge in vehicular networks some studies like: *An efficient authentication scheme for vehicular communications* by Chenxi et al. (2008) and auditable and privacy preserving authentication in vehicular network by Kim et al. uses MAC chain and k-anonymity for privacy preservation, pseudonym change at frequent intervals and maintains a revoke list.
- **Secure Localization:** The location information of a vehicle is vital in a vehicular network for applications like traffic congestion notification, emergency warning, cooperative driving, traffic reports and entertainment. Different technologies are adopted for localization namely the radar, cooperative localization of a vehicle by the traces of previous vehicles, Geographical Positioning System (GPS) and Radio Frequency Identification (RFID). Tang et al. (2006) proposed a secure relative localization method in vehicular network to locate the correct location of a vehicle by the relative location of the connected neighbor vehicles.
- **Data Security:** The vehicular cloud environment provides an efficient way to share the resources of vehicles in the cloud. The data stored in the vehicles can be modified by the intruders. Hence, the data stored in the cloud should be encrypted. When a parked vehicle is used as a data center, it is essential to move the data to other storages once when the vehicle leaves the location.
- **Access Control:** Since multiple vehicles are involved in a vehicular network, it is a challenging task to identify a valid user in the network. The identification of the user has to be verified before providing access to the vehicular cloud.
- **Communication:** The VCC relies on the basic infrastructure of VANET and other communication networks. The incorporation of emerging technologies like 4G and 5G will support higher end services in vehicular cloud.
- **Trust Relationship Establishment:** The vehicles communicating and sharing the resources in a Vehicular Cloud should establish a trust among themselves before sharing the information. This enables a trustful communication between the vehicles and the vehicle and the infrastructure in a vehicular cloud.
- **Heterogeneity of Vehicles:** In a vehicular cloud the resources of varied heterogeneous vehicles are involved in resource sharing and allocation. The computation involved in sharing of resources varies based on the vehicles. The computational power of higher end vehicles will be more compared to lower end vehicles. It is necessary to balance this in a vehicular cloud to maintain a stable service.
- **Key Distribution and Management:** In a vehicular cloud environment, security is an important factor which should be considered. The safer key distribution and management helps the vehicles communicate among themselves and the infrastructure in a genuine manner without any suspicion on security.
- **Security Cost:** The cost of the security parameters maters more when concerned with the Quality of Service (QoS) and bandwidth. The QoS is concerned with reliability and availability of the resources. The bandwidth required to provide the resources should be effective to provide the resources on time for the vehicles in the vehicular cloud. In addition, the cost incurred by the system should be minimal.

Further, the challenge involved is the establishment of the vehicular cloud. The vehicular cloud networks are very dynamic and the number of resource providers change over time. During the time of travel, a vehicle can obtain the resources from its neighbors or the RSU which is a member of cloud. The stability of the cloud may vary based on the velocity of the vehicles, obstacles and unpredictable trajectory of the vehicles. When the vehicles are static, for example in a parking slot their resources can be shared. And this mimics the conventional cloud. However, this cloud is not permanent. When the vehicles leave the parking lot and if the resources are used by some other vehicles this creates problem. According to the arrival and departure of the vehicles there will be change in the list of available resources.

So maintaining the control policies is very important to maintain the up-to date knowledge for the cloud members and to avoid the resource fluctuations. When the vehicles are on the move and if its resources are utilized by some other vehicles it should be conserved until the mission is accomplished. So it is essential to maintain the cloud stability and accordingly the controller selection has to be done. There should be a good recovery strategy to keep the vehicle cloud stable.

The second important challenge is on resource management. Based on the different layers the issues on resource management are listed. In the Medium Access Layer (MAC), the resource allocation, scheduling, multi-channel allocation and assignment has to be sorted out for efficient resource allocation. In a vehicular cloud, there is a possibility of high packet loss due to collision caused by multiple factors namely, mobility, less signal strength and low quality link. The enormous number of vehicles using the cloud makes this a bigger challenge. In Network Layer (NL), data dissemination and load balancing are the challenges to be addressed. In a vehicular cloud, data related to various applications are transferred between the vehicles and between vehicle and the infrastructure. Transferring the messages from the conventional cloud based on the type of applications is challenging. In addition, frequent link interruptions can also be addressed by suitable protocols. The other main criterion in NL is load balancing. Suitable path selection algorithms have to be selected for lossless transmission of data and congestion avoidance. There is a possibility for congestion if the entire data is forwarded through the same path. The traffic paths have to be selected based on the application or the type of the traffic. In Application Layer (AL), estimation of the future resources, resource allocation and the types of services and applications are the challenges to be considered. Estimating the available resources for the vehicular cloud and allocating them to the needed vehicles is a challenge which should be addressed.

2.3 Security and Privacy Threats

Since the vehicular cloud environment is dynamic, it is difficult to identify the attacker information since it closely represents like original information. Hence, it is mandatory to categorize the security and privacy threats involved in vehicular communication. Some of the major threats include the following.

- **ID Spoofing:** To obtain the information from the cloud the intruder pretends to be another user like man-in-the-middle attack. Both Alice and Bob will be unaware that they are communicating with the intruder.
- **Information Disclosure:** The intruder acquires the personal information of the users namely the ID, residence of the user, medical statements, geographical history, finance and legal issues.
- **Denial of Service:** The intruder acquires the system resources and makes it unable to respond by unusually popping up with more requests and responses.

- **Repudiation:** The intruder is active in manipulating the ID of the user and the operations performed by the user.
- **Tampering of Data:** The intruder modifies the data of a genuine user.
- **Sybil Attack:** The intruder misuses the ID of other vehicle user and pretends to be multiple users. One example is formulating congestion on roads when there is actually no congestion.
- **Data Analysis Attack:** The data from the end user can be analyzed by the computing unit can eavesdrop the transmission of messages and analyze them.
- **Arbitrary Results Attack:** It is possible for the computing unit to return the results to the end user and can return the incorrect results.
- **Elevation of Privilege:** The bugs, design flaws, configuration mistake system leakage in an operating system are considered by the attacker to have an elevated access privilege to the protected data or resources.

3. STATE OF THE ART

The security and privacy issues in Vehicular Networking are widely addressed in the literatures. Golan and Meinel (2002) proposed the use of Global Positioning System (GPS) along with digital signatures to securely position and identify vehicles, and provide new schemes for both the public and private sectors. Based on the perspective of positioning of vehicles leads to privacy loss, Hubaux et al. (2004) suggested the use of electronic license plates to identify vehicles. Though the approaches of Hubaux et al. (2004) recognize the importance of conditional privacy, no specific solution is suggested.

Using this framework, Zarki et al. (2002), Raya et al. (2005), Sampigethaya et al. (2005 a), and Choi et al. (2005) propose to preload each vehicle with a pool of certified aliases called pseudonym generated by the TA during vehicle registration and/or renewal. The pseudonyms are used to hide a vehicle's unique ID. When a vehicle needs to report an event, it randomly picks one pseudonym and signs the message with it, using Public Key Cryptography (PKC). The generation of random pseudo ID makes it difficult to track a vehicle simply by observing the pseudonym it uses, and thus privacy is preserved. In trying to preserve privacy, these schemes have been shown to be susceptible to Sybil attacks. This is because a malicious vehicle may broadcast multiple messages, each signed with a different pseudonym selected from the given pool. Since other vehicles and RSUs would not know the pseudonym pool for each vehicle, they will be not being able to recognize that the messages are from one vehicle.

Raya et al. (2005) proposed the use of pseudonyms to preserve the privacy of the vehicle users. In this scheme, a set of pseudonyms are preloaded in the vehicle's On Board Unit (OBU) and is changed frequently while communication. This scheme resulted in the launch of Sybil attacks. Sha et al. (2006) proposed an adaptive privacy preserving authentication scheme based on group signatures. In this scheme, a user can choose the degree of privacy according to the resources supported by the OBU. Lower privacy level costs less and incurs less communication, computation and memory overhead, while a higher privacy level incurs more communication, computation and memory overhead. In this scheme, the OBU is loaded with the public keys required for group communication. A binary tree structure is generated with public keys at the end and each sub root is expressed by a binary number. All the RSUs are loaded with this binary tree, which is used for verification process.

The authentication process is initiated by the RSU, which broadcasts its certificate signed by TA. Before an authentication request is send by the OBU, the OBU verifies the identity of the RSU using the public key of TA. During the process of authentication request, along with session key and time stamp, the index of the sub tree root is send to indicate the users desire level of privacy. This adaptivity provides mutual authentication between OBU and RSU since both shares the same tree of public keys. However, this scheme is not scalable and has practical difficulties on implementation. This is because each OBU should maintain the public keys of all the other vehicles in the group. Since the group size varies very frequently in high speed VANETs, the OBU should update the public keys of all its group members often, which are inefficient in real time.

Calandriello et al. (2007) combines the traditional public key infrastructure with group signatures. Each vehicle holds an individual private key with a common group public key and the OBU in each vehicle generates and certifies its own pseudonyms without communication with TA. Each vehicle uses the group private key to sign the public key pairs. In the receiver side, first verification is made whether the message is delivered by a trusted group member. And, second, the message is checked for integrity.

A scalable privacy preserving method proposed by Xi et al. (2007) uses symmetric random key set in VANETs. In this scheme, a random key set is chosen from the central pool and is shared among multiple vehicles in the group. This makes the RSU difficult to trace a vehicle in its communication range. To prevent correlation attack from RSU, this scheme uses a predefined authentication period for each key set chosen from the key ring. The ID of the vehicle can be traced with the involvement of the key distribution center. However, the limitation in this scheme includes the compromise of the RSU.

Later Raya et al. (2007) introduced a scheme of pseudonyms with expiry time. During the time of change of pseudonyms, it is obtained from the nearest RSU. Wasef and Shen (2013) proposed a scheme called Expedite message authentication protocol in VANET which uses public key based authentication. Feiri et al. (2015) proposed an authentication scheme, which pre-loads the certificates for broadcast authentication. In this approach, omission schemes are employed to reduce the cryptographic packet loss. Two types of pre distribution methods are employed namely geographic and temporal pre-distribution. However, this scheme is limited to one hop dissemination. Though the certificates are pre-loaded, the overhead incurred is high and privacy is not addressed.

To avoid unlinkability among the changing of pseudonyms, Freudiger (2007) to some extent proposed another scheme of changing the pseudonyms in social spots called mix zones. He has proposed that the pseudo IDs can be changed at fixed locations called mix zones so that when all the vehicles in a mix zone changes their pseudo IDs at one point of time, it becomes impossible for the adversary to track the vehicle. Gerlach et al. (2007) considered the number of nodes in its proximity as a criterion to define the mix zones. When the neighboring vehicles of a particular source have similar direction then such places are considered to be mix zones where the pseudo IDs are changed. These mix zones are said to be context mix zones. Though, how the vehicles in the neighborhood are detected and how the nearby vehicles are guaranteed to similarly react is not included in this scheme. Ren et al. (2008) introduced user revocation for privacy aware PKI.

Buttyan (2009) proposed a scheme where the vehicles are allowed to change the pseudo IDs when the speed of the vehicle drops below 30 km/hr. This period is known as silent period where the vehicle does not transmit safety messages and just changes the pseudo ID. But, the occurrence of an emergency event in the silent period when not reported immediately may cause a serious issue. Active and passive location security algorithms were proposed by Yan et al. (2008). In this paper, radar is used as a "virtual eye," and onboard radar can detect the location of vehicles. GeoEncrypt in VANETs has been proposed by Yan et al.(2009). The idea is to generate a secret key using the geographic location of a vehicle. The secret key is used to encode the messages. The receiving vehicles must be physically present in the geographic region specified by the sender to be able to decrypt the message.

A privacy preserving pseudonym based Sybil attack detection scheme known as P^2DAP proposed by Tong Zhou et al. (2011) is a light weight and scalable protocol. By means of passive overhearing by the fixed units namely the RSUs Sybil attacks can be overheard. This scheme employs a two level hashing scheme where the TA generates a coarse level hash group and from the coarse level hash groups, each group generates a set of multiple fine grained hash groups. Each fine grained hash group is assigned to a vehicle and is loaded in its OBU. The pseudonyms are changed within fixed time intervals during the report of an event.

Rongxing et al. (2012) suggests that changing pseudo IDs at fixed intervals makes the adversary to trace the original ID of the vehicle and has proposed a more similar approach where the pseudo IDs are changed in social spots like traffic signals and parking lots where the quantity of vehicles arriving is quite high. In addition to the criteria proposed by Gerlach et al. (2007), Stubing (2013) has considered the adversary's tracking capabilities as a measure for defining inappropriate time when a pseudo ID can be changed.

Dong et al. (2011) proposed a lite-TA based public key cryptography and on path onion encryption scheme which withstands traffic tracing attack. Though this scheme eliminates deployment complexity, it is dependent completely on TA for certificate generation.

Another scheme is proposed by Sun et al. (2010), which addresses the issue of handling revocation of certificates. An OBU resigns the certificate obtained from TA with the RSU keys. However, RSUs should be deployed in regular intervals on roads and is subject to compromise attack. Also, no solution is provided for Sybil attacks. Implementing such a scheme leads to malicious activities and would be expensive.

Shim (2012) has devised an efficient conditional privacy preserving scheme. In this scheme, each message sent from the vehicle is mapped to a distinct pseudo identity. It supports the fastest batch verification by verifying 750 signatures simultaneously in less than 300 milli-seconds. This scheme uses a tamper proof device for storing the master secret and the Map To Point function used by Zhang et al. (2008 a) is replaced. Though, the master secret is not stored in the tamper proof device for a long time, it is known that such tamper proof devices can be easily cracked by side channel attacks. The cracking of proof devices becomes a major drawback for this approach.

Zhang et al. (2008) uses an approach where original ID is used to generate the pseudo IDs. This scheme heavily relies on the Tamper Proof Device (TPD) for storing all the pseudo IDs generated by the trusted authority during vehicle registration. Though, it is proved to be conditionally traceable it is required to store the set of generated pseudo IDs in the TPD and each time a message is send a different pseudo ID is utilized. Kyung-Ah-Shim (2012) has inducted the scheme of Zhang et al. (2008 c) and has included the lifetime for pseudo IDs. Based on the lifetime, the pseudo IDs are changed. This scheme is also proved to be conditionally traceable. These schemes heavily rely on the TPD to store all the generated pseudo IDs and these pseudo IDs are alive for a lifetime. And, does not provide any solution for efficiently changing the pseudo IDs such that it becomes difficult for the adversary to trace the original ID of the vehicle.

Gerlach et al. (2007) considered the number of nodes in its proximity as criteria to define the mix zones. When the neighboring vehicles of a particular source have similar direction then such places are considered to be mix zones where the pseudo IDs are changed. These mix zones are said to be context mix zones. However, the way the vehicles in the neighborhood are detected and the nearby vehicles are guaranteed to react is not included in this scheme. In addition to the above criteria, Stubing (2013) has considered the adversary's tracking capabilities as a measure for defining an appropriate time when a pseudo ID can be changed. Buttyan et al. (2009) proposed a scheme where the vehicles are allowed to change the pseudo IDs when the speed of the vehicle drops below 30 km/hr. This period is known as silent period where the vehicle does not transmit safety messages and just changes the pseudo ID. But, the occurrence of an emergency event in the silent period may cause a serious issue if not reported immediately.

Subir Biswas et al. (2011) advised to use a common pseudo ID for all the vehicles in a particular communication range. And, this pseudo ID is selected from the most significant bits of the Geographical Positioning System (GPS) coordinates, so that all vehicles within a communication range hold the same pseudo identity information. This method provides a complete privacy for the user, but does not provide any solution for conditional traceability.

However in a vehicular cloud, the communication, storage and computing resources are not properly addressed. The idea of a vehicular cloud was proposed by Weigle (2009) and the paper emphasized on the research challenges and applications of vehicular cloud. Md Whaiduzzaman et. al (2013) in the paper *A Survey on Vehicular Computing* has proposed a key management scheme for securing the vehicular cloud and provided a detailed survey on the various challenges, threats, requirements and usage of a vehicular cloud.

To summarize, all these issues from existing works indicate that, storing the generated pseudo IDs in TPD and using them at regular intervals or changing it at fixed intervals or in mix zones becomes more frantic and even an adversary can trace the ID of the vehicle when it is changed. To solve these issues, the proposed system model shows the generation of secured and dynamic pseudo ID by the vehicle instantly on its travel during the report of an event.

4. SYSTEM MODEL

This section briefly describes the proposed scheme of secured and dynamic privacy preservation for vehicular cloud. Safety applications in VANET need to be handled with utmost care so that no adversary is allowed to intrude the privacy of the user. Replacement of original ID of the vehicle with pseudo ID makes this possible. Trusted Authority (TA) generates a set of pseudo IDs and stores them in the Tamper Proof Device (TPD) of each vehicle and changes them at fixed intervals. This paves way for an adversary to trace the original ID of the vehicle. Also, the space required for storing the generated pseudo IDs in the TPD of each vehicle is large. The secured and dynamic pseudo ID generation scheme addresses the generation of secured and dynamic pseudo ID by the vehicle instantly on its travel during the report of an event by the origin.

The TA generates password for driver authentication and issues to vehicle user during registration. To improve the security of the generated passwords, instead of storing the passwords in the TA, passwords are split by adopting Shamir's secret sharing scheme proposed by Dan et al. (2007) and distributed to the TMS located in various zones. The pseudo ID is dynamically generated based on the current location of the origin of the message, which is used for communication between the vehicle and the RSU. The cluster head verifies the pseudo ID, in case of any dispute, forwards it to the TMS to take necessary action.

The functionality of the proposed system is categorized into three sections namely the original ID generation, secured and dynamic pseudo ID generation, conditional traceability.

4.1 Original ID Generation

During the time of vehicle registration in the transport authority, based on the vehicle registration number, the user is provided with a unique ID and password for authentication. The first part is the vehicle ID that represents the state, second part represents the district, third part represents the sector of the vehicle and fourth part is a 4 digit unique number. Table 1 shows the format of the original ID. The vehicles OBU is preloaded with the password *p*, the public key of the authority T_{pub} and the hashed value of the password $K=H(m)$.

4.2 Secured and Dynamic Pseudo ID Generation

Once the authentication of the driver is confirmed, the pseudo IDs are generated by the OBU in the vehicle dynamically, based on the location when an event is to be reported. Since GPS position information represents a broader range, each vehicle in the network will not have unique positioning information by, which the pseudo ID generated will not be of a random value. To alleviate this issue, an extra parameter from the format of GPS data representation GPGGA (Global Positioning System Fix Data) as referred by National Marine Electronics Association (NMEA) standard can be incorporated during this process as given in Table 2. Along with the parameters latitude and longitude, time parameter is also included for unique positioning. Using this location information pseudo ID generation is done.

For example, the GPS coordinates of Sathyabama University, Sholinganallur, India is considered. Approximately for a region of 1 km, the GPS coordinate in terms of latitude and longitude is $12^0 52'$ and $80^0 13'$ of degrees and minutes and there is minor change in the seconds, which varies from $2''$ to $32''$. For processing, this location information in Degrees, Minutes, Seconds (DMS) form is converted into decimal degrees using equation (1).

$$Dec_d = d + \frac{m}{60} + \frac{s}{3600} \tag{1}$$

where, Dec_d represents decimal degrees, *d* represents the degree, *m* represents minutes and *s* represents seconds. So, by applying equation (1) the latitude and longitude for $12^0 52' 26''$ and $80^0 13' 6''$ is 12.87 and 80.21 in terms of decimal degrees. Thus the GPS coordinates are mapped to their corresponding Cartesian coordinates for experimentation. Generation of pseudo ID is done dynamically during the report of an event. Table 3 shows the format of an event report.

The system parameters namely the password, master secret and public key of TA is loaded in the OBU of the vehicle. Let *m* be the master secret of the TA. The TA generates the password for driver authentication as follows and issues it to the driver to keep it secret. Let, *p* be the password and is computed by multiplying the master secret *m* and a random number *r* as given in equation (2). The public key of the trusted authority is computed as given in equation (3) where p is a random prime number.

$$PWD = m * q \tag{2}$$

$$T_{pub} = m * p \tag{3}$$

The master secret *m* is hashed and stored in the TPD as shown in equation (4). The values obtained from equation (2) (3) and (4) along with parameter *p* is loaded in the TPD of the OBU.

$$Q = H(m) \tag{4}$$

Algorithm 1 shows the dynamic pseudo ID generation based on the location information during the report of an event to the vehicle.

4.2.1 Conditional Traceability

Conditional traceability is defined as the ability to trace the original ID of the vehicle if it is identified to be malicious. When a vehicle is identified as malicious, it is required to trace its original ID for taking necessary action by legal authorities. Once when a malicious event is identified by the RSU, the location of the received malicious event is forwarded to the TMS, which further forwards to the TA for tracing the original ID. The proposed model supports conditional traceability of a malicious vehicle by the TA. The process of conditional traceability is described using a sequence diagram shown in Figure 2.

Given a pseudo ID P_{ID}, the TA computes a function by using the public parameters Q, T_{pub}, password *PWD* and the location as given in equation (5), (6), (7) and (8).

Algorithm 1. Dynamic pseudo ID generation

Aim: To generate pseudo ID dynamically on receiving an event report
Input: System parameters, Event report
Output: Pseudo ID
Steps:
1. Event = {event ID, location, time}
2. Let L be the location coordinates with x and y.
3. Let m be the master secret.
4. Let T_{pub}=m*p be the public key of TA.
5. Let p be a random prime number.
6. Assign Q = $H(m)$ where $H(.)$ is a SHA-1 function.
7. Compute the hash value for x and y as,
$n_1 = H(x,y,t)$ where, $H(.)$ is a SHA-1 function.
8. Compute s with the system parameters Q, T_{pub}, and k as
9. $n_2 = Q * T_{pub}$
10. s = $n_1 + n_2$
11. Compute the pseudo ID as, $P_{ID} = p \oplus s$

$$n_1 = H\left(x, y, t\right) \tag{5}$$

$$n_2 = Q * T_{pub} \tag{6}$$

$$n = n_1 + n_2 \tag{7}$$

$$d = n \oplus P_{ID} \tag{8}$$

where $H(x,y)$ represents the SHA-1 hash function applied upon the location coordinates (x,y) and time t. T_{pub} is the public key of the trusted authority as shown in equation (4). Q is the hash value of the master secret m as shown in equation (5). The value n is the additive of n_1,n_2.

5. EXPERIMENTAL ANALYSIS

5.1 Original ID Generation

Table 4 shows the sample list of vehicle numbers in India and their corresponding original IDs generated. The first part in the vehicle ID represents the state, second part represents the district, third part represents the sector of the vehicle and fourth part is a 4 digit unique number.

The pseudo ID generation is tested for vehicles from 5 States namely TN, KL, KA, AP and MP. These States are assigned values 1, 2, 3, 4, 5 during original ID generation respectively. Three categories namely private, Governmental and commercial vehicles are considered and their values include 1, 2 and 3 respectively.

5.2 Pseudo ID Generation

As pseudo ID generation depends upon the location, the GPS coordinates of the vehicle location is mapped to its corresponding Cartesian coordinates. As mentioned before, the GPS location is coarse for an area of 1 km. If pseudo ID is generated with this coarse location, the difference in pseudo IDs generated will be minimal, which results in easier traceability of original ID by a malicious user. Table 5 shows the GPS to Cartesian coordinate mapping for the sample data. The Cartesian coordinates obtained are used for pseudo ID generation as given in Table 6.

A hash function is applied to the x and y coordinates of the location retrieved and the first 24 bits are extracted and stored in n_1 as given in equation (5). Then equations (6) and (7) are applied to these parameters, which when XORed with the password p provides the pseudo ID. Table 6 shows the pseudo IDs generated for a sample list of vehicles at a time instance t using only coordinate values without involving time factor in the computation. Since pseudo ID generation solely depends upon the location of an event occurrence, the GPS location of the vehicle reporting the event has to be a fine grained to show the difference in the pseudo ID among neighbor vehicles. To incorporate this, time parameter is then included for finding a fine grained location. The time parameter considered is in terms of *hours:minutes:seconds.* This time parameter is added with the (x, y) location coordinate to obtain a fine grained location.

Table 7 shows the pseudo IDs for the sample list of original IDs at time instance *t* with (x,y) location attribute and with time parameter.

5.3 Conditional Traceability

As an example on conditional traceability, a sample list of original IDs and their corresponding passwords generated is shown in Table 8. Table 9 shows the sample calculation of n_1, n_2, n and p from a pseudo ID.

6. PERFORMANCE ANALYSIS

The level of security and privacy has to be analyzed to show the performance of the secured and dynamic pseudo ID generation. The analysis is based on the following.

1. To show that the adversary is uncertain about the original ID of the vehicle.
2. To prove that two pseudo IDs generated from an original ID is unlikable with each other.
3. To show that the original ID of a malicious vehicle is conditionally traceable.

Even though, pseudo IDs are dynamically generated by the OBU of the vehicle itself, it is mandatory to trace the original ID of the vehicle when required, so that the entire network can be secured from adversaries. This traceability of the original ID of the vehicle is a condition-based and it should be carried out by the TA.

6.1. Privacy Analysis of Pseudo ID

The level of privacy provided can be measured using anonymity and unlinkability. The quantitative measures such as anonymity set size and Pearson's correlation coefficient are used.

6.1.1 Anonymity

Anonymity as defined by Pfitzmann and Hansen (2009) states that it is the state of being not identifiable within a set of objects namely the anonymity set. The level of anonymity achieved is investigated in terms of a quantitative measure called anonymity set size. All the vehicles reporting the event will generate pseudo ID based on the current location. Let $V=\{v_1,v_2,\ldots,v_n\}$ be the set of vehicles reporting an event. So, the anonymity set size is $A_s = V_n$. If the number of vehicles reporting the event increases, greater will be the anonymity set size. An analytical model is introduced to theoretically investigate the anonymity. Let V_a be the vehicle arrival during the occurrence of an event and let it be a Poisson process. Let t_a be the inter arrival time of vehicles and is considered to be exponential with mean $\frac{1}{-}$. Let $T_a = t$, where t=30 and 60 seconds be the waiting time of the vehicles until the path is cleared.

Let V be the random variable of vehicles arriving near the occurrence of an event during a time period T_a. So, the probability of a vehicle v arriving at time t can be represented as given in equation (9).

Suppose a message M is communicated by a source vehicle v using its P_{ID} and let V be the total number of vehicles waiting for clearance of path. If an attacker has obtained a pseudo ID of a particular vehicle with no further knowledge, the anonymity set size for the adversary A with respect to time t is shown in equation (10) and (11).

$$P\left[V = v \mid T_a = t\right] = \frac{\left(t\right)^v}{v!} e^{-t} \tag{9}$$

$$E\left[V \mid T_a = t\right] = t \tag{10}$$

$$A_s = t \tag{11}$$

Figure 3 shows the anonymity set size with respect to the inter arrival time of vehicles.

6.1.2 Unlinkability

From the attackers perspective, unlinkability of two or more pseudo ID is defined as the disability of the attacker to map two or more pseudo ID received from the same source.

Making an adversary impossible to map two or more pseudo IDs received from the same vehicle or a pseudo ID to its corresponding original ID is essential. This feature of unlinkability should exist for pseudo IDs received within a short period or a long period. In this system, the IDs are represented using alphanumeric codes. For analysis, the ASCII values of the corresponding alphanumeric codes are calculated and are represented using numeric values.

Case 1: Unlinkability Between Original ID and Pseudo ID of a Vehicle

Let $r_1, r_2, \ldots, r_n$ be the original IDs and $p_1, p_2, \ldots, p_n$ be their corresponding pseudo IDs. In this case, unlinkability can be expressed as the probability that an adversary can use the received pseudo IDs to successfully map their corresponding original IDs.

Let R be a random variable that denotes the set of original IDs of vehicles $\{r_1, r_2, \ldots. r_n\}$. Let P be another random variable that denotes the set of pseudo IDs generated for the corresponding original IDs. The correlation between the two random variables R and P can be measured by the Pearson's correlation coefficient, r as given in equation (12).

$$r = \frac{n \sum PS - \sum P \sum S}{\sqrt{n \sum P^2 - \left(\left(\sum P\right)^2\right)\left(n \sum S^2 - \left(\sum S\right)^2\right)}}. \tag{12}$$

where $\sum R$ represents the original IDs varying from 1 to n and $\sum P$ represents the pseudo IDs varying from 1 to n. When the value of *r* is close to zero, it means that the dispersion is more and the variables are uncorrelated. For analysis, a sample set of data shown in Table 4 and 7 is considered. For the given sample values, the correlation coefficient *r* is 0.05, which is significant. This ensures that by knowing the pseudo ID, the adversary finds it difficult to trace the original ID.

Let *r* be the original ID of a vehicle and p_1, p_2,...p_n be the corresponding pseudo IDs. Equation (12) is used to show that the pseudo IDs are unlinkable with each other.

The correlation between the pseudo IDs generated for a single original ID is measured. The correlation coefficient *r* is zero, which proves that the pseudo IDs are unlinkable with each other. This shows that when an adversary captures two pseudo IDs generated from a single original ID, it becomes difficult for an adversary to link both pseudo IDs to map an original ID. For analysis, the number of pseudo IDs are varied from 10 to 100 and the correlation coefficient is fixed to a probability value of 0 to 1. From Figure 4 it is evident that the correlation coefficient for the dynamically generated pseudo IDs is comparatively less when compared to static pseudo IDs.

Case 2: Unlinkability Between the Pseudo IDs for the Same Vehicle

Let P ={p_1,p_2..., p_n} be the set of pseudo IDs received from the same vehicle. To represent any of the items in the set *is related* an equivalence relation is used, which is represented by, $\sim_{r(P)}$. Let p_i and p_j be two randomly selected pseudo IDs from the list. For a random variable X, equation (13) shows the posterior probability of the attacker, given two items p_i and p_j.

$$P\left(p_i \sim_{r(P)} p_j\right) = P\left(X = \left(p_i \sim_{r(P)} p_j\right)\right) \tag{13}$$

and, $P\left(p_i \nsim_{r(P)} p_j\right)$ denotes the probability that p_i and p_j are not related to each other as shown in equation (14):

$$P\left(p_i \sim_{r(P)} p_j\right) + P\left(p_i \nsim_{r(P)} p_j\right) = 1 \tag{14}$$

$\forall p_i, p_j \in P$.

To measure the unlinkability of two received pseudo IDs, the attacker's entropy can be used as shown in equation (15):

$$H\left(i, j\right) = H\left(X\right) \tag{15}$$

Entropy is a mathematical function that quantifies the expected value of information contained in a received message. It measures the uncertainty on the information sent by the source. In VANET, each vehicle reports the occurrence of an event. During this process, the pseudo ID is generated to represent the source vehicle. When the next transmitted pseudo ID is unpredictable then it is more uncertain and this leads to maximum entropy. So, if an adversary is uncertain about the next possible pseudo ID, then the entropy is high. Entropy is computed as given in equation (16):

$$H\left(X\right) = -P\left(x_i\right)\log_b P\left(x_i\right) \tag{16}$$

The degree of unlinkability can be achieved as given in equation (17):

$$d\left(i,j\right) = H\left(i,j\right) \tag{17}$$

When equation (16) is substituted in equation (17), the entropy between two pseudo IDs can be computed as shown in equation (18):

$$d\left(i,j\right) = -P\left(p_i \sim_{r(P)} p_j\right)\log_2\left(P\left(p_i \sim_{r(P)} p_j\right)\right) - P\left(p_i \nsim_{r(P)} p_j\right)\log_2 P\left(p_i \nsim_{r(P)} p_j\right) \tag{18}$$

If *d(i,j)=0*, it leads to minimum entropy and *d(i,j)=1* leads to maximum entropy.

For the sample list of pseudo IDs given in Table 7, the entropy is computed. For the pseudo IDs obtained by the adversary, p_i and p_j are related if both the values are equal in their corresponding bit positions and is not related if the values are different in their corresponding bit positions. Since there are two possibilities, the maximum entropy is $log_2(2)=1$. Table 10 shows the entropy computed for two pseudo IDs generated at two different time instances for a single original ID using equation (18).

From Table 10 it can be inferred that for all the samples the entropy is closer to 1, which means it is closer to maximum. So, it can be assured that pseudo IDs generated dynamically are not able to be linked with each other.

When there is an occurrence of a malicious event, the conditional traceability of original ID of the vehicle is not possible in traditional schemes. To alleviate this limitation, Tong et al. (2011) proposed a scheme of static generation of pseudo IDs for a vehicle. In this scheme, a set of pseudo IDs are generated based on one way hash function and that belongs to the same pool. Table 11 shows the sample list selected from statically generated pseudo IDs.

Figure 5 compares the inability to be linked feature of static pseudo IDs proposed by Tong et al. (2011) and dynamic pseudo IDs proposed by this research work. From Figure 5, it can be inferred that the uncertainty incurred by dynamic pseudo IDs is nearer to one. Also, compared to statically generated pseudo IDs, the entropy obtained by dynamic pseudo IDs is 0.2 to 0.6 bits higher. So, it can be found that the dynamic pseudo IDs provides more privacy compared to statically generated pseudo IDs.

7. CONCLUSION

Privacy of the pseudo ID is analyzed based on anonymity and unlinkability. According to the analysis, it can be inferred that greater the inter-arrival time, lesser will be the vehicles near the event. So, lesser the anonymity set size, lesser will be the privacy. Unnlinkability is measured in terms of two scenarios. In scenario 1, Pearson's correlation coefficient is used to measure the correlation between the original ID and the pseudo IDs. As an average, the correlation of dynamic pseudo IDs is lesser than the static pseudo IDs. In scenario 2, unlinkability between two pseudo IDs of the same vehicle is analyzed based on entropy. Inference shows that dynamic pseudo IDs have higher entropy compared to statically generated pseudo IDs. This system provides a better way of generating pseudo IDs which are conditionally traceable in need.

REFERENCES

Bevish Jinila, Y. (2015). *Anonymization based location privacy preservation in Vehicular ad hoc networks.* Academic Press.

Bevish Jinila, Y., & Komathy, K. (2013). A privacy preserving authentication framework for safety messages in vanet. *4th International Conference on Sustainable Energy and Intelligent System (SEISCON 2013)*, 456-461.

Bevish Jinila, Y., & Komathy, K. (2014). Distributed and secured dynamic pseudo ID generation for privacy preservation in Vehicular ad hoc networks. *Journal of Applied and Theoretical Information Technology, 66*(1), 126 – 134.

Biswas, S., & Misic, J. (2013). A Cross-Layer Approach to Privacy Preserving Authentication in WAVE-Enabled VANETs. *IEEE Transactions on Vehicular Technology, 62*(5), 2182–2192. doi:10.1109/TVT.2013.2238566

Buttyan, L., Holczer, T., Weimerskirch, A., & Whyte, W. (2009). Slow: A practical pseudonym changing scheme for location privacy in vanets. *IEEE Vehicular Networking Conference (VNC)*, 1-8. 10.1109/VNC.2009.5416380

Calandriello, G., Papadimitratos, P., Hubaux, J. P., & Lioy, A. (2007). Efficient and Robust Pseudonymous Authentication in VANET. *Proc. 4th ACM Int. Workshop VANET*, 19-28. 10.1145/1287748.1287752

Choi, J. Y., Jakobsson, M., & Wetzel, S. (2005). Balancing auditability and privacy in vehicular networks. *Proceedings of the 1st ACM international workshop on Quality of service & security in wireless and mobile networks*, 79-87.

CintraM. O. (2014). Retrieved from http://blogchicao.tripod.com/custocarro.html

Dong, X., Wei, L., Zhu, H., Cao, Z., & Wang, L. (2011). An Efficient Privacy-Preserving Data-Forwarding Scheme for Service-Oriented Vehicular Ad Hoc Networks. Vehicular Technology. *IEEE Transactions on, 60*(2), 580–591.

El Zarki, Mehrotra, Tsudik, & Venkatasubramanian. (2002). Security issues in a future vehicular network. *European Wireless*.

Feiri, M., Pielage, R., Petit, J., Zannone, N., & Kargl, F. (2015). Pre-distribution of certificates for pseudonymous broadcast authentication in VANET. *IEEE 81st Vehicular Technology Conference (VTC Spring),* 1-5. 10.1109/VTCSpring.2015.7146029

Gerla, M. (2012). Vehicular Cloud Computing. *Workshop on Vehicular Communications and Applications*.

Gerlach, M., & Guttler, F. (2007). Privacy in VANETs using changing pseudonyms-ideal and real. In *Vehicular Technology Conference, 2007. VTC2007-Spring*. IEEE. 10.1109/VETECS.2007.519

Gollan & Meinel. (2002). Digital signatures for automobiles. In *Systemics, Cybernetics and Informatics*. SCI.

Hubaux, J.-P., Capkun, S., & Luo, J. (2004). The security and privacy of smart vehicles. *IEEE Security and Privacy Magazine*, *2*(3), 49–55. doi:10.1109/MSP.2004.26

Kim, S.H., Kim, B.H., Kim, Y.K., & Lee, D.H. (n.d.). *Auditable and privacy preserving authentication in vehicular networks*. IEEE.

Lu, R. (2012). Pseudonym changing at social spots: An effective strategy for location privacy in vanets. *IEEE Transactions on Vehicular Technology*, *61*(1), 86–96. doi:10.1109/TVT.2011.2162864

Mell, P., & Grace, T. (2011). *NIST definition of cloud computing*. National Institute of Standards and Technology. doi:10.6028/NIST.SP.800-145

Neha Roy, Y., & Jinila, B. (2015). A survey on security challenges and malicious vehicle detection in Vehicular ad hoc networks. *Contemporary Engineering Sciences, 8*(5-8), 235-240.

Raya, M., & Hubaux, J.-P. (2005). A security of vehicular ad hoc networks. *Proc. SASN*. 10.1145/1102219.1102223

Raya, M., & Hubaux, J.-P. (2007). Securing vehicular ad hoc networks. *Vehicular Technology. IEEE Transactions on*, *15*(1), 39–68.

Sam Mathews, Y., & Jinila, B. (2014). An effective strategy for pseudonym generation and changing scheme with privacy preservation for VANET. *International Conference on Electronics and Communication Systems (ICECS) 2014*, 1-6.

Sampigethaya, K., Huang, L., Li, M., Poovendran, R., Matsuura, K., & Sezaki, K. (2005a). *CARAVAN: Providing location privacy for VANET*. Washington Univ Seattle Dept of Electrical Engineering.

Sampigethaya, K. M. Y., Huang, L. P., & Poovendran, R. (2005). Amoeba: Robust Location Privacy Scheme for VANET. *IEEE Journal on Selected Areas in Communications*, *25*(8), 1569–1589. doi:10.1109/JSAC.2007.071007

Sha, K., Xi, Y., Shi, W., Schwiebert, L., & Zhang, T. (2006). Adaptive Privacy-Preserving Authentication in Vehicular Networks. *Proceedings of theFirst International Conference on Communications and Networking in China,* 1–8. 10.1109/CHINACOM.2006.344746

Shim, K. A. (2012). An Efficient Conditional Privacy-Preserving Authentication Scheme for Vehicular Sensor Networks. *Vehicular Technology. IEEE Transactions on*, *61*(4), 1874–1883.

Stubing. (2013). *Multilayered security and privacy protection in car-to-X networks: Solution from application down to physical layer*. Springer.

Sun, Y., Lu, R., Lin, X., Shen, X., & Su, J. (2010). An efficient pseudonymous authentication scheme with strong privacy preservation for vehicular communications. *Vehicular Technology. IEEE Transactions on*, *59*(7), 3589–3603.

Tang, L., Hong, X., & Bradford, P. (2006). Secure relative location determination in vehicular network. *Mobile Ad hoc and sensor networks*, 543-554.

Wasef, A., & Shen, X. (2013). EMAP: Expedite message authentication protocol for vehicular ad hoc networks. *IEEE Transactions on Mobile Computing*, *12*(1), 78–89. doi:10.1109/TMC.2011.246

Whaiduzzaman, Sookhak, Gani, & Buyya. (2013). A Survey on Vehicular cloud computing. *Journal of Network and Computer Applications*.

Xi, Y., Sha, K., Shi, W., Schwiebert, L., & Zhang, T. (2007). Enforcing Privacy UsingSymmetric Random Key-Set in Vehicular Networks. *Proceedings of EighthInternational Symposium on Autonomous Decentralized Systems, ISADS'07*, 344–351. 10.1109/ISADS.2007.37

Yan, G., Olariu, S., & Weigle, M. (2009). Providing location security in Vehicular ad hoc networks. *IEEE Wireless Communications*, *16*(6), 48–55. doi:10.1109/MWC.2009.5361178

Yan, G., Olariu, S., & Weigle, M. C. (2008, July). Providing VANET security through active position detection. *Computer Communications*, *31*(12), 2883–2897. doi:10.1016/j.comcom.2008.01.009

Zhang, Lin, & Lu, Ho, & Xuemin. (2008, November). An efficient message authentication scheme for vehicular communications. *IEEE Transactions on Vehicular Technology*, *57*(6).

Zhang, Lin, Lu, & Ho. (2008). An efficient message authentication scheme for vehicular communications. *IEEE Transactions on Vehicular Technology, 57*, 3357–3368.

Zhou, Choudhury, Ning, & Chakrabarty. (2011). P2DAP – Sybil Attacks Detection in Vehicular Ad Hoc Networks. *IEEE Journal on Selected Areas in Communications, 29*(3), 582 – 594.

Zingirian, N., & Valenti, C. (2012). Sensor clouds for intelligent truck monitoring. *Proceedings of the IEEE Intelligent Vehicles Symposium*.

KEY TERMS AND DEFINITIONS

Anonymity: The state of being not identifiable within a set of objects namely the anonymity set.

Conditional Traceability: Conditional traceability is the ability to trace the original ID of a user based on a specified condition.

Entropy: Entropy is a mathematical function which quantifies the expected value of information contained in a received message.

Unlinkability: The incapability of relating two items is known as unlinkability.

APPENDIX

Tables and Figures

Table 1. Original ID format

State	District	Sector	Unique number

Table 2. GPGGA data representation

Sl. No.	Parameters
1	Time
2	Latitude
3	Longitude
4	Fix quality
5	Number of satellites
6	Horizontal dilution of precision
7	Altitude
8	Height
9	Time since last DGPS update
10	DGPS reference station ID
11	Checksum

Table 3. Format of event report

Event ID	Location (x,y)	Time (t)

Table 4. Sample list of vehicle ID and their original ID

Vehicle ID		Original ID
Tamil Nadu	TN22BR4995	12214995
	TN21AG1510	12121510
	TN18CA1212	11831212
Kerala	KL02MD1200	20211200
	KL11CA1000	21131000
	KL13AG1312	21321312
Karnataka	KA13AG2100	31322100
	KA10MA1490	31011490
	KA09CA7000	30937000
Andra Pradesh	AP01AG2300	40122300
	AP12CA4325	41234325
	AP14PM1100	41411100
Madhya Pradesh	MP15CA2434	51532434
	MP13AG1890	51321890
	MP05GD8065	50518065

Table 5. GPS to Cartesian coordinate mapping

Sl. No.	GPS Coordinate	Cartesian Coordinate
1	$12^{0}52^{'}26'' \, and \, 80^{0}13^{'}6''$	(12.873, 80.218)
2	$12^{0}52^{'}28'' \, and \, 80^{0}13^{'}7''$	(12.874, 80.218)
3	$12^{0}52^{'}32'' \, and \, 80^{0}13^{'}11''$	(12.875, 80.219)
4	$12^{0}52^{'}38'' \, and \, 80^{0}13^{'}17''$	(12.877, 80.221)
5	$12^{0}52^{'}42'' \, and \, 80^{0}13^{'}20''$	(12.878, 80.222)

Table 6. Sample list of generated pseudo IDs without time parameter

Sl. No.	Cartesian Coordinates	Pseudo ID
1	(12.873, 80.218)	16212224
2	(12.874, 80.218)	16212223
3	(12.875, 80.219)	16212048
4	(12.877, 80.221)	16212003
5	(12.878, 80.222)	16948186

Table 7. Sample list of pseudo IDs generated with time parameter

Sl. No.	Cartesian Coordinates	Time	Pseudo ID
1	(12.873, 80.218)	12:11:10	17231989
2	(12.874, 80.218)	12:11:11	10999306
3	(12.875, 80.219)	12:11:12	13121922
4	(12.877, 80.221)	12:11:13	09655713
5	(12.878, 80.222)	12:11:14	12167282

Table 8. Original IDs and their passwords

Original ID	Password
12214995	1234
20211200	1325
31322100	1410
40122300	1521
51532434	1610

Table 9. Traceability of the original ID for the sample pseudo ID

Pseudo ID	n_1	n_2	N	PWD	Original ID
17231989	16140895	1092168	17233063	1234	12214995

Table 10. Entropy

Pseudo ID	Entropy
17231989	0.92
10999306	0.97
13121922	0.92
09655713	0.95
12167282	0.95

Table 11. Sample list of statically generated pseudo IDs

Time Instance	Statically Generated Pseudo IDs	Entropy
t_1	33554431 33552383	0.45
t_2	33548287 33549311	0.53
t_3	33550248 33550271	0.62
t_4	33550239 33550271	0.45
t_5	33553919 33553663	0.75

Figure 1. Taxonomy of vehicular clouds

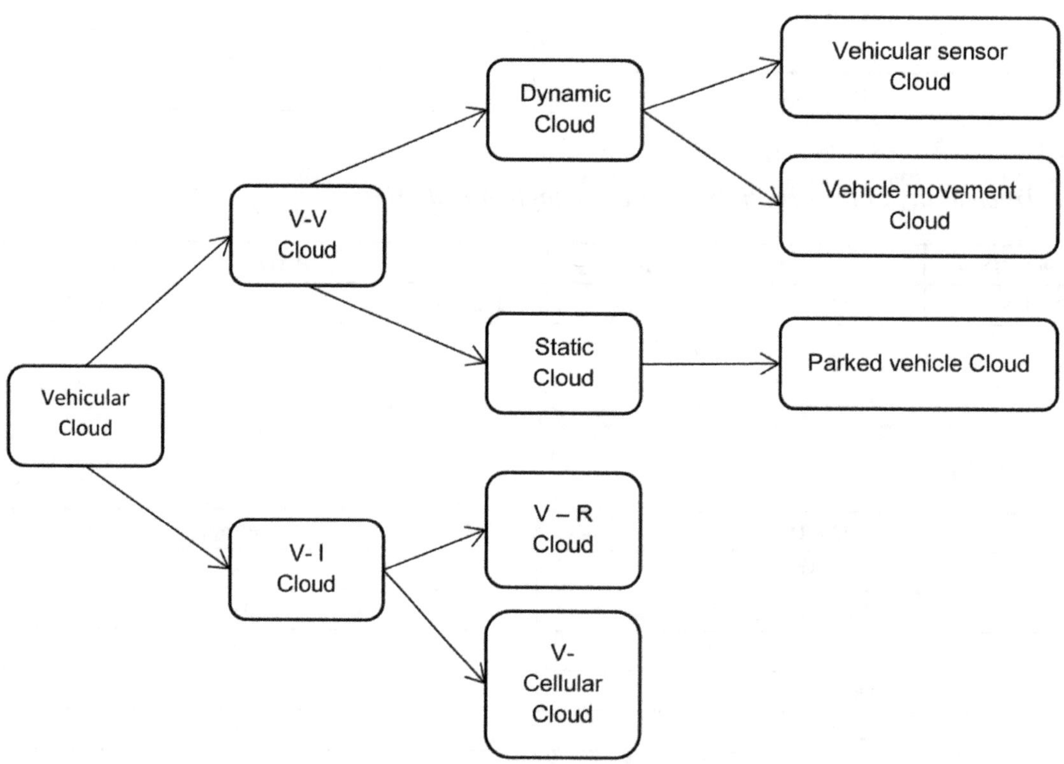

Figure 2. Sequence diagram for conditional traceability

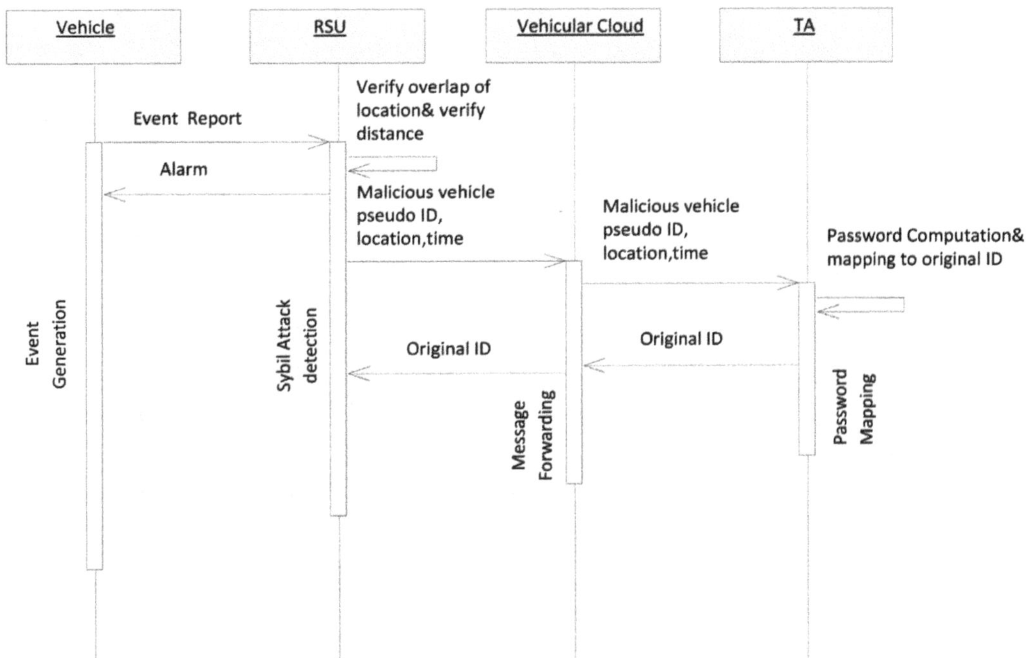

Figure 3. Anonymity set size

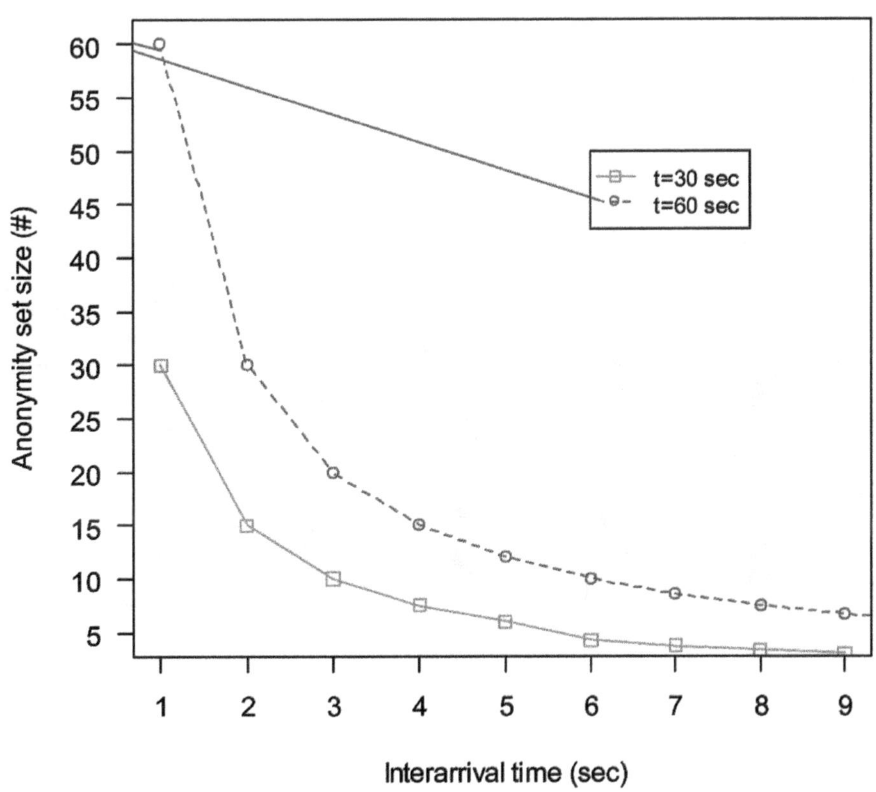

Figure 4. Correlation coefficient vs. pseudo ID

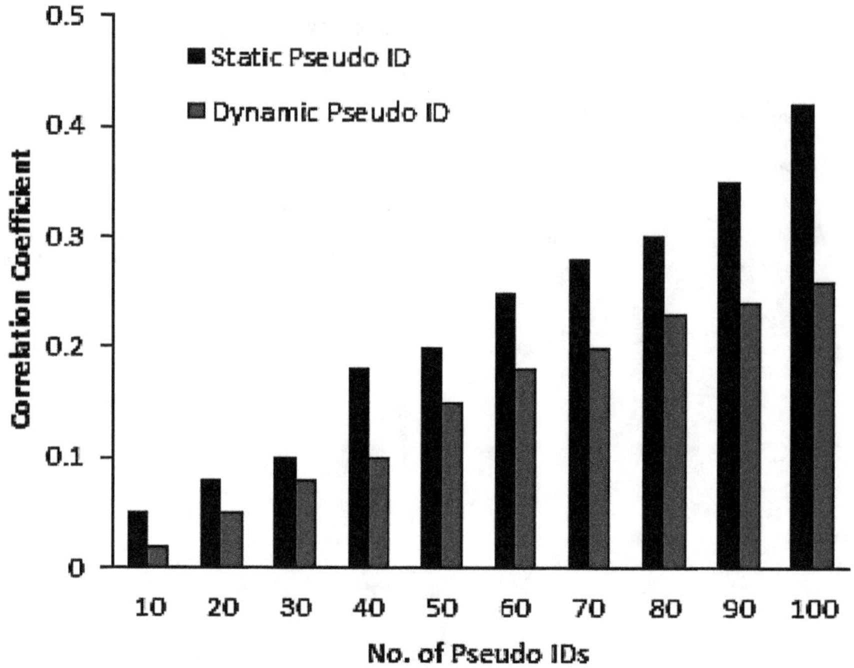

Figure 5. Entropy of static vs. dynamic pseudo IDs

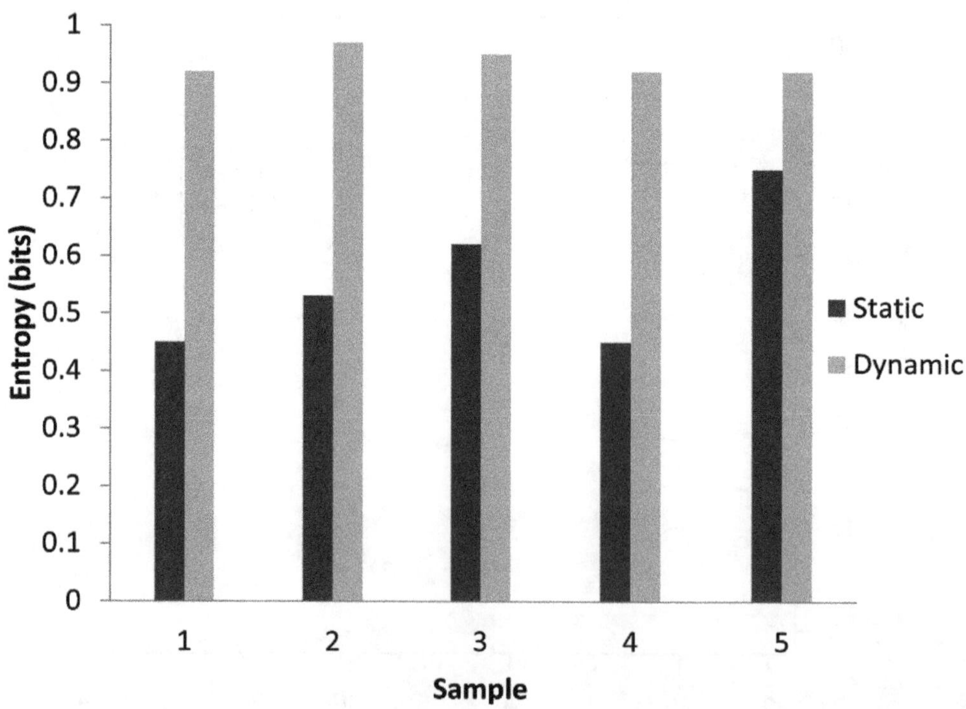

Chapter 6
V2V Communication Protocols in Cloud-Assisted Vehicular Networks

Amal Ahmed Eltahir
Sudan University of Science and Technology, Sudan

Rashid A. Saeed
Sudan University of Science and Technology, Sudan

ABSTRACT

Integration of vehicular ad-hoc network (VANET) and cellular network is a promising architecture for future machine-to-machine applications. This integration helps the vehicles have steady internet connection through cellular network (i.e., LTE), and at same time communicate with other vehicles. However, dead spot areas and unsuccessful handoff processes due to the high speed of vehicles that can disrupt the implementation of this kind of architecture. In this chapter, simplified cluster-based gateway selection (SCGS) scheme for multi-hop relay in VANET network is proposed. The scheme is achieved by utilizing a new routing protocol called an enhanced hybrid wireless mesh protocol (E-HWMP). The simulations results show that SCGS scheme through E-HWMP protocol performed better than ad-hoc on demand distance vector (AODV) routing protocol. Furthermore, SCGS scheme through E-HWMP is compared with other cluster-based gateway selections used in the previous works; the result shows that SCGS scheme through E-HWMP protocol outperforms the other cluster-based gateway selections schemes.

INTRODUCTION TO V2V

Along with the ongoing advances in dedicated short-range communication and wireless technologies, inter-vehicular communication and road–vehicle communication have become possible, giving birth to a new network-type called Vehicular Ad-hoc Network (VANET). VANETs aim to improve road safety and transportation efficiency, as well as to reduce the impact of transportation on environment. Vehicle communication networks are formed by connecting devices inside public and private vehicles with each other (vehicle–to–vehicle communications) and with fixed communication infrastructure (vehicle–to–

DOI: 10.4018/978-1-5225-3981-0.ch006

infrastructure communication) (Eltahir, & Saeed, 2015; Elathir, & Saeed, 2016). Providing internet access to devices (end users) in vehicles enables various applications in safety and emergency warning, traffic management and entertainment. The key role that VANETs can play in the realization of intelligent transport systems has attracted the attention of major car manufacturers. Heterogeneous wireless network is the one of the hot research topics in wireless domain. Recent research in wireless networking has been focusing upon the heterogeneous integration of IEEE 802.11-based wireless ad hoc networks with 3GPP cellular networks (Liu, Jia, Wang, Lu, & Wu, 2015; Saeed, Alawi, & Hassan, 2011; Eltahir, Saeed, & Mokhtar, 2014). The 3GPP cellular networks such as Universal Mobile Telecommunications System (UMTS), are predominantly used for wide-area wireless data and voice services via access to a Base Station Transceiver (BST), also referred to as UMTS Node B. On the other hand, VANETs are used for short range, high-speed communication among nearby vehicles, and between vehicles and roadside infrastructure units. Vehicle to- Vehicle (V2V) communication supports services such as car collision avoidance and road safety by exchanging warning messages across vehicles (Eltahir, Saeed, Mekherjee, & Kamrul Hasan, 2016; Eltahir, Saeed, & Alawi, 2013).

The major weakness of IEEE 802.11 radio is its very limited transmission range which is typically 200-300 meters (Jiau, 2015; Wan, Tang, & Wolff, 2008). In order to cover the wide area, a large number of access point is needed which results in high deployment cost. From this it is understood that the better solution is to use a small number of access points (Aps) along the roadside and form an ad-hoc network among vehicles and apply a relay mechanism to forward the packets for vehicles outside of the AP range or to integrate VANET to the network that can provide high coverage range (i.e. cellular network) in order to reduce the number of access points and the switching process from one access point to another.

The unique characteristics of VANET are the high mobility and rapidly changing network topology caused by the high travelling speed of the nodes; the constrained pattern due to the restricted roads; limitations of bandwidth due to the absence of a central coordinator that controls and manages communications between nodes; disconnection problems owing to the frequent fragmentation in the networks and signal fading caused by objects that form obstacles between the communicating nodes.

Consequently, the main challenges facing VANET are to decide upon the routing protocol that should be used to control the process of forwarding packets through nodes on the network; determining how to select the next-hop node to use to forward packets to their final destination, particularly in a sparse environment, depending on the presence of the unique characteristics of VANET. Therefore, providing a robust routing protocol is considered to be the most crucial solution in VANET (Matthew, 2012). This involves using new parameters to take the decisions regarding selecting the next-hop node to enhance the efficiency of the routing process and increase performance. As a result, it is anticipated that designing an efficient routing protocol will aid in accomplishing the task of delivering packets to their destinations in VANET via a more realistic method, which promises to apply road safety efficiently.

VANET is a specific circumstance of MANET. Many routing techniques have been designed in MANET to tackle the limitations with transmission packet delivery delay, involving packets being dropped, wasting bandwidth, mobility and security. These techniques could not be appended to VANET owing to its particular characteristics, such as the restricted mobility pattern.

It is observed that frequencies network disruption is one of the main problems in VANET network. Many researchers in this area have proposed some solutions in order to eradicate this problem. However, since VANET involves vehicles moving in a very high speed with limited coverage area of IEEE 802.11, lack of high technology makes it difficult for such kind of network to have stable connection especially in the rural area where by the network coverage is quite low.

The main aim of this chapter is to enhance integration of VANET and cellular network in a way that extends the coverage in dead spot areas, reduces the amount of overhead cost, reduces delay metrics and increases the packet delivery ratio.

The specific objectives of this chapter are:

1. To design "end to end" multi-hop scheme to extend the coverage of VANET network.
2. To develop a new cluster-based gateway (cluster head) selection algorithm for Vehicular communication.
3. To design an Enhanced Hybrid Wireless Mesh Protocol (E-HWMP) Protocol for cluster-based multi-hop Vehicular Communications.
4. To evaluate the proposed scheme under multi-metric design parameters for overall throughput.

VANET-UMTS Heterogeneous Wireless Network

The Intelligent Transportation System (ITS) is moving significantly toward advanced active safety in the recent years, and various kinds of research and efforts has carried out in academia and in industry. One the promising technology, which has improved transportation safety, is using wireless communication between cars on the road. This wireless network between vehicles is known as vehicular ad hoc network. Vehicles, which use wireless communication, need to follow same regulation and standard in order to be able to exchange information. The regulation is a law which is suggested by a governmental agency, for instance Federal Communications Commission (FCC) determines the wireless frequency spectrum which is allowed to be transmitted by vehicles. On the other hand, the standard defines protocols or interfaces which advance interoperability between vehicles developed by professional organization, for

Figure 1. VANET-UMTS heterogeneous wireless network architecture

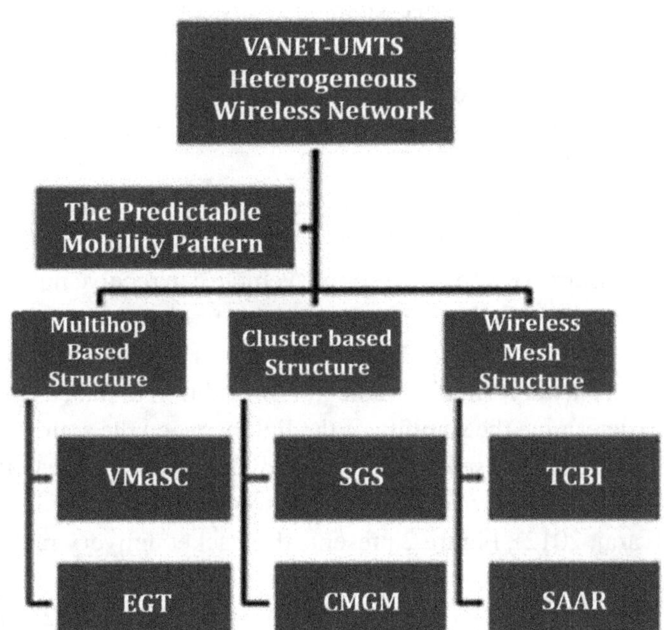

example IEEE. In 1997, the United States Federal Communications Commission (FCC) allocated 75MHZ of Dedicated Short Range Communications (DSRC) spectrum at 5.9 GHZ to be used for vehicular ad hoc network. The DSRC is a radio services which receives 75 MHZ of spectrum in the 5.85-5.925 GHZ range. Furthermore, in 2008, the European Union (EU), which is a confederacy of 27 European countries, allocated 30 MHz of spectrum, 5.875–5.905 GHz, for wireless communication between vehicles (Saeed, Alawi, & Hassan, 2011; Benslimane, Taleb, & Sivaraj, 2012). VANET-UMTS Heterogeneous Wireless Network Architecture is presented in Figure 1.

Cluster Based Structure

In this section some schemes and protocols that are closely related to the intended proposed scheme are reviewed and their strengths and weaknesses are highlighted and at the end suggestions have been given as what should be done to improve the current schemes. In addition to that, from this review it is observed that two recently scheme proposed by (Saeed, Alawi, & Hassan, 2011; Bensalimane, Taleb, & Sivaraj, 2012) has got some weaknesses where at the end they have been selected and their weakness are improved in VANET –cellular networks scenario. In (Zhioua, Tabbane, Labiod, & Tabbane, 2014), the authors proposed gateway selection method considering traffic priority first, with the purpose of ensuring the Quality of Service (QoS) of different traffic data. Nevertheless, we note that most of the existing studies focus on the uplink strategy which indicates how to select a suitable gateway by source vehicle to upload message to a server, while the downlink message dissemination strategy is seldom considered. In (Saeed, Alawi, & Hassan, 2011), a simplified gateway selection (SGS) was proposed to enable vehicles to continue connecting to the internet or other mobile service (i.e. location map download). Also assumed that each vehicle located inside the coverage zone has two interfaces one to connect to infrastructure mode which is UMTS UTRAN interface and the other is to connect to the ad-hoc mode (IEEE802.11), both interfaces must be active all the time and vehicles located outside the coverage zone contained only one interface. This scenario involves vehicles moving on the highway into two opposite directions. All vehicles access the UMTS network via UMTS Node B over the UMTS radio interface (UTRAN). The vehicles will experience different signal strength (RSS) from Node B. All vehicles connect directly to Node B if they are inside of the coverage (using UMTS UTRAN interface) and once UMTS RSS fall to certain predefined threshold the vehicle will use it IEEE802.11 interface to find the partner to act as relay to connect to the UMTS Node B.

However, it should be noted that this solution is less reliable in terms of availability and complex in terms of cost of creating. In addition, the delay between broadcast Relay Request messages (RReq), receiving Relay Respond messages (RRep) (The source vehicle can receive numbers of RRep but choose the best one) is not taken into consideration. Therefore, in this work a new gateway selection mechanism is proposed by formulating a cluster only at the LTE cell edges. This is done by using received signal strength (RSS) threshold as a cluster formation indicator. In addition to that, a hybrid metrics of the Link Expiration Time (LET) to determine the stability of the link between the source vehicle and the GW vehicle along with the best route capacity selection to avoid overload and hence bottleneck in one gateway.

The use of multi metrics for gateway selection was also studied in (Saeed, Alwai, & Hassan, 2011; Benslimane, Taleb, & Sivaraj, 2012). Figure 2 presents the Packet delivery ratio of SGS scheme on top of Reactive and proactive routing protocol with variation of mobility speed of vehicles (Saeed, Alwai, & Hassan, 2011) and Figure 3 presents the DPDR for different average IEEE 802.11p wireless transmission ranges (Benslimane, Taleb, & Sivaraj, 2012).

Figure 2. Packet delivery ratio of SGS scheme on top of reactive and proactive routing protocol with variation of mobility speed of vehicles

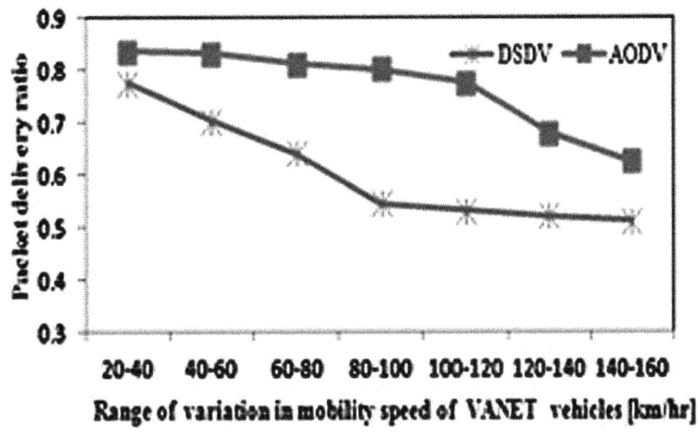

Figure 3. DPDR for different average IEEE 802.11p wireless transmission ranges

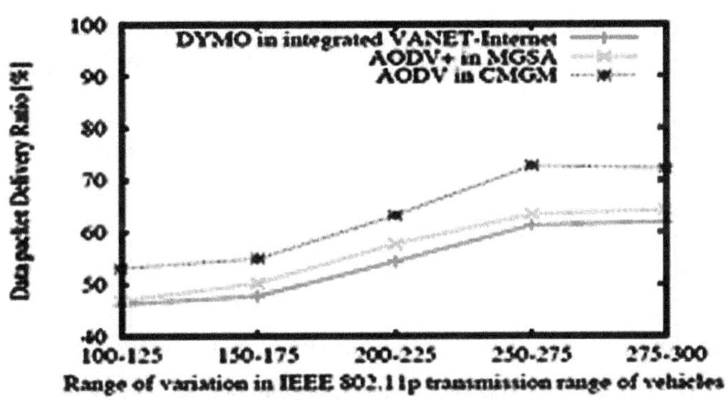

The Predictable Mobility Pattern

In (Namboodiri & Gao, 2007), a prediction-based routing (PBR) protocol is proposed specifically for mobile gateway. This protocol takes the advantage of the predictable mobility pattern of the vehicles on highways. The protocol allows the vehicles which have both WLAN and WWAN interfaces to act as mobile gateway and the normal vehicles contained only WLAN interface to use that mobile gateway to connect to Internet. The work assumes the vehicles moving into two opposite directions in the highway and source vehicle uses the link lifetime between the source vehicles and the mobile gateway to select the optimal mobile gateway. However, it can be noted that in this work different approaches have been used to determine the link lifetime. However, only velocity and location of the vehicles in relation to the mobile gateway were used in calculating the route lifetime.

The minimum of the link lifetime along the route is used as the route life time utilized by the source vehicle in selecting the route. Two PBR variants are also proposed in this work: PBR-S and PBR-M in case when an old route breaks. A normal PBR during the event of the route break, the new nearest gateway is selected even though the route can be established using the same gateway. In PBR-S the

Figure 4. Multi-head election

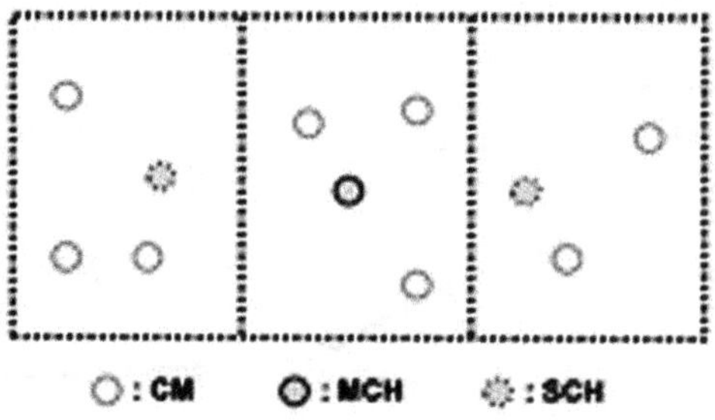

same gateway is selected during the route failure until no route can be constructed to it. This mechanism reduces the gateway switching which has an impact on network performance. PBR-M selects the gateway among all those within a certain number of hops with largest predicted route lifetime whereas during the route breakage PBR-M selects a new gateway with the same largest predicted lifetime criteria. In (Lo, Shou-Chih, Yi-Jen Lin, & Gao, 2013), a Clustering technique is proposed to group nodes into several clusters. Each node in the cluster structure plays one of three roles: Cluster Head (SCH), Cluster Gateway (MCH), and Cluster Member (CM). The cluster head in a cluster plays the roles as coordinator and backbone. It is in charge of all the communications inside a cluster, managing medium access and allocating the resource to cluster members. A MCH is a border node of a cluster that can communicate nodes belonging to different clusters. Figure 4 presents the Multi-head election (Lo, Shou-Chih, Yi-Jen Lin, & Gao, 2013).

Multi-Hop Based Structure

In (Benslimane, Taleb, & Sivaraj, 2012), a Clustering-based Multi-metric Adaptive Mobile Gateway Management Mechanism (CMGM) is proposed. Here, two types of the vehicles were assumed: the vehicles which have two interfaces, UTRAN interface and IEEE 802.11p interface which are referred to as gateway candidates (GWC) and normal vehicles containing only one interface (IEEE 802.11p). The scheme allows minimum number of vehicles to connect to UMTS network by allowing normal vehicles to use the GWC to communicate with the UMTS network. Clustering mechanism is used to cluster the gateway candidates which depend on direction of movement, UMTS-RSS (Received Signal Strength), and IEEE 802.11p transmission ranges. GWCs closed to the cluster center are always selected as Cluster head. Normal vehicles use cluster heads to communicate to UMTS network. Multi metrics Mobile selection algorithm based on Simple Adaptive Weighting (SAW) technique (Setiawan, Bouk, & Sasase, 2008) is used as well in (Benslimane, Taleb, & Sivaraj, 2012) .

Three metrics are used for mobile gateway selection; mobility speed, UMTS-RSS and link stability. The paper also discusses handoff mechanism and gateway management in VANET-UMTS integration. However, the scheme proposed in (Benslimane, Taleb, & Sivaraj, 2012) is complex in terms of cost of creating and maintaining the clusters for rapid environment as in VANET. In addition to that, the cluster head election process was based only on the geographic position of the vehicles for a lower TTL value. This could result in the cluster heads not possessing the essential credentials to manage the VANET clusters and to enable them communicate with the UMTS also the signaling traffic and time that is used for clustering formation and cluster head selection is larger than the time of data traffic exchange.

In (Ucar, Seyhan, Coleri Ergen, & Ozkasap, 2013), the authors introduce VMaSC: Vehicular Multi-hop algorithm for Stable Clustering. VMaSC is a novel clustering technique based on choosing the node with the least mobility calculated as a function of the speed difference between neighboring nodes as the cluster head through multiple hops. In (Anjum Khan, 2015), a two-layer novel Evolutionary Game Theoretic (EGT) framework is presented to solve the problem of in-stable clustering in VANETs. The aim of this research is to model the interactions of vehicular nodes in VANETs, to retain a stable clustering state of the network with evolutionary equilibrium as the solution of this game. A stable clustering scenario in VANETs is modelled with a reinforcement learning approach to reach the solution of an evolutionary equilibrium. Performance of the proposed \evolutionary game based clustering algorithm "is empirically investigated in different cases and he simulation results show that the system retains cluster stability.

In (Matthew, 2012), dynamic clustering and gateway management in vehicular ad-hoc networks are proposed. This method consider two types of the vehicles: first, gateway candidates (GWC) referred to the vehicles that have two interfaces, UTRAN interface and IEEE 802.11p interface and second, normal vehicles that include single interface (IEEE 802.11p). By means of the decided on VANET mobile gateways, VANET is connected to UMTS using the Universal Terrestrial Radio Access Network (UTRAN) interface. Dynamic clustering mechanism is used to organize the gateway candidates d into clusters. A minimum number of gateways, per time instance, are employed to attach the regular vehicles with the UMTS network. Route stability, mobility features, and signal strength of vehicles are all taken into consideration when clustering vehicles and selecting vehicle gateways. A Multi-metric Mobile Gateway Selection Algorithm (MMGSA) is proposed, it is employed upon the available CHs of the GWC sub-cluster. An optimal gateway is then selected by the source vehicle using the MMGSA mechanism.

Wireless Mesh Structure

In (Kathirvel, & Loyd, 2014), an optimized Hybrid Wireless Mesh Protocol had been suggested to construct a routing protocol over VANET. A real-time packet loss estimation was utilized as an additional parameter to the parameters used in (Benslimane, Taleb, & Sivaraj, 2012) (UMTS received signal strength, link stability and available relay capacity) to elect an efficient mobile gateway node for the transfer of packets without losing any of them.

In (Meng, 2016) the authors proposed spatial reusability-aware single-path routing (SASR) and any path routing (SAAR) protocols, and compare them with existing single-path routing and any path routing protocols, respectively. The evaluation results show that our protocols significantly improve the end-to-end throughput compared with existing protocols. Specifically, for single-path routing, the median throughput gain is up to 60 percent, and for each source-destination pair, the throughput gain is as high as 5.3x; for any path routing, the maximum per-flow throughput gain is 71.6%, while the median gain is up to 13.2%. The proposed system architecture is shown in Figure 5 (Meng, 2016).

In (Li, 2014) the authors propose a reliable multicast protocol, called CodePipe, with energy-efficiency, high throughput and fairness in lossy wireless networks. Building upon opportunistic routing and random linear network coding, CodePipe can not only eliminate coordination between nodes, but also improve the multicast throughput significantly by exploiting both intra-batch and inter-batch coding opportunities. In particular, four key techniques, namely, LP-based opportunistic routing structure, opportunistic feeding, fast batch moving and inter-batch coding, are proposed to offer significant improvement in throughput, energy-efficiency and fairness. Moreover, the authors design an efficient online extension of CodePipe such that it can work in a dynamic network where nodes join and leave the network as time progresses.

Figure 5. System architecture

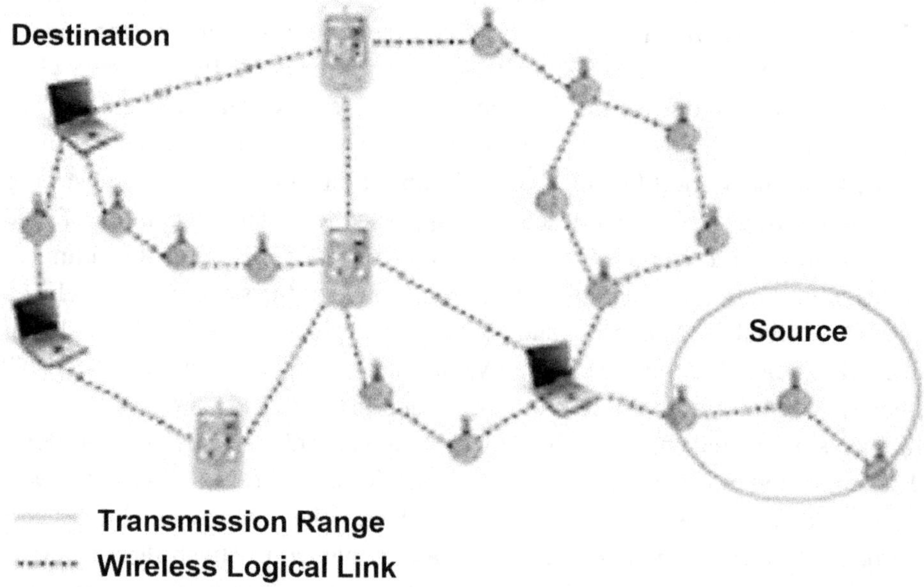

The proposed protocol evaluated on ns2 simulator by comparing with other two state-of-art multicast protocols, MORE and Pacifier. Simulation results show that CodePipe significantly outperforms both of them.

In (Zhou, 2015) the authors proposed a novel threshold credit-based incentive mechanism (TCBI) based on the modified model of population dynamics to efficiently resist the node compromise attacks, stimulate the cooperation among intermediate nodes, maximize vehicular nodes' interest, and realize the fairness of possessing the same opportunity of transmitting packets for credits. Then, a TCBI-based privacy-preserving packet forwarding protocol was proposed to solve the open problem of resisting layer-adding attack by outsourcing the privacy-preserving aggregated transmission evidence generation for multiple resource-constrained vehicles to the cloud side from performing any one-way trapdoor function only once. The vehicle privacy is well protected from both the cloud and transportation manager. Finally, formal security proof and the extensive simulation show the effectiveness of their proposed TCBI in resisting the sophisticated attacks and the efficiency in terms of high reliability, high delivery ratio, and low average delay in cloud-assisted vehicular DTNs.

In (Viriyasitavat, 2015) the authors classify and describe the most relevant vehicular propagation and channel models, with a particular focus on the usability of the models for the evaluation of protocols and applications. First they classify the models based on the propagation mechanisms they employ and their implementation approach. They also classify the models based on the channel properties they implement and pay special attention to the usability of the models, including the complexity of implementation, scalability, and the input requirements (e.g., geographical data input). They also discuss the less-explored aspects in vehicular channel modelling, including modelling specific environments (e.g., tunnels, over-passes, and parking lots) and types of communicating vehicles (e.g., scooters and public transportation vehicles). They conclude their paper by identifying the under researched aspects of vehicular propagation and channel modelling that require further modelling and measurement studies.

Gateway Mechanism

The topology of the proposed architecture consider a scenario of two different tracks over a particular road (e.g. Highway), with two lanes for each direction, involves vehicles moving on the highway into two opposite directions. Two types of the vehicles were assumed, vehicles inside the coverage zone have two interfaces: E-UTRAN interface and IEEE802.11p interface are called Gateway Candidates (GWCs), GWCs that are moving in the same direction are grouped into one cluster using a dynamic clustering mechanism, all GWCs having their 3GPP E-UTRAN interfaces activated and used it to connect direct to LTE network other type of vehicles occupied with only one interface (IEEE802.11p) are called ordinary vehicles, those vehicles connect to LTE network through gateway)GW), GW is the cluster head (CH) of the cluster. For position all vehicles have built-in GPS device, connect to different LTE signal strength along the road. To reduce the delay of handovers between gateways, a new gateway elected when serving gateway near to lose its optimality. Communication over the VANET network is multi-hop and the nodes of VANET communicate with each other on a peer- to-peer basis. In the proposed architecture, the IEEE802.11p WLAN mesh, forms the standard for wireless access for vehicular environment (WAVE), which utilizes vehicular communication among vehicles (V2V). The standard does not support vehicle to infrastructure (V2I) communication. An LTE eNode B is used to provide connectivity of internet to the V2V network.

An enhanced hybrid wireless mesh protocol (E-HWMP) assists any source vehicle to deliver its packet to eNodeB through GW in multi-hop manner. The main components of the LTE network are:

- **Base Station Transceiver eNodeB:** Communication with the GW and GWc via E-UTRAN interface.
- **Serving Gateway (S-GW):** Serving gateway and forward between base station and P-GW.
- **Packet Data Network Gateway (P-GW):** Through P-GW, LTE network is connected to the external IP networks.

E-HWMP is standard for wireless access to vehicular environments (WAVE) that employs a vehicular connection in vehicles (V2V). A Vehicle to infrastructure (V2I) connection is not supported by the standard for wireless access. The Evolved Universal Terrestrial Radio Access Network (E-UTRAN) interface is utilized by the VANET to connect to the internet through eNodeB. Gateway numbers are required to be lower. The proposed scheme is cluster based. The clusters are formulated only at a certain point of the 3GPP LTE cell's edges when the received signal strength (RSS) crumbles to less than assured threshold.

This scheme adopts a hybrid (Saeed, Alawi, & Hassan, 2011) that is a combination of both proactive and the reactive approaches. Firstly, the gateway broadcasts the Gateway Advertisement message (GANN) proactively to an area confined to a limited number of hops which is known as the TTL value of the message. The TTL is set to advertisement zone so that the advertisement message can be forwarded only up to this maximal number of hops through the ad hoc network. This area is known as the proactive zone. The nodes that are outside the proactive zone establish connection to the gateway reactively, On-demand Broadcast of Gateway Request message (GREQ), by Active Sources in VANET. In the hybrid approach the gateway does always send a GWADV message every 5 seconds. The main purpose of this approach (hybrid gateway discovery mechanism) is to the control overhead due to the reactive discovery scheme when there are many source nodes in the VANET. This approach also aims to reduce the overhead of the proactive protocols when the number of gateways increases. Architecture of t proposed mechanism implemented using E-HWMP protocol. E-HWMP is a combination of IEEE802.11p (AODV Routing Protocol) and IEEE802.11s (HWMP Routing Protocol. Proposed protocol provides a tradeoff between reactive and proactive gateway discovery solutions and bypasses the drawbacks of previous hybrid protocols, which consists of adjusting the scope for the gateway advertisement for proposed protocol. AODV was chosen over HWMP as the routing protocol to be modified. The basic idea behind hybrid routing protocols is to use proactive routing mechanisms inside the coverage zone and reactive routing on demand to be able to connect to the network through gateway in multi-hop manner. The layer-3 approach (E-HWMP) implements mesh functionality at network layer.

E-HWMP: Proactive Routing Protocol

E-HWMP is a proactive routing protocol for VANETs that periodically broadcast information about a route to connect to active GW with all nodes inside the coverage zone (GWCs). At each GWC when a route information is received, a route to the GW is created and updates its E-HWMP routing table.

E-HWMP: Reactive Routing Protocol

E-HWMP is a reactive routing protocol for VANETs that creates road-based paths (or routes) on-demand, this happens when the vehicle is out of coverage zone and need to continuo connect to the network, it uses a reactive mechanism to find a route in multi-hop manner through intermediate node.

Message exchanges of the E-HWMP protocol are:

1. **Gateway Announcement (GANN) (Broadcast):** Use to inform about the presence of Gateway (GW). Active GW periodically broadcast GANN to all Gateway candidates (GWCs) in the cluster. The information contained in the GANN is used to disseminate route metrics to the GW. At each GWC when GANN is received a route to the GW is created and updates its E-HWMP routing table.

```
IF (route entry for GANN node exists(
update route entry
ELSE
create route entry for GANN node
ENDIF
```

2. **Gate Request (GREQ):** When a source node needs to connect to a gateway node,E-HWMP is designed to assist source vehicle to connect to the gateway in multi-hop manner. The source vehicle starts to sends Gateway Request message (GREQ). Gateway Requests message are initially sent with small Time-to-Live (TTL) field, the time which a SV wait before sending another GREQ to another nearest node,. Gateway Discovery Mechanism illustrate later in this chapter.

```
IF (the neighbor vehicle is a GWC(
SV send multi-hop cluster request
```

```
IF (receives OK(
send their packets
Else
IF (the neighbour vehicle within multi-hop cluster(
updates the HWMP table and sends multi-hop cluster request
IF (receives OK(
send their packets
Else
IF (the neighbour vehicle is not GWc cluster nor within multi-hop cluster and
reply time out)
SV search for hop to LTE
 Else
SV buffer packet for available hop to LTE
ENDIF
ENDIF
ENDIF
ENDIF
ENDIF
```

3. **Gate Replay (GREP) (ACK):** Upon receiving the GREQ by the nearest node, GREP message sent by the nearest node of the source vehicle if its GWC cluster or it's not GWC cluster but within multi-hop cluster. If a GREP is not received by the source within a short time limit after it sends GREQ, the source sends GREQ to other nearest node.

```
IF ((nearest vehicle is GWC or within multi hop cluster)
nearest vehicle sends ok to the SV
ELSE
SV send GREQ to other nearest node.
ENDIF
```

When the source vehicle receives the GREP it will update the E-HWMP routing table and start to send its packet.

4. **Message Error (MERR):** Notify the source of the broken link, it is propagated to all the affected destinations. E-HWMP works depend on three main metrics for cluster and three algorithms for gateway management. The three metrics are:

    a. Direction of Movement (θ).
    b. LTE Signal Strength.
    c. IEEE 802.11p wireless transmission range.

The three algorithms are:

a. Multi-metric Mobile Gateway Selection.
b. Gateway Discovery/Advertisement.
c. Multi-metric Mobile Gateway Handover.

E-HWMP is designed to assist source vehicle to connect to the gateway in multi-hop manner. The destination sequence numbers are used by E-HWMP in order to detect stale routing information. Newly received routing information with a smaller sequence number than the sequence number of the corresponding information already known to the node will be discarded, because it is outdated, this avoids the creation of routing loops. The E-HWMP uses the route discovery process difference from AODV. The time to live field (TTL) defines the scope of the GANN in number of hops.

Gateway Discovery Mechanism

This algorithm shown in algorithm (1) is employed to allow source vehicle to connect to LTE network through Gateway. In the proposed Gateway Discovery mechanism there is no need to broadcast the packet to all the neighboring nodes, as this has an impact on reducing the overheads generated. A peer to peer mode can be utilized as a communication technique between vehicles. The proposed algorithm done in the following steps:

1. When the LTE receive signal strength (RSS) is degrades below the threshold, it means that the vehicle is nearly to move outside current LTE cell area. This will be followed by handoff to next cell or enter to the dead spot zone. The source vehicle (SV) connects to the LTE network through multi-hop and with the help of cluster head. SV sends a message for multi-hop cluster. For scalability reasons, the message includes time to live (TTL) value, which defines the scope of the GREQ in number of hops that needed to connect with the GW that can forward the source vehicle to the LTE network, and to mitigate the broadcast storm.
2. If the nearest vehicle is a GWc cluster then the SV send multi-hop cluster gateway request message (GREQ).
3. If the nearest nodes send back GREP message to the source to inform about the route to the GW then SV send their packets, if not repeat steps 1 to 3.
4. If its nearest vehicle is not a GWc cluster but within multi-hop cluster, the source vehicle updates the HWMP table and sends GREQ (each node maintains a route table in which next-hop routing information for destination nodes is stored. Each routing table entry has an associated lifetime value. If a route is not utilized within the lifetime period, the route is expired otherwise, each time the route is used, and the lifetime period is updated so that the route is not prematurely deleted).
5. If nearest node sends back GREP message to the source then SV send their packets, if not it repeats the steps 1 to 3. If nearest vehicle is not GWc cluster nor within multi-hop cluster and reply time out, source vehicle search for hop to LTE.
6. If its nearest vehicle is neither a GWc cluster nor within a multi-hop cluster but the reply time does not out, the source vehicle socket information buffered for available hop to LTE) the idea is to buffer packets for which there is no path to the destination and attempt to deliver these buffered packets when an alternative route is found or to pass them to neighbors who might eventually be able to establish a route to the destination).

This mechanism reduces the amount of time required to find the route since all the nodes inside the cluster have already know the route to the GW.

Gateway Selection Parameters

Simplified Cluster -Based Gateway Selection (SGS) employed upon the available GWcs. Multi metric gateway selection algorithm is proposed (Algorithm 2) which based on the three metrics, LTE receive signal strength, available route capacity and link stability.

LTE Received Signal Strength

In VANET, the nodes are moving in high speed which cause the frequently change LTE received signal strength. Normally, when the vehicle moves toward enodeB, its RSS is increases and if it moves away, its RSS decreases. However, the faster a vehicle moves towards or away from the enodeB, the faster the increase or decrease in its received LTE signal strength.

The LTE- RSS of the vehicles at the time t is computed as in (Benslimane, Taleb, & Sivaraj, 2012). Suppose that the velocity of the vehicle at time t is v_t and at time t-1 is v_{t-1}. The current vehicle RSS_t for LTE is calculated as shown in Equation 1:

$$RSS_t = RSS_{t-1} \pm \left(1 + e^{\frac{|v_t - v_{t-1}|}{\alpha}}\right) \tag{1}$$

where:

v_t and v_{t-1}: are the vehicle speed at time t and t -1.

α: is constant defines the rate of variation of the mobility speed in a particular movement direction relative to position of the LTE Base Station Transceiver (BST).

$\left(1 + e^{\frac{|v_t - v_{t-1}|}{\alpha}}\right)$: denotes the changes in LTE-RSS by the corresponding changes in the mobility speed of vehicles.

RSS_{t-1}: is the received signal strength at the time t-1.

Available Route Capacity (AVC)

In multi hop scenario, a multiple route has vehicles common in the route from the source vehicle to relay or gateway vehicles. This causes bottleneck problem on route or on the relay node buffer. To overcome these problems, the source vehicle (SV) selects the route based on available route capacity λ. Available route capacity (λ) can be defined as the minimum available load capacity at any node including intermediate nodes and the gateway node, in that route, is computed similar to the process in (Safda & Sasase, 2009). Suppose the maximum load capacity of node m is $C_{\max}$ and the current traffic load handled by m is TL_m then available load capacity of the node m is λ_A, it is computed using Equation

2, and the current traffic load can be calculated using Equation 3,where TL_m is the current traffic load on the node m and p_{r_j} , p_{k_j} denote the probability of the average packet arrival rate and the probability of the average packet size of the traffic from source j, respectively.

$$\lambda_A = C_{\max} - TL_m \tag{2}$$

$$TL_m = \sum_{j=1}^{s} p_{r_j} * p_{k_j} \tag{3}$$

The overall residual load capacity λ_i of route j is the minimum available load capacity at any vehicle. λ_i is computed in as Equation 4, where λ_j denotes the residual capacity of the intermediate nodes in the route including gateway node.

$$\lambda_i = \min \sum_{j=0}^{n} \lambda_j \tag{4}$$

where j denotes vehicles in a route including GW vehicle.

Link Stability(LS)

Link Stability (LS) which depends on the link Expiration time (LET) and the Route Expiration Time (RET) between the source and the CH. LET is the link lifetime between the two adjacent vehicles and the minimum lifetime between the vehicles to select gateway which is referred as (RET). Intuitively the larger the value of the LET, the higher is the stability of the link LET and RET are calculated as Equation 5 and Equation 6 similar to the process in (Setiawan, Bouk, & Sasase, 2008).

$$LET_{ij} = \frac{-\left(ab + cd\right) + \sqrt{\left(a^2 + c^2\right) R^2 - \left(ad - bc\right)^2}}{a^2 + c^2} \tag{5}$$

where:

$a = v_i cos\theta_i - v_j cos\theta_j$

$b = x_i - x_j$

$c = v_i sin\theta_i - v_j sin\theta_j$

$d = y_i - y_j$

$$RET_{n-1} = \min\left\{LET_{i,i+1}\right\}, i = 1,2,3,...,n-1. \tag{6}$$

(x_j, y_j) and (x_i, y_i) denote the Cartesian coordinates of two neighbor vehicles i and j, moving at speeds v_i and v_j, along two roads inclined at θ_i and θ_j $\left(0 < \theta_i, \theta_j < 2\pi\right)$ with respect to the axis, respectively. R denotes the maximum wireless transmission range.

Multi-Metric Gateway Selection Mechanism

Simplified Cluster -Based Gateway Selection (SCGS) algorithm is shown in Algorithm 2. Similar to (Saeed, Alawi, & Hassan, 2011; Benslimane, Taleb, & Sivaraj, 2012) all vehicles within the LTE coverage zone communicate through LTE E-UTRAN interface. When the LTE radio signal strength (RSS) deteriorates below the threshold, indicating that the vehicle starts to move outside the 4G coverage. Vehicles formulate cluster and select a cluster head (CH) to be a gateway for the cluster. The delay between the actual relay breakage, notification and new found are avoided (the new gateway is already elected.

It is employed upon the available GWcs (vehicles inside the converge zone and equipped with dual interface) cluster. The algorithm is based on the Simple Additive Weighting (SAW) technique (Saeed, Alawi, & Hassan, 2011), is used as well to outrank the optimum gateway node. Three metrics are used for mobile gateway selection: LTE RSS, available route capacity and link stability. Given the fact that, the hybrid gateway discovery mechanism is employed. The metric information of each CH lies with the GWcs of the cluster.

(SAW) technique as in (Saeed, Alawi, & Hassan, 2011) is used to calculate the overall metric score as the weighted sum of all metric values. There are three fundamental steps of SAW method calculations:

1. Scaling the comparable value.
2. Applying the weighting factors.
3. Summing up the weight values of each metric.

1. Scaling Method

Each metric has different range and value unit. Hence; we firstly have to scale the metrics to bring their values into non-dimensional values. This scaling has two ways, the positive criteria and the negative criteria (Saeed, Alawi, & Hassan, 2011). Whereas in this case LTE-RSS, link stability and available route capacity are all metrics with positive criterion (i.e., more optimality with increase in values). The best score of any criterion is 1, and the worst one is 0 .Metric with positive criterion is calculated as Equation 7 and Metric with negative criterion is calculated as Equation 8.

$$M_i = \frac{Z_{\max} - Z_i}{Z_{\max} - Z_{\min}} \tag{8}$$

where:

M_i is scaled metric value of themetric.

Z_{max} is maximum value of the metric.
Z_{min} is the minimum value of the metric.
Z_i is the value of the metric.

2. Weighting Factor

After normalizing each metric to the comparable scale, scaled values are then weighted based on the priority given to each metric. Weight is the priority given to each metric out of 1 and hence the sum of weights is always equal to 1, is computed similar to the process in (Eltahir, Saeed, Mukherjee, & Kamruk Hasan, 2016) as shown in Equation 9. There are some ways to determine these weightn factors, such as Eigen vector method (Saeed, Alawi, & Hassan, 2011), direct specification (Benslimane, Taleb, & Sivaraj, 2012), and pair comparison (Setiawan, Bouk, & Sasase, 2008). Direct specification method is used to define the weight factors. Weight factor assignment is flexible where every node can select weight factors according to their priority. Below is the prerequisite of weight factor assignment where PF is a priority factor selected by the Source Vehicle (SV), i is number of metrics, PF_1, PF_2, PF_3 are weight factors of all metrics, respectively.

$$\sum_{i=1}^{3} PF_i = 1. \tag{9}$$

The last step is to calculate the final weight value (W) for each gateway candidate, are computed similar to the process in (Benslimane, Taleb, & Sivaraj, 2012) as shown in Equation 10.

$$W_{GWc} = \sum_{i=1}^{3} \left(Z_i [PF] * M_i \right) \tag{10}$$

where PF is priority factor selected by the source node.

The GWc with the maximum weight is selected as the gateway(CH) of the cluster.

Multi-Metric Gateway Handover Mechanism

To perform handover of the serving gateway to one or more optimal gateways when the serving gateway loses its optimality.

The algorithm shown in Algorithm 3 when LTE RSS of the gateway goes below the signal strength threshold and/or if the RET of the gateway with the source vehicle goes below its pre-determined threshold, migrate the responsibilities of the currently serving gateway to the ready new gateway, selected by SGSS. Handover, thus, aims to sustain the inter-connectivity of the integrated network to pursue the data transaction. Of course, there is no common neighbor GWc between the two different clusters.

Algorithm 1. Gateway discovery algorithm

```
Begin Algorithm 1
1.  When the source vehicle LTE receive signal strength (RSS) is degrades be-
low the threshold, the source vehicle (SV) connects to the LTE network through
multi-hop and with the help of cluster head.
2. SV connect to the multi hop and send GREQ message for multi hop cluster.
3. When receiving GREQ by the nearest vehicle:
if (Nearest vehicle _ Type == GWc) then
3.1. Source vehicle send GREQ message for multi hop cluster
if (Source vehicle receive GREP message) then
3.1.1Send their packet.
else
3.1.2Repeat step 2 to 3 for selecting a new nearest vehicle.
end if
else
if (Nearest vehicle _ Type != GWc and in multi hop cluster then
3.2Source vehicle update its HWMP table and send GREQ message for multi hop
cluster.
if  (Source vehicle receive GREP message) then
3.2.1Send their packet.
else
3.2.2Repeat step 2 to 3 for selecting a new nearest vehicle.
end if
else
4. Look to its reply time out .
if (reply time out>1) then
4.1Socket information buffering.
else
4.2Source vehicle search for hop to LTE.
end if
end if
end if
End of Algorithm 1
```

Vehicular Ad Hoc Network and Cloud Computing

Vehicular ad hoc network are different from other ad hoc networks because of their hybrid architecture and node movement. Though ad hoc network, wirelesses LAN, cellular technology for intelligent transport system – all work together in VANET, address routing is the most important of all. VANET can employ vehicle-to-vehicle (V2V) and vehicle-to-infrastructure (V2I) (Eltoweissy, Olariu, & Younis, 2010b).

Recently (Eltoweissy, Olariu, & Younis, 2010b) introduced the concept of a Vehicular Cloud (VC) that leverages the on board resources in participating cars. Some vehicles are parked for long times while

Algorithm 2. Multi-metric gateway selection algorithm

Begin Algorithm 2
1. Dynamic cluster initiation when serving GW observed its LTE-RSS near to fall below specific threshold (γ).
2. When the LTE radio signal strength (RSS) for serving Gateway near to fall below the specific threshold:
if (LTE-RSS[GW] == LTE-RSS$_{th}$) then
    2.1 Calculate the weight of each GWc.
    2.1.1 The scaled metric M_i must be calculated as fallow.
For each metric Z_i of the GWc, where 1≤ i ≤ 3, **do**
 if (Z_i [CRITERION] is POSITIVE) then

$$M_i = \frac{Z_i - Z_{\min}}{Z_{\max} - Z_{\min}}$$

else
   if (Z_i [CRITERION] is NEGATIVE) then

$$M_i = \frac{Z_{\max} - Z_i}{Z_{\max} - Z_{\min}}.$$

        end if
   end if
end for
2.1.2 $W_{GWc} = \sum_{i=1}^{3}(Z_i[PF] * M_i)$.
2.2 **if** ($W_{GWc} == W_{\max}$) then
The GWc with the maximum weight elected as a (CH) GW
else
2.3 **if** ($W_{GWc} < W_{\max}$) then
    Go to the next GWc.
    Repeat Steps 2.1 to 2.3 for electing a new *GW.*
end if
 end if
end if
3. When serving GW loses its optimality the new elected GW takes place.
End of Algorithm 2

others are stuck in congested traffic and move slowly, altering their position in some wireless network. Finally, our cars spend significant time on the road and may face dynamically fluctuating locations. In this case, the vehicles will help local consultants resolve traffic incidents in a timely fashion which is not possible with the municipal traffic management centers alone due to the lack of adequate computational resources. We expect that, the vehicles are capable of solving problems in many situations that may require an indefinite time for a centralized system. Ultimately, by using self-organized autonomous

Algorithm 3. Multi-metric Gateway Handover Algorithm

```
Begin Algorithm 3
Migrate the responsibility of the currently serving gateway to the new elected
one with respect to its sources.
if (LTE-RSS[GW]<LTE-RSSth) or (RET[GW] <RETth) then
the new elected GW will take place.
Forward source vehicle data to the new elected GW.
GWc gives information about the new elected GW by using a hybrid gateway dis-
covery and advertisement mechanism.
end if
Elected GW starts serving GW and the source vehicle starts to forward the data
through the new elected GW in multi hop manner.
End of Algorithm 3
```

resources, vehicles will serve on demand in real time to resolve large, serious problems of unexpected occurrences. The new vehicular clouds will help resolve technical challenges and contribute to complex transportation systems with their evolving behavior. The Vehicular Cloud Computing can be defined as follows (Eltoweissy, Olariu, & Younis, 2010b; Olariu, Hristov, & Yan, 2013):

A group of largely autonomous vehicles whose corporate computing, sensing, communication and physical resources can be coordinated and dynamically allocated to authorized users.

The VC concept is a further step to assemble the computational and situational consciousness of drivers in public and the greater portion of the population. The ultimate focus of the VC is to offer on demand solutions for unpredictable events in a proactive fashion (Olariu, Hristov, & Yan, 2013).

V2I Simulation and Modeling

NS2.35 and VanetMobiSim-1.1 are the tools which is used in the simulation. Both are open source applications and are descripted in more detail below.

VanetMobiSim

VanetMobiSim is an extension to CanuMobisim (CANU Research Group, 2001) which is a general purpose mobility simulator.Mobility model is a framework which describes the vehicular movement pattern and it categorized into two main classes (Harri, Filali, & Bonnet, 2009):

- **Macroscopic Mobility Model:** This model defines the macroscopic characteristic of the vehicular traffic, for instance vehicles motion constraint, road topology, velocity of vehicles, and number of lanes.
- **Microscopic Mobility Model:** This model defines the personal actions of each driver, for example it defines the behavior of the each driver in traffic jam or car queues.

VanetMobiSim supports both micro-mobility and macro- mobility in simulating the realistic movement in the road. Thus, this application was chooses to make the simulation be more realistic and cover up as more as possible the behavior of vehicular movement in the road. After design the mobility model in VanetMobism and generate the trace file which can be imported in NS2 as mobility file. Simulation scenarios for VanetMobiSim are defined in XML format.

NS2.35

Network Simulator (Version 2), widely known as NS2, is simply an event driven simulation tool that has proved useful in studying the dynamic nature of communication networks. Simulation of wired as well as wireless network functions and protocols (e.g., routing algorithms, TCP, UDP) can be done using NS2.The network simulator (ns2) contains all commonly used IP protocols. The network animator (NAM) is used to visualize the simulations. NS-2 (Nsam, 2014) fully simulates a layered network from the physical radio transmission channel to high-level applications. After simulation, simulation produces two output file (Nsam, 2014):

- **Trace File:** This file provide a full details about the events that take place at the network.
- **NAM File:** It provides network animation.

VanetMobiSim was used in the simulation because it supports both micro-mobility and macro-mobility models (Haerri, Fiore, Fethi, Bonnet, 2006; Alsaqour, Alsaqour, Alahdal, Saeed, & Al-Hubaishi, 2015; CANU Research Group; NS, 2015). VanetMobiSim was used to simulate the mobility model while NS2.35 was used to measure the network performance of the proposed mechanism. This makes our simulation more realistic and covers up, as much as possible, the behavior of vehicular movement on the road. The result of the simulation is a trace file which can be imported in NS2 as a mobility file. The Simulation Architecture is shown in Figure 6.

In order to have a realistic vehicular movement pattern, it is necessary to define the attributes of this model, for instance the velocity of the vehicles, trip motion, and road topology. VanetMobiSim for scenario is shown in Figure 7 comprises with the creation of the highway with two lanes for each direction whereby the vehicles are moving in opposite and base station which is connected to the wired network. But this base station and other part of wired network are configured in network simulator and topology size of our scenario is kept to be 8000 to 1000. The scenario is defined in xml format.

Furthermore, it is possible in VanetMobiSim to change the number of vehicles on the road. Figure 7, shows only 30 vehicles in the roads but we varies the number of vehicles for different scenarios, starting from 10 to 50 vehicles From this it is understood that we change the number of vehicles in this order for the purpose of getting much details of the network behavior upon increasing of the number of vehicles.

Figure 6. Simulation architecture

Figure 7. Vehicles mobility model for 30 vehicles on the highway

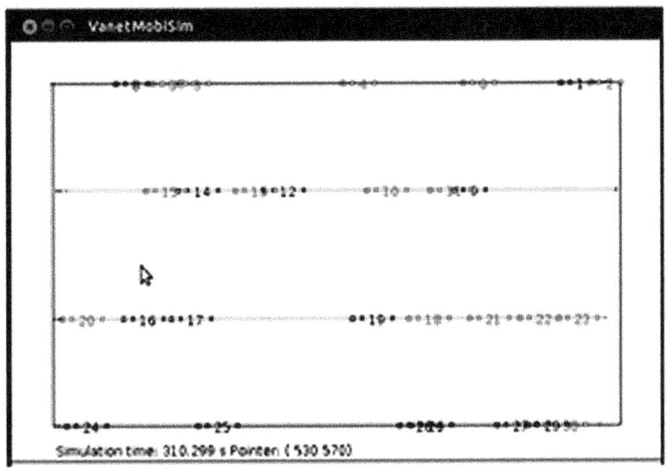

Moreover, it is possible in VanetMobiSim to change the speed of vehicles, so we study performance of the scheme in different vehicle speeds because speed of the vehicles can affect the network performance. For the simulation, vehicle speed is change from 18 to 180 km/hr and the stop time is between 1 to 3 second. the simulation is set to be 500s second in both VanetMobiSim and NS2 which is long enough to precisely observe the network performance.

After design the mobility model in VanetMobism, a trace file is generated which can be imported in NS2 as mobility file.

A screenshot of the simulation scenario is shown in Figure 8, consider a scenario of two different tracks over a particular road (e.g. Highway), with two lanes for each direction and 30 vehicles moving in these four lanes. Vehicles are assumed to be equipped with GPS devices, connect to different LTE signal strength along the road. The vehicles is designed to move in a three dimensional topology. However the third dimension (Z) is not used. That is the mobile node is assumed to move always on a flat terrain with Z always equal to 0. Thus the mobile node has X, Y, Z (=0) co-ordinates that is continually adjusted as the node moves. (x_1,y_1) = initial position. Speed limit for each lane is5 m/s and 50 m/s. 18 km/hr and 180 km/h were used respectively as the minimum and maximum speeds for the vehicles. The distance between each lane in the same direction is 50 meters and the distance between the two directions is 60 meters. The number of the vehicles was 30. The source vehicle is constant bit rate traffic sources. 500 seconds is considered as the simulation time. Simulations run twenty-five times.

The simulation involves one Base station (E-Node B) which is connected to cellular network and the number of vehicles which use E-Node B to connect to the cellular network. This study uses only one base station in the simulation since the proposed model does not allow handover process between base stations. This has been done as one of the objectives that intends to study how multi hop communication can be used to increase the network coverage in the area whereby no coverage. However, handover process can take place between the selected gateway vehicle to another gateway vehicle which are used to forward the packet to E- Node B when the formal one loss their optimality. When the simulation starts all the vehicles connect directly through the E-Node B, the transmission range of E-Node B is set to 10km. When the vehicles are in transmission range no clustering are formed and all the vehicles use their EUTRAN interfaces to connect to E- Node B. When the vehicle moves away from the E-Node

Figure 8. Simulation scenario with thirty vehicles

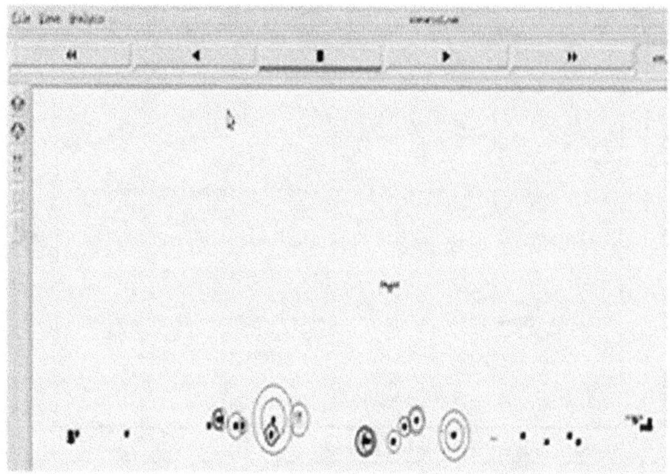

B and found that its LTESRSS falls below the Threshold($_\gamma$), this is the point where the vehicles start to form clusters by using multi hop communication to connect to the E-Node B. Formation of clusters inside the LTE cell was eradicated in order to reduce the overhead and to simplify the gateway selection for connection to the cellular network.

Each of the selected vehicles will initiate FTP application and send data to cellular network through E-Node B. Vehicles which located inside the coverage zone starts to send the packet to E-Node B which is part of wired network and when the vehicle observe that its LTE-RSS from the E-Node B falls below the predefined value, the vehicle will start to initiate the searching of the mobile gateway by sending the GREQ with the TTL value 10 and waiting for the reply. If no reply for about 5s which is sending time interval suggested in (Bilal & Umar, 2010) the vehicle increases the TTL value by 10 and send again the GREQ.

It will continue to increase the TTL value and send the GREQ until it received the GREP. After getting the GREP, the vehicle will start to connect to E-Node B using multi hop communication. The migration of one gateway to another will take place if the received signal strength (RSS) drops below 25% and priority factors for three metrics in Algorithm 2 are assigned equal values (i.eMi = 1/3 € for i = 1,2,3) as in (Benslimane, Taleb, & Sivaraj, 2012).

Through this simulation five performance metrics are analyzed: Packet Delivery Ratio, End-to-End Delay, Control Packet Overhead, Throughput and Packet Loss. The definitions of network performance metrics are written based on these papers: (Ali Hamidian, 2003; Liu, Jiang, Nishiyama, Kato, & Xuemin, 2012; Wang, Yang, & An, 2009).

- **Packet Delivery Ratio:** The number of received packets compared to the number of generated packets is defined as the packet delivery ratio. This parameter determines the protocol's performance in terms of delivery of packets, while speed, acceleration or payloads are variables. This property can be defined as:
  Packet delivery ratio = (received packets / generated packets)
  Note that packet drop ratio can be calculated as: 1 - (packet delivery ratio).

- **Delay Metric:** This refers to time elapse since the Source Vehicle (SV) send the GReq until the gateway is selected.
- **Control Packet Overhead:** This measures the ratio of the total number of control packets to the total number of packets generated within the VANET- cellular network integration.
- **Throughput:** This is defined as the total amount of data a receiver receives from the sender divided by the time between receiving the first packet and last packet.

CONCLUSION

Vehicular Ad-hoc Network (VANET) and cellular network integration are promising architectures for future machine-to-machine application. In this research, a new network architecture is proposed for integration VANET - cellular network to enhance the heterogeneity in vehicular communication. The proposed scheme is implemented using the proposed E-HWMP protocol. The implementation of E-HWMP mechanism has led to enhancement of the integration between VANET and cellular network, such as LTE, and extended the coverage in the area where there is no coverage, and reduced the amount of overhead, and reduced delay metrics, and increased the packet delivery ratio, and also reduced the dead spot in the multi-hop VANET- 3GPP LTE network. The mechanism involves a multi-hop relay communication for the VANET over wireless mesh networks to connect to 3GPP LTE network when the vehicle is nearly to be outside the coverage zone.

The processes involved in E-HWMP scheme are: gateway management, gateway selection and gateway handoff, all performed in a dynamic way using different metrics.

An integrated simulation environment combining VanetMobiSim and NS2 was developed to evaluate the performance of the overall architecture. Two types of comparison were made to validate the performance of the proposed scheme. The first study aimed to determine which routing protocol (E-HWMP or AODV) proposed scheme is to be implemented. The result shows that the E-HWMP protocol performed better than Ad-hoc on Demand Distance Vector (AODV) routing protocol. The second study compared E-HWMP with other cluster-based gateway selection that used in previous works. The result shows that E-HWMP protocol (proposed by this study) outperforms the other cluster-based gateway selections schemes in terms of connection delay, control packet overhead, packet delivery ratio and overall throughput.

The most important point which should be considered is that the nature of vehicular communication is based on the movement. Therefore, it is necessary to implement a realistic vehicular movement in the vehicular network simulation. In other words, all the important parameters should be implemented accurately in the VANET simulation.

REFERENCES

Alsaqour, O. (2015). *A Comparative Study of Simulation Based Performance Evaluation of Routing Protocol for Ad-Hoc Networks*. Springer International Publishing.

Benslimane, A., Taleb, T., & Sivaraj, R. (2012, March). Dynamic clustering-based adaptive mobile gateway management in integrated VANET - 3G heterogeneous wireless networks. *IEEE Journal on Selected Areas in Communications*, *29*(3), 559–570. doi:10.1109/JSAC.2011.110306

Bilal, M., & Umar, W. (2010, June). *Issues of Routing in VANET* (Master Computer Science Thesis no: MCS-2010-20). Blekinge Institute of Technology, Sweden.

CANU Research Group. CanuMobiSim (Mobility Simulation Environment). (2001). Institute of Parallel and Distributed Systems (IPVS). University of Stuttgart. Retrieved from http://canu.informatik.uni-stuttgart.de/

CANU Research Group (Stuttgart University). (n.d.). Retrieved from http://canu.informatik.uni-stuttgart.de/mobisim/

Eltahir, A. A., Saeed, R. A., Mukherjee, A., & Hasan, M. K. (2016). Evaluation and analysis of an enhanced hybrid wireless mesh protocol for vehicular ad hoc network. *EURASIP Journal on Wireless Communications and Networking*, *2016*(169), 1–11. doi:10.118613638-016-0666-5

Eltahir & Saeed. (2015). Performance Evaluation of an Enhanced Hybrid Wireless Mesh Protocol (E-HWMP) Protocol for VANET. *International Conference On Computing, Control, Networking, Electronics And Embedded Systems Engineering (ICCNEEE 2015)*, 95 - 100.

Eltahir, Saeed, & Alawi. (2013). An Enhanced Hybrid Wireless Mesh Protocol (E-HWMP) Protocol for Multi-hop Vehicular Communications. *International Conference On Computing, Electrical And Electronic Engineering (ICCEEE 2013)*, 1 - 8.

Eltahir, Saeed, & Mokhtar. (2014). Vehicular Communication and Cellular Network Integration: Gateway Selection Perspective. *International Conference On Computer & Communication Engineering (ICCCE 2014)*, 64 - 67.

Eltoweissy, M., Olariu, S., & Younis, M. (2010b). Towards autonomous vehicular clouds. In J. Zheng, D. Simplot-Ryl, & V. M. Leung (Eds.), *Ad hoc networks* (Vol. 49, pp. 1–16). Berlin: Springer. doi:10.1007/978-3-642-17994-5_1

Haerri, J., Fiore, M., Fethi, F., & Bonnet, C. (2006). *VanetMobiSim: generating realistic mobility patterns for VANETs*. Institut Eurécom and Politecnico Di Torino.

Hamidian. (2003). *A Study of Internet Connectivity for Mobile Ad Hoc Networks in NS 2*. Department of Communication Systems Lund Institute of Technology, Lund University.

Harri, J., Filali, F., & Bonnet, C. (2009). Mobility models for vehicular ad hoc networks: A survey and taxonomy. *IEEE Communications Surveys and Tutorials*, *11*(4), 19–41. doi:10.1109/SURV.2009.090403

Jiau, M.-K. (2015). Multimedia Services in Cloud-Based Vehicular Networks. *IEEE Intell. Transport. Syst. Mag.*, *7*(3), 62–79.

Kathirvel & Loyd. (2014). Optimized Hybrid Wireless Mesh Protocol using Estimation of Paket Loss Rate Algorithm for VANET. *International Journal of Computer Science and Mobile Applications, 2*(3), 117-123.

Khan, A. A. (2015). *An Evolutionary Game Theoretic Approach for Stable Clustering in Vehicular Ad hoc Networks (VANETs)* (Master of Engineering thesis). University of Technology.

Li, P., Guo, S., Yu, S., & Vasilakos, A. V. (2014, December). Reliable Multicast with Pipelined Network Coding Using Opportunistic Feeding and Routing. *IEEE Transactions on Parallel and Distributed Systems, 25*(12), 3264–3273. doi:10.1109/TPDS.2013.2297105

Liu, Jiang, Nishiyama, Kato, & Shen. (2012). End-to-End Delay in Mobile Ad Hoc Networks with Generalized Transmission Range and Limited Packet Redundancy. *IEEE Wireless Communications and Networ king Conference (WCNC 2012)*, 1731-1736.

Liu, B., Jia, D., Wang, J., Lu, K., & Wu, L. (2015). Cloud-Assisted Safety Message Dissemination in VANET-Cellular Heterogeneous Wireless Network. *IEEE System Journal*, 1–12.

Lo, S.-C., Lin, Y.-J., & Gao, J.-S. (2013). A multi-head clustering algorithm in vehicular ad hoc networks. *International Journal of Computer Theory and Engineering*, *5*(2), 242–247. doi:10.7763/IJCTE.2013.V5.686

Meng, T. (2016). Spatial Reusability-Aware Routing in Multi-Hop Wireless Networks. IEEE TC, 65(1), 244-255.

Namboodiri & Gao. (2007). Prediction-based routingfor vehicular ad hoc networks. *Vehicular Technology, IEEE Transactions on, 56*(4), 2332–2345.

ns. (n.d.). *ns2*. Retrieved from http://www.isi.edu/nsnam/ns/

nsnam. (2014). *Publications*. Retrieved from http://www.nsnam.org/overview/publications/

Olariu, S., Hristov, T., & Yan, G. (2013). The next paradigm shift: from vehicular networks to vehicular clouds. In Mobile ad hoc networking: cutting edge directions (2nd ed.). John Wiley & Sons, Inc. doi:10.1002/9781118511305.ch19

Saeed, Alawi, & Hassan. (2011). Simplified Gateway Selection in Integrated VANET-UMTS Heterogeneous Wireless Network. IST-Africa 2012 Conference & Exhibition.

Safda, H., & Sasase, I. (2009). Multiple End- to-End QoS Metrics Gateway Selection Scheme in Mobile Ad hoc Networks. *International Conference on Emerging Technologies.*

F. P. Setiawan, S. H. Bouk, & I. Sasase (Eds.). (2008). An optimum multiple metrics gateway selection mechanism in MANET and infrastructured networks integration. *IEEE Wireless Communications and Networking Conference (WCNC 2008)*. 10.1109/WCNC.2008.394

F. P. Setiawan, S. H. Bouk, & I. Sasase (Eds.). (2008). An optimum multiple metrics gateway selection mechanism in MANET and infrastructured networks integration. *IEEE Wireless Communications and Networking Conference (WCNC 2008*. 10.1109/WCNC.2008.394

Shobin, (2012). Dynamic Clustering and Gateway Management in Vehicular Ad-Hoc Networks. *International Journal of Communications and Engineering, 3*(1).

Ucar, Ergen, & Ozkasap. (2013). VMaSC: Vehicular multi-hop algorithm for stable clustering in Vehicular Ad Hoc Networks. Wireless Communications and Networking Conference (WCNC).

Viriyasitavat, W., Boban, M., Hsin-Mu Tsai, & Vasilakos, A. (2015). Vehicular Communications. *Survey and Challenges of Channel and Propagation Models Vehicular Technology Magazine, IEEE, 10*(2), 55–66. doi:10.1109/MVT.2015.2410341

Wan, S., Tang, J., & Wolff, R. (2008). Reliable routing for road side to vehicle Communications in rural areas. *Proceedings of IEEE International Conference on Communications*, 3017-3021.

Xiao-bo, W. A. N. G., Yu-liang, Y. A. N. G., & Jian-wei, A. N. (2009). *Multi-Metric Routing Decisions in VANET*. Beijing, China: Department of Communication Engineering, University of Science and Technology Beijing.

Zhioua, G. (2014). A fuzzy multi-metric qos-balancing gateway selection algorithmin a clustered vanet to lte advanced hybrid cellular network. *Vehicular Technology. IEEE Transactions on, 64*(2), 804–817.

Zhou, J. (2015). Seure and privacy preserving protocol for cloud-based vehicular DTNs. *IEEE Transactions on Information Forensics and Security, 10*(6), 1299–1314. doi:10.1109/TIFS.2015.2407326

Chapter 7
On-Demand Routing Protocols for Vehicular Cloud Computing

Ramesh C. Poonia
Amity University Jaipur, India

Linesh Raja
Amity University Jaipur, India

ABSTRACT

In vehicular cloud computing systems (VCC), the overlapping transmission range of each vehicle ensures a unified and common channel for communication among the vehicles. The flexibility of VCC systems opens the door to myriad applications that contribute to the safety and comfort of the passengers. They are distributed, self-organizing communication networks built up by moving vehicles, and are thus characterized by very high node mobility and limited degrees of freedom in mobility patterns. Such particular features often make the standard networking protocol inefficient or unusable, hence the growing effort in the development of communication protocols, which are specific to vehicular networks. Routing protocols should be selected carefully after carrying out literature review. This chapter has investigated different on-demand routing protocols and focused to identify the efficient on-demand routing protocol that can give better performance in realistic environments of vehicular cloud computing systems.

1. INTRODUCTION

In recent times, the number of vehicles is increasing at rapid speed. Vehicles can communicate with each other and to the road side unit with the help of some wireless technology. Inter-communication of vehicles and to the road side unit is considered as Vehicular Ad-hoc Networks (VANETs).

In order to establish communication between vehicles and roadside units (RSU's), every vehicle must contain some kind of radio interface or onboard unit (OBU) that would be responsible for short range communication in Vehicular Ad-hoc Networks. For smooth communication, all vehicles should be equipped with GPS system. Various protocols are responsible for this network.

DOI: 10.4018/978-1-5225-3981-0.ch007

Intelligent Transport System (ITS) is emerging field in research which is used in inter vehicular and vehicle to RSU communication to get updated information about surrounding environment like automatic road enforcement, traffic light sequence and collision avoidance system. For information exchange, ITS used various protocols and accurate positing System (Balon, 2006). For vehicle communication, various standards i.e. WAVE 1609, IEEE 802.11p, IEEE 802.11a, and IEEE 802.11e are used that supports wireless communication between V2V and V2I (Li, Mirhashemi, Laurent & Gao, 2010; Morgan, 2010).

The various factors associated with technology, business, regulation and social behavior, naturally and logically speak in favor of wireless ad-hoc networking. The wireless data communication, which is advancing both in terms of technology and usage/penetration, is a driving force, thanks to the internet and the success of second generation cellular systems (Frodigh, Johansson & Larsson, 2000). In terms of price, portability and usability and in the context of an ad-hoc network, many computing and communication devices such as PDAs and mobile phones, already possess the attributes that are desirable.

An Ad-hoc network is "infrastructure-less "network that does not have any base station. It is a collection of multiple nodes having wireless networking capabilities. It also supports anytime and anywhere computing. There are two types of topologies formed in Ad-hoc network:

- Heterogeneous, which have differences in capabilities.
- Homogeneous or fully symmetric, in which all nodes have identical capabilities and responsibilities.

Adaptive and self-organization are the main feature of ad-hoc network that allow prompt formation and deformation of network. Each node in the network behaves as a router and supports peer-to-peer and peer-to-remote communications. It is cost effective and easy to deploy too. One of the fastest growing domains of interest in ad-hoc networks is the safety, where transmissions are exchanged in order to improve the passengers and driver behavior in Indian automotive networks. In this chapter the basic overview of mobile and vehicular networks is discusses. Its forms the basic idea behind the safety to passengers and drivers. Author discussed their characteristics, significance and performance related to ad-hoc networks.

In the similar manner, the world is witnessing a rapid growth in the cloud computing and their prospective applications in terms of data, storage and processing. The computer and various communication devices can utilize these resources, software and information, on-demand over the internet. The cloud helps us to develop, test and deploy a network centric application for the scalable environment at low cost computing in terms of IP networks (Whaiduzzaman, Sookhak, Gani & Buyya, 2014). The various characteristics of cloud computing is listed below:

- It provides large scale computing resources,
- They are highly scalable in nature,
- They shared a common resource pool,
- It provides dynamic resource scheduling, and
- They are used for general purpose computing.

The cloud system provides their services at three different layers. Figure 1 shows the different layers of the cloud system.

Figure 1. Cloud layers

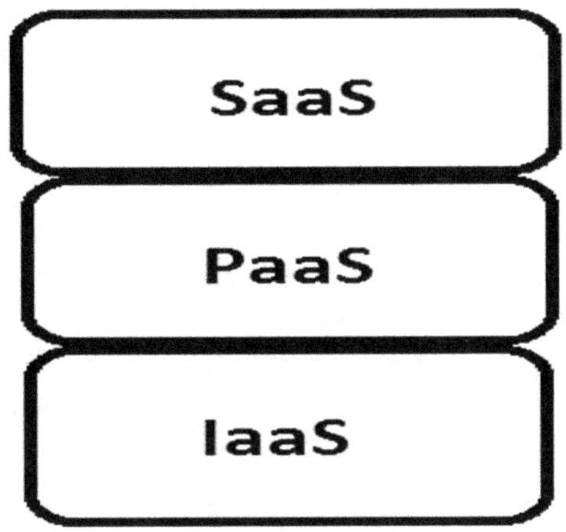

Infrastructure as a Service (IaaS) provides computation communication resources as a service. It includes hardware, storage and server. Their examples are Amazon Elastic Cloud Computing and Simple Storage Services. The user application can be build, tested and deployed at Platform as a Service (PaaS). It mostly includes software testing, delivery and managing software applications. Their examples are Google Apps Engine and Amazon Maps. The users can share and access various applications remotely via Internet at Software as a Service (SaaS). It includes hosting and managing software applications; software up-gradation and security patching (Aloqaily, Kantarci, & Mouftah, 2015; Tyagi & Niladhuri, 2016). Their examples are Microsoft Office 365 etc. Table 1 shows the service provides along with their services offered.

MANET is an autonomous collection of mobile users that communicate over relatively bandwidth constrained wireless links. Since the nodes are mobile, the network topology may change rapidly and unpredictably over time. The network is decentralized, where all network activity including discovering the topology and delivering messages must be executed by the nodes themselves i.e., routing functionality will be incorporated into mobile nodes (Loo, Mauri & Ortiz, 2016). The set of applications for MANET

Table 1. Service provider along with their services offered

Cloud Service Provide	IaaS	PaaS	SaaS
Amazon	Yes	Yes	No
Century Link	Yes	Yes	No
Google	Yes	Yes	Yes
IBM	Yes	Yes	Yes
Microsoft	Yes	Yes	Yes
Rackspace	Yes	Yes	No
Salesforce.com	No	Yes	Yes
SAP	Yes	Yes	Yes
Verizon Terremark	Yes	Yes	No

is diverse, ranging from small, static networks that are constrained by power sources, to large-scale, mobile, highly dynamic networks. The design of network protocols for these networks is a complex issue. Regardless of the application, MANETs need efficient distributed algorithms to determine network organization, link scheduling, and routing. However, determining viable routing paths and delivering messages in a decentralized environment where network topology fluctuates is not a well-defined problem. While the shortest path (based on a given cost function) from a source to a destination in a static network is usually the optimal route, this idea is not easily extended to MANETs. The network should be able to adaptively alter the routing paths to alleviate any of these effects. Moreover, in a military environment, preservation of security, latency, reliability, intentional jamming, and recovery from failure are significant concerns. Military networks are designed to maintain a low probability of intercept and/or a low probability of detection. Hence, nodes prefer to radiate as little power as necessary and transmit as infrequently as possible, thus decreasing the probability of detection or interception. A lapse in any of these requirements may degrade the performance and dependability of the network (Wang & Lin, 2008).

Mobile Cloud Computing (MCC) is one of the promising models where mobile devices can be considered as client as well as server (in terms of data, storage and processing) (Hassanein Abdelhamid & Elgazzar, 2015). The number of mobile users in the world is more than 2.1 billion. In recent times Smart phone are the most popular way for internet access. It helps us to enquiry about our surroundings (e.g. ATM, Hospital, Restaurant, Taxi Services, and Entertainment etc.) and mobile agents (people and vehicles). These mobile agents are considered as Mobile Clouds (MC) (Gerla, 2012).

One of the significant contributions of Mobile Cloud is in vehicular networking. The vehicles are challenged to bring more communication systems, high processing and increased power. Figure 2 shows the vehicular environment.

The Vehicular Cloud Computing (VCC) is one solution to the proposed challenge. This also uses Infrastructure, Platform and Software as a basic service set to accomplish a particular task (Aloqaily, Kantarci, & Mouftah, 2015).

Figure 2. Vehicular environment

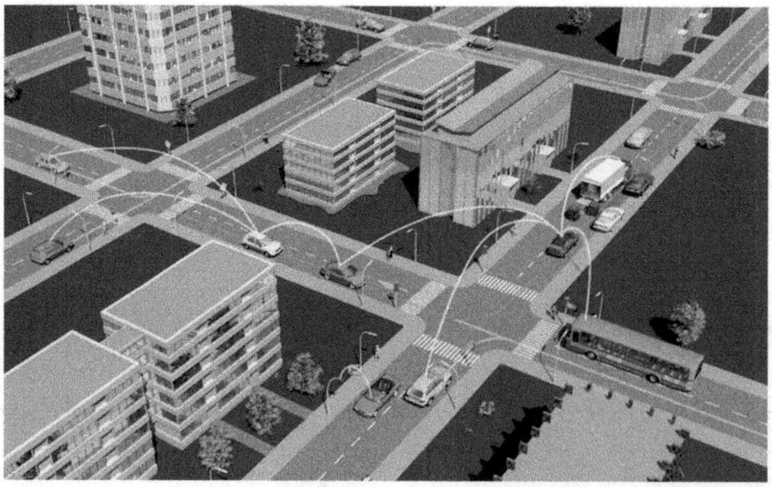

A Vehicular Ad-hoc Network (VANETs) is a technology that uses moving cars as nodes in a network (mobile network). In this each vehicle can receive and transmit others messages through the wireless network. VANETs turn every participating car into a wireless router or node, allowing cars approximately 100 to 500 meters of each other to connect and, in turn, create a network with a wide range. As cars fall out of the signal range and drop out of the network, other cars can join in, connecting vehicles to one another so that a mobile Internet is created. The vehicles are equipped with advanced wireless communication devices without any base stations (Pesel & Maslouh, 2011). The main characteristic of the VANETs is the infrastructure absence, such as access point or base stations, existing in the Wi-Fi, WiMax, GSM or UMTS. The communication between nodes that are beyond the reach of transmission of the radio is made in multi hops through the intermediate nodes contribution. However, the topology of the networks can move dynamically. On the other side, the media without wire, i.e. in the absence of infrastructures and the multi hops routing transforms these networks in potential targets of diverse types of attacks that go since simple eavesdropping (espionages) passive of the messages until active interferences with the creation, modification and destruction of the messages. This type of networks can provide wide variety of services. The safety application is one of the most important. For example, if a vehicle detects road accident, it will inform other neighboring vehicles about this road accident. The safety messages must be delivered to each neighboring node with almost no delays. A single delayed or lost event-driven safety messages could result in loss of life (Darus & Bakar, 2011). VANETs can be used to communicate premonitions, notifications of emergencies and warnings about traffic conditions, but also for distributing information about road conditions and maintenance, weather forecasts or other relevant data distribution requirements between vehicles.

It is estimated that the first systems that will integrate this technology are police and fire vehicles to communicate with each other for safety purposes. VANETs enable the use of Advanced Driver Assistance Systems (ADAS) and use ad-hoc communications for performing efficient driver assistance and car safety. The communications include data from the roadside and from other cars. VANETs research aims to supply drivers with information regarding obstacles on the road and emergency events, mainly due to line-of-sight limitations and large processing delays (Pesel & Maslouh, 2011). A vehicular ad-hoc network is an active research area. The various standards are developed, which has the tremendous potential to improve vehicle and road safety, traffic efficiency and convenience as well as comfort to passengers and drivers.

It works on the various layers of the transmission system. Few of the commonly used layers for VANETs (in case of Dedicated Short Range Communication – DSRC) are the Physical and Mac layer. Recent research efforts have placed a strong emphasis on novel VANETs design architectures and implementations. In short Communication VANETs uses the IEEE 802.11 Protocol set. The IEEE.11 is a set of standards for implementing wireless local area network computer communication in the 2.4, 3.6 and 5 GHz frequency bands. The IEEE 1604 family protocol are also used for the VANETs.

The remaining sub sections of the chapter are set out as follows. Section 2 discusses the vehicular cloud computing. The architecture of vehicular cloud computing is formulated in Section 3. The routing protocols have been discussed in Section 4. Finally, a conclusion is drawn in Section 5.

2. VEHICULAR CLOUD COMPUTING

In recent times the improvement in computer (software and hardware) and their communication technology leads to the design and implementation of various types of networks. These networks can be deployed to various environments. One of the promising network environments is the Vehicular Ad-hoc networks (VANETs). VANETs contains moving vehicle as a nodes in a wireless communication. It applies the various communication technologies to provide the services to driver-passengers safety and traffic management etc. At the same time VANETs contributes to various solutions in areas, e.g. driver-passengers road safety, traffic management and Intelligent Transport System (ITS). It also faces an uphill task of uploading real-time communication information on the cloud such as traffic status or emergency situation etc. As it is costly and time consuming (Alam, Saini & Saddik, 2015).

Vehicular cloud computing is the extension of the cloud computing with several new dimensions which is used in intelligent transport system. They are considered for providing safety to driver and passengers. They utilize vehicular resources such as computing, storage and internet for communication and smart decision. We are motivated because as on date the vehicular resources are underutilized. So once these resources are utilized to their full extent then it can help the society in terms of communication and smart decision. The pooling of some other vehicular resources such as computing power, storage facility and internet connection among drivers are similar to the present conventional cloud resources. The same can be rented to the customer. There are four different parameters for the VCC application. They are related to active safety, public services, improved driving and business and entertainment. Based on these parameters the contents type of the vehicular cloud computing is listed below:

- Vehicles software updates availability for the users,
- Accidents alerts at road side units,
- Road condition sharing during disaster,
- Safety application for life critical scenario such as collision avoidance, security threats etc,
- Driver receives traffic status report, i.e. Traffic Navigation,
- Intelligent parking system
- Commercial advertisement and
- Multimedia files sharing.

The lifetime of the above applications varies from a fraction of second to all days. At the same time their validity ranges 100 m to 5 km. Some must be applicable to all the vehicles and very few applications are subscription based (Lee, Gerla, & Oh, 2014; Sharma & Kaur, 2015).

Intelligent transport system in vehicular ad-hoc networks support the growing number of wireless products that can be used in different vehicles (De Fuentes, González-Manzano, González-Tablas, & Blasco, 2014). In ITS, each vehicle takes on the role of sender, receiver and router to broadcast information to the vehicular networks or transportation agency, which then uses the information to ensure safe, free flow of traffic (Li, Wu, Lin, Lan, Chou & Hsu, 2010; Morgan, 2010).

Intelligent Transportation System (ITS) applies advanced technologies of electronics, communications, computers, control and sensing and detecting in all kinds of transportation system in order to improve safety, efficiency and service, and traffic situation through transmitting real-time information (Racehorse, Zhang, Sussman, Jian, & Parker, 2014).

Their objectives are given below:

- Improve traffic safety
- Relieve traffic congestion
- Improve transportation efficiency
- Reduce air pollution
- Increase the energy efficiency
- Promote the development of related industries

Interest in Intelligent Transport System (ITS) comes from the problems caused by traffic congestion and a synergy of new information technology for *simulation, real-time control, and communications networks*. Traffic congestion has been increasing worldwide as a result of increased motorization, urbanization, population growth, and changes in population density. Congestion reduces efficiency of transportation infrastructure and increases travel time, air pollution, and fuel consumption (Pesel, 2011). The various applications of Intelligent Transport System are given below (Li, Wu, Lin, Lan, Chou, & Hsu, 2011; Fujise, Kato, Sato & Harada, 2002):

- **Automatic Fare Collection:**
    - ETMs (Electronic Ticketing Machines)
    - Smart Card based Ticketing and Pass
    - Mobile based Ticketing (NFC, RFID)
    - Access gates and turnstiles
    - Fare Collection Modules
- **Handheld Solutions:**
    - ETMs
    - eChallan and Enforcement
    - Parking payments using Credit/Debit Cards
    - Smart Card Validator
    - Billing Applications
    - Field Surveys
- **Identity Management:**
    - Smart Card based Driving License, RC, Permits for Commercial Vehicles
    - Learner License
    - Public Service Vehicle Badge
    - Biometrics
- **Intelligent Signaling:**
    - Queue detection
    - Automatic number plate recognition
    - Signaling cycle standards
    - Signal light standards
    - Vehicle detection standards for traffic density
    - Camera
    - Radar
    - RFID tags

- **Highway Traffic Management:**
  - Vehicle detection technologies
  - Automatic traffic counters and classification
  - Control center
  - Emergency call
  - Surveillance
- **Location Based Technologies:**
  - Vehicle tracking
  - Fleet Management (Public transport, Railway wagons, Cabs, Trucks)
  - Mobile phone based tracking
  - Passenger Information Displays
  - Traffic information
  - Intelligent Routing
  - GIS

3. ARCHITECTURE

This section provides the brief idea of the Vehicular Cloud Computing (VCC) environment and their architecture. But before that it is required to understand the basic VANETs environment. Figure 3 shows the VANETs environment which significantly provides the communications and routing services to the connecting nodes.

In the similar fashion, Figure 4 provides the VANET cloud environment. At the same time it also offers *content routing, security privacy, monitoring, and virtualization services*. Collectively such environment is considered as Vehicular Cloud Computing (VCC). It is easier to build application on VCC (Yu, Zhang, Gjessing, Xia & Yang, 2013).

Figure 3. VANETs environment

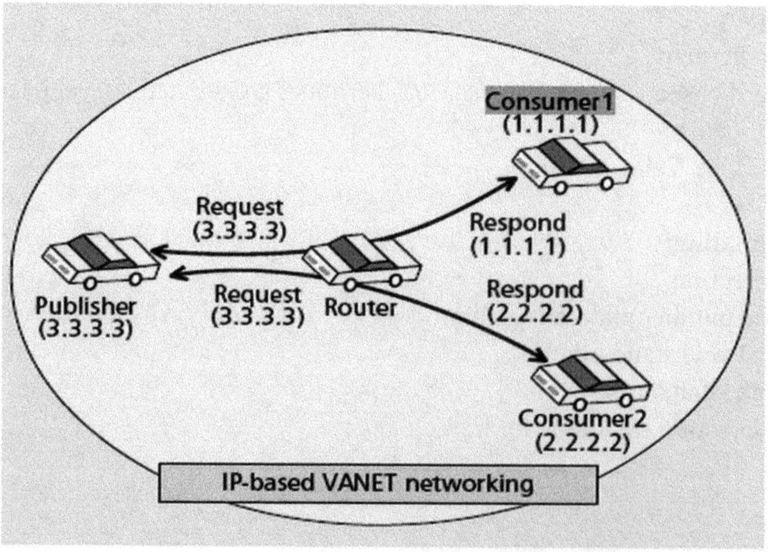

Figure 4. VANETs cloud environment

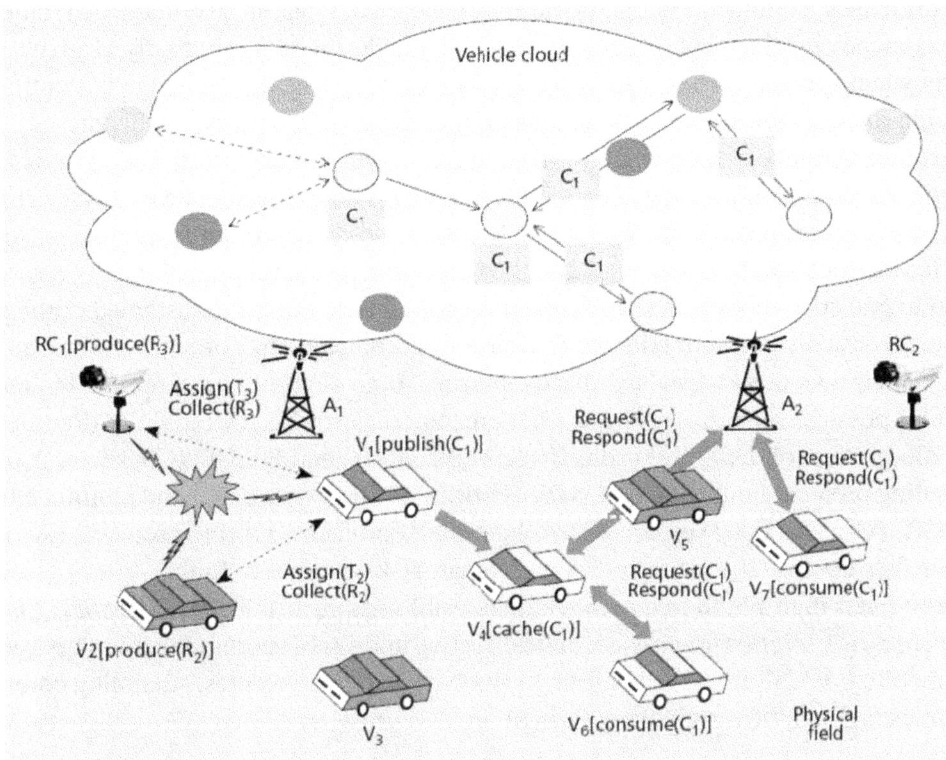

In order to monitor and select the appropriate schemes, the vehicular cloud computing is integrated with various networking services. As a part of the monitoring services VCC can monitor group mobility.

Once considered for group mobility, a VCC can track group mobility as part of its monitoring service. Vehicles determine that they are traveling in a group via crowd sourcing, a VCN service. The VCN can then exploit group mobility and offer it as a property to applications. For instance, intelligent transport can exploit group mobility in the rerouting of traffic by rerouting groups of cars going to the same destination rather than rerouting individual cars. This helps scalability (Sharma & Mahmoud, 2015; Yu, Zhang, Gjessing, Xia, & Yang, 2013).

VANET offer challenging network environment in which finding, maintaining and using the routes for communications and decision making is not easy for computing, storage and internet. There are so many issues that prevent a routing problem to perform its best like short unpredictable node density, little or no path redundancy coupled with strict application requirements and communication opportunity etc. As Vehicular Cloud Computing (VCC) is direct offshoot of Mobile Cloud Computing (MCC), therefore it is appropriate to study and analyze the performance of routing protocols developed for MCC (Sallam, 2015; Li & Wang 2007; De Oliveira Barros & De Morais Gomes, 2013).

In vehicular environment, routing protocols are used to calculate efficient routes. These protocols are divided into two main categories with respect to their routing behavior. They are On Demand (Reactive) and Table Driven (Proactive). In this sub section these two classes of routing protocols are defined.

In proactive routing protocols the routing table gets continuously update, even there is no communication. Due to this reason, sustained routing overhead is generated. It has an advantage that route discovery process has no latency involvement. Therefore, routes to the destinations are easily available and can be used promptly when application has data to send. It is based on periodic exchange of control messages and maintaining routing tables, for the same reason, this protocol is known as table driver protocol.

Reactive routing protocols experience initial route discovery latency with the help of flooding to discover routes, which is their main overhead. These routing protocols calculate routes for destination in the network, when it is needed therefore these are known as on-demand routing protocols. Applications have to wait for the routes to be discovered and then use them for communication.

In contrast to proactive routing protocols, reactive protocols do not make sustained routing overhead. Reactive protocols usually take more time to find a route as compared to a proactive protocol. These two protocols have their own advantages and disadvantages, so no single routing protocol is able to work efficiently in all possible scenarios. Ad-hoc On demand Distance Vector (AODV) (Perkins, Belding-Royer, Das, 2003) protocol, DSR (Johnson, Hu & Maltz, 2001) and DYMO (Chakeres, 2008) belongs to reactive routing protocol family, while DSDV (Perkins, & Bhagwat, 1994) and Optimized Link State Routing (OLSR) (Clausen, 2003) protocol is an example of proactive routing protocols (Azouqa, 2013; Shastri, Dadhich & Poonia, 2011; Singh, Singh, Kumar, 2011; Kumar & Singla, 2011).

VANET integrates multiple ad-hoc networking technologies such as *WiFi IEEE 802.11 b/g, WiMAX 802.16, Bluetooth, IRA, ZigBee* for easy accurate effective and simple communication between vehicles on dynamic mobility. VANET helps in defining safety measures in vehicles, streaming communication between vehicles, infotainment and telematics.

4. ROUTING PROTOCOLS

Various routing protocols have been developed to compete with sudden changes that may arise due to nature of the networks. Route discovery, route maintenance and sudden change in the topology are the major barriers for routing protocols in MANET. Due to these problems several routing protocol to have been developed that can meet the dynamic nature of ah-hoc network. These routing protocols are categorized into different types: *Proactive Routing Protocol, Reactive Routing Protocol and Hybrid Routing Protocol*, which are described below and shown in Figure 5.

These protocols are characterized on the basis of area/application where they are most suitable. Proactive routing maintains routes to all nodes, including nodes to which no packets are sent. Such methods react to topology changes, even if no traffic is affected by the changes. They are also called table-driven methods.

Reactive routing is based on demand for data transmission. Routes between hosts are determined only when they are explicitly needed to forward packets. Reactive routing are also called on-demand methods.

They can significantly reduce routing overhead when the traffic is lightweight and the topology changes less dramatically, since they do not need to update route information periodically and do not need to find and maintain routes on which there is no traffic. Hybrid routing combine proactive and reactive methods to find efficient routes, without much control overhead.

Figure 5. Types of routing protocols

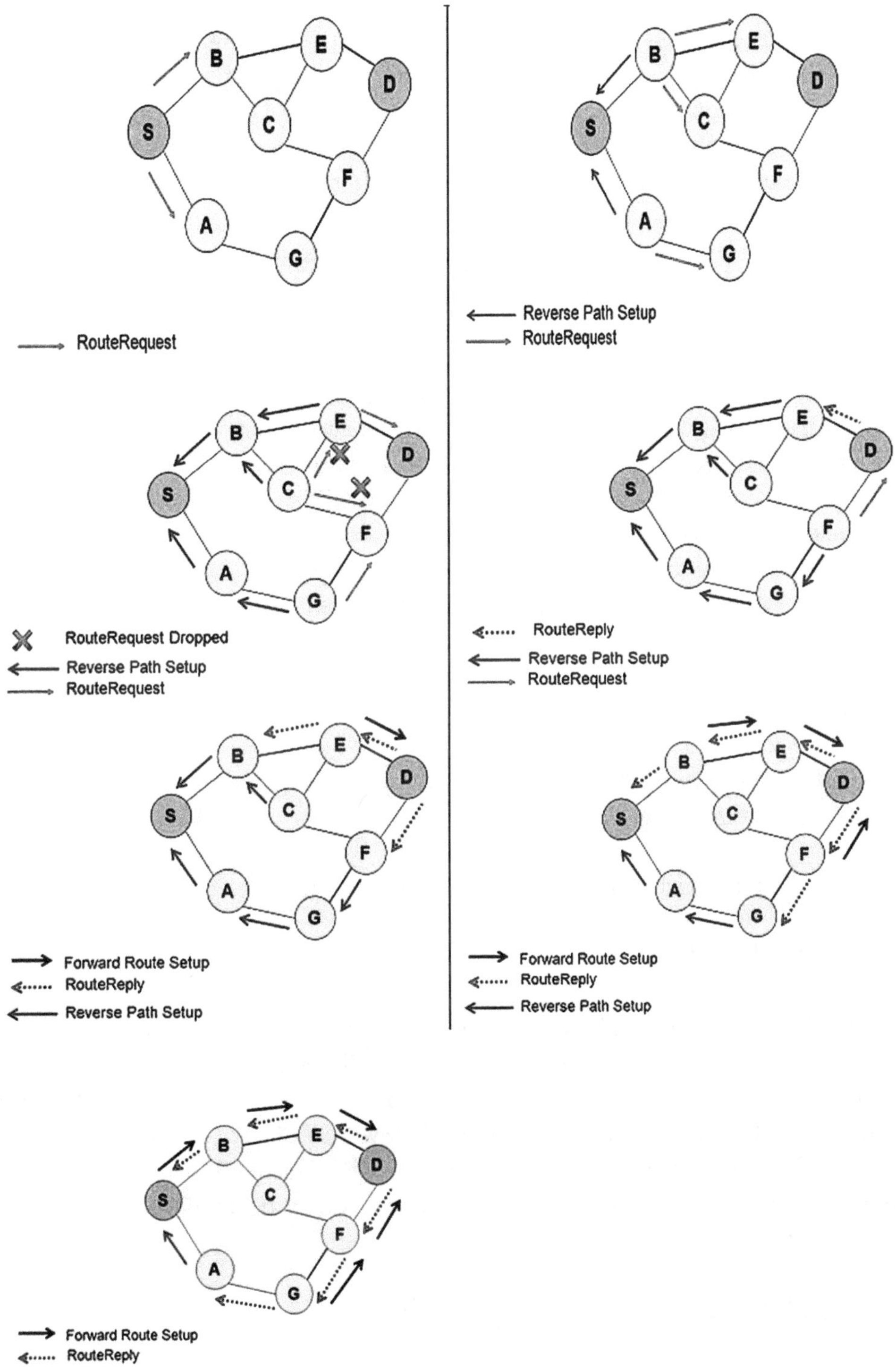

4.1 Proactive Routing Protocols

These routing protocols act just like wired networks. In proactive routing, each node keeps tables that have the updated information of the routing of nodes in the network. In this protocol, each row keeps the information about the next hop for reaching a node subnet and the cost of this route. Various table-driven protocols differ in the way the information about a change in topology is propagated through all nodes in the network (Vidhale & Dorle, 2011). There are some differences among these protocols depending upon the routing information being updated in each routing table. Additionally, these routing protocols also build various tables. These proactive protocols are not suitable for larger networks, because they need to maintain node entries for each node in the routing table of every node. This ends up with more overhead in the routing table and consumption of more bandwidth. For example, the conventional routing schemes are Destination Sequenced Distance Vector (DSDV) and Optimized Link State Routing (OLSR). Destination-Sequenced Distance-Vector Routing (DSDV) and Optimized Link State Routing (OLSR) is a table-driven routing protocol for ad-hoc mobile networks (Azouqa, 2013).

Proactive routing protocols are based on shortest path algorithms. They keep information of all connected nodes in form of tables because these protocols are tabling based (Paul, 2011). Furthermore, these tables are also shared with their neighbors. Whenever any change occurs in network topology, every node updates its routing table.

Proactive Routing Protocols are mainly categorized into five types. They are as listed below.

- *Fisheye State Routing* (FSR),
- *Link State Routing* (e.g. - OLSR),
- *Distance Vector Routing* (e.g. DSDV),
- *Wireless Routing Protocol* (WRP), and
- *Source Tree Adaptive Routing* (STAR).

The working details for proactive routing protocols are as follows:

4.1.1 Fisheye State Routing

FSR (Gerla, Hong & Pei, 2002) is similar to Link State Routing Protocol that each node maintain topology map, but the difference is in the process of information dissemination.

Fisheye State Routing table (Gerla, Pei & Lee, 1999) is mainly designed for mitigate the routing overhead in large network. This technique is based on *Fisheye Technique*. In Link State protocol, packets are flooded whenever there is change in topology, but in FSR, updated information is maintained in a table by each node, and this table is periodically exchanged among the neighbor nodes. Packets are not flooded in FSR as done in Link State protocol.

4.1.2 Destination Sequence Distance Vector Routing

DSDV (Perkins and Bhagwat, 1994), (Abolhasan, Wysocki & Dutkiewicz, 2004) use Distance Vector shortest path routing algorithm, it provides loop free single path to the destination. DSDV sends two types of packets "full dump" and "incremental".

(Lee, Gerla & Toh, 1999) DSDV is the enhanced version of Bellman-Ford Routing algorithm that is table driven based routing algorithm. DSDV uses bidirectional links and has a new attribute named as *Sequence Number* in each routing table. This sequence number avoids the loop formation in the network. As the behavior of convention routing algorithms, all nodes in the DSDV also maintain a routing table which contains information regarding metrics, destination, next hop and sequence number. When a change is detected (Joining of node in the network or leaving of a node in the network) each node advertises the updated routing information to its neighbor nodes with the help of broadcasting *update packet*. This update packet starts with *one* in the metric to all neighbor nodes. After receiving the *update packet*, each node updates its routing table and one is incremented from metric, and again this update packet is forwarded to next neighbor. This same process is followed until all the nodes in the network receive the *update packet* with corresponding metric value. In case, a node receives more than one update packet, route having the more sequence number is prefer. Even if there is clash between sequence numbers of update packets for same node, update packet having the smallest metric will be preferred and rest all routes will be discarded. However, DSDV faces the problem of Dump Fluctuations because old sequence number is discarded when new sequence number is generated. Moreover, knowledge Asymmetry and sink Un-reachable problems are also encountered due to unidirectional links in the ad-hoc network. Multi Path routing is also not supported in DSDV and it suffers overhead because of periodical updates.

4.1.3 Optimized Link State Routing

OLSR (Abolhasan, 2004; Clausen, 2003) maintains routing information by sending link state information. After each change in the topology, every node sends updates to selective nodes. By doing so, every node in the network receive updates only once. Unselected packets cannot retransmit updates; they can only read updated information.

(Jacquet, Muhlethaler, Clausen, Laouiti, Qayyum, & Viennot, 2001) OLSR is a proactive routing protocol that inherits the feature of link state routing algorithm. It has two main benefits, first is size of control packet are reduced and second is, it mitigates the flooding of control packets using *multipoint relay*. In OLSR, hop by hop routing is performed that means most recent information is used by node in the network for routing. Multipoint Relay is the backbone of OLSR that is used to reduce the flooding of packets in the ad-hoc network. It is possible to do because each node selects set of nodes from its neighbor, which also retransmit the packets. The set of this selected node is known as multipoint relay (MPRs) of that node. Nodes which are not the part of multipoint relay do not transmit the broadcast packet, they only process and the read the packet.

4.1.4 Wireless Routing Protocol

WRP described in (Murthy & Garcia-Luna-Aceves, 1996) that is a table-based protocol with the goal of maintaining routing information among all nodes in the network. Each node in the network is responsible for maintaining four tables, i.e.: (a) distance table; (b) routing table; (c) link-cost table; and (d) message retransmission list (MRL) table. It is a path finding algorithm, specially designed for mitigate the loops occur in the network. WRP is based on Radio connectivity among the nodes.

4.1.5 Source Tree Adaptive Routing

STAR is another link State protocol. In STAR, (Abolhasan, 2004; Jiang & Garcia-Luna-Aceves, 2001) preferred routes to every destination are saved in each router. It reduces overhead on the network by eliminating periodic updates. There is no need of sending updates unless any event occurs. This protocol can be suitable for large-scale networks, but it needs large memory and processing because it has to maintain large trees for whole network (Steenstrup, 1995). Proactive based routing protocols may not be suitable for high mobility nodes because distance vector routing takes much bandwidth to share routing information with neighbors. Furthermore, size of the table is also quite big while discussing about large networks and in case of link state routing a lot of memory and processing may also be required.

As in VANET, nodes (vehicles) have high mobility and moves with high speed. Proactive based routing is not suitable for it. Proactive based routing protocols may fail in VANET due to consumption of more bandwidth and large table information.

4.2 Reactive Routing Protocols

Reactive routing is also known as on-demand routing protocol since they don't store routing information or routing updates at the network nodes if there is no communication. These protocols take a different approach to routing and route according to present condition. They do not store or continuously update their route tables with the new route topology. According to demand if a node wants to send a packet to another node then this protocol searches for the best possible route and establishes the connection to transmit and receive the packet. The route discovery usually done by flooding the route request packets all over the network. Examples of reactive routing protocols are the Dynamic Source Routing (DSR) and Ad-hoc On Demand Distance Vector Routing (AODV) (Belding-Royer, & Das, 2003). Cloud computing is a way of enabling on-demand network access to resources which can be provided and released with minimal management effort. It has many advantages: it reduces the cost of capital expenditure; it eases the maintenance and provides high performance with less effort. In VANET, cloud can be viewed in three ways, Infrastructure-as-a-Service (IaaS), either as Network-as-a-service (NaaS) or storage-as-a-service (SaaS). In NaaS, the vehicles can share their network access with other vehicles, while in SaaS, the vehicles can share their storage with other vehicles or any other computing services (Bitam, Mellouk & Zeadally, 2015).

Reactive routing protocols were designed to overcome the overhead that was created by proactive routing protocols. Reactive routing opens the route only when it is necessary for a node to communicate with each other. Reactive routing consists of route discovery phase in which the query packets are flooded through the network for the path search, and this phase completes when the route is found (Nagaraj & Dhamal, 2012).

Route discovery in reactive routing can be done by sending RREQ (*Route Request*) from a node when it requires a route to send the data to a particular destination. After sending RREQ, node then waits for the RREP (*Route Reply*) and if it does not receive any RREP within a given time period, source node assumes that either route is not available or route expired. When RREQ reaches the particular destination and if the source node receives RREP then by using unicasting, information is forwarded to the source node in order to ensure that route is available for communication.

Reactive routing protocols have been mainly categorized into three types. They are as listed below:

- *Ad-hoc On Demand Distance Vector Routing* (AODV),
- *Dynamic Source Routing* (DSR), *and*
- *Temporally Ordered Routing Algorithm* (TORA).

4.2.1 Ad-Hoc On-Demand Distance Vector Routing

AODV (Perkins, 2003) is an example of pure reactive routing protocol. AODV belongs to multi-hop type of reactive routing. AODV routing protocol works purely on the demand basis when it is required by network, which is fulfilled by nodes within the network. Route discovery and route maintenance is also carried out on the demand basis even if only two nodes need to communicate with each other. AODV cuts down the need of nodes in order to always remain active and to continuously update routing information at each node. In other words, AODV maintains and discovers routes only when there is a need of communication among different nodes.

- **AODV Route Discovery:** Route discovery is one of the most important characteristics of any protocol in wireless communication. The need for basic route discovery arises when a source node wants to communicate with any particular destination node in order to forward data packet. AODV uses route discovery by broadcasting RREQ to all its neighboring nodes. The broadcasted RREQ contains addresses of source and destination nodes in order identify those particular nodes for which route have been demanded.
- **AODV Route Table Management:** Routing table management in AODV is needed to avoid those entries of nodes that do not exist in the route from source to destination. Managing routing table information in AODV is handled with the destination sequence numbers. The need for routing table management is important to make the communication loop free. The following are characteristics to maintain the route table for each node (Tsirigos & Haas, 2001).
    - IP address of the particular destination
    - Total number of hops to the destination
    - **Next Hop:** It contains information of those nodes that are used to forward data packets by using the current route.
    - Destination sequence numbers
    - **Active Neighbors:** Those nodes that currently using the active route.
    - **Expiration Time:** It contains information for the total time that route is being valid.
- **AODV Route Maintenance:** When nodes in the network detects that a route is not valid anymore for communication it deletes all the related entries from the routing table for those invalid routes. And sends the RREP to current active neighboring nodes that route is not valid anymore for communication. AODV maintains only the loop free routes, when the source node receives the link failure notification, it either start the process of rebroadcasting RREQ or the source node stop sending data through invalid route. Moreover, AODV uses the active neighbor's information to keep tracking of currently used route.
- **AODV Features:** It is a distance vector routing protocol when a node wants to establish new communication with another node, it searches for an available path to the destination node, in its local routing table (Perkins, Belding-Royer, & Das, 2003). AODV protocol is generally used when one node wishes to communicate with another node which is not in the range, in this case it will use intermediate node for communication. AODV works on two phases:

- **Route Discovery:** Route Discovery is process in which a node floods a Request packet to its neighbor, until the route is not established of Destination Node.
- **Route Maintenance:** Route Maintenance is the second phase after the Route Discovery, after the establishment of the route. There are three advantages of route maintenance: 1.) it provides the scalability to the route; 2.) it provides the response of the links; and 3.) reduces the cost and time of discovering new route.

AODV protocol works on four types of messages RREQ (Route Request), RREP (Route Reply), RERR (Route Error), HELLO Packet. RREQ and RREP are used in Route Discovery and RERR and HELLO used in route maintenance.

To understand in better way, see the figure number 8, in which Node A is Source Node and H is Destination. So, Node A generates and floods RREQ message to its neighbors i.e. Node C and Node B. It will initial TTL with 1. This RREQ contains IP address of Node A, IP address of Node H, Hop Count and Sequence Number. When Node B and C will receive RREQ and as they are not the destination nodes they will again broadcast or flood the RREQ message until the Node G receive RREQ. When Node G receive RREQ message, it will send RREP message to the Node A. During forward the RREQ message, Node C and Node B store the reverse count back to the Node A. When there will be multiple reverse route it will pick the smallest Hop Count to identify the shortest path. Sequence Number is maintained to avoid the loops. RERR is an error message is used when changed its position or connection is lost or crashed, route become invalid or not able to communicate with its neighbor. If a node gets a RERR message that it deletes all the routes related with error message. HELLO message is used to ensure that all neighbors are still connected among each other. Every node sends HELLO packet after regular time intervals. If a node does not receives a HELLO packet from its neighbor from a fixed time interval, it is assumed as node is no longer exits. However, since AODV depends upon flooding the RREQ message so latency is very high and also it is depending upon RREQ message protocol so it is not efficient route maintenance strategy (Chakeres, & Belding-Royer, 2004; Perkins, 2001; Royer & Perkins, 2000; Sanzgiri, Dahill, Levine, Shields & Belding-Royer, 2002).

The message formats for AODV contains four different message formats. This is provided in Tables 2-5 respectively.

Figure 6. AODV route discovery

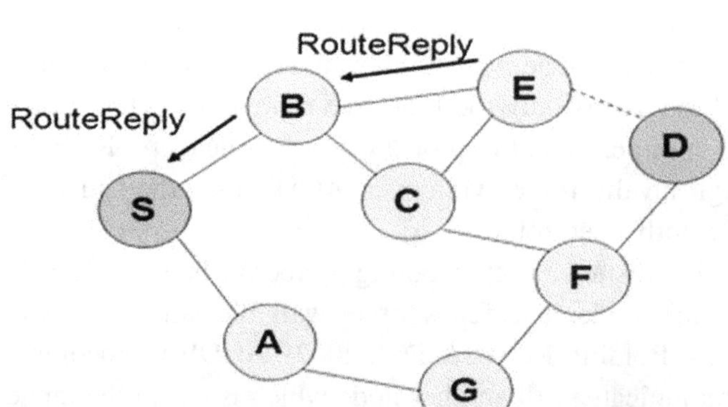

Figure 7. AODV route maintenance

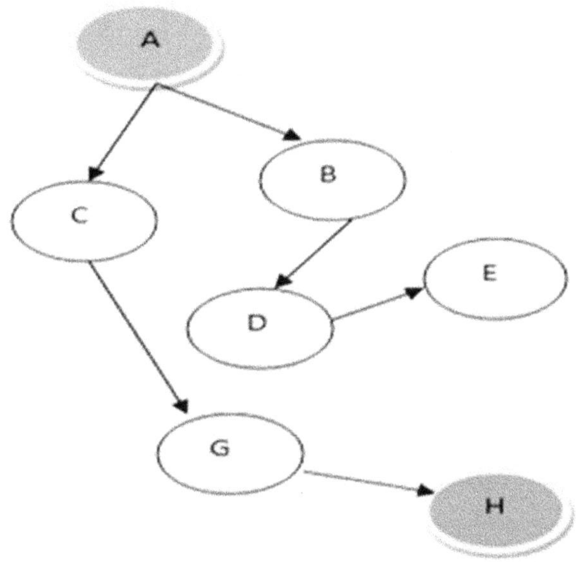

Figure 8. Route discovery in AODV

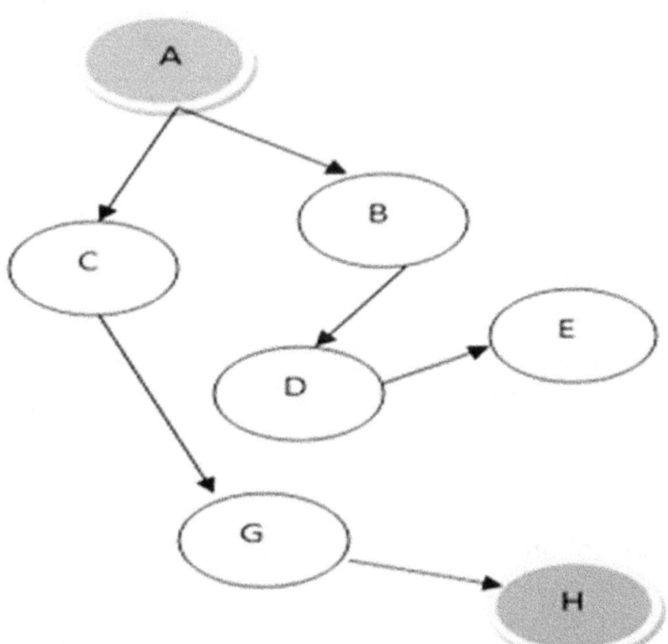

Table 2. Route Request (Route_Req) message

Type
Route_Req_ID
Destination_IP_Address
Destination_Sequence_Number
Originator_IP_Address
Originator_Sequence_Number

Table 3. Route Reply (Route_Rep) message

Type
Destination_IP_Address
Destination_Sequence_Number
Originator_IP_Address
Lifetime

Table 4. Route Error (Route_Err) message

Type
Unreachable_Destination_IP_Address
Unreachable_Destination_Sequence_Number
Additional_Unreachable_Destination_IP_Addresses
Additional_Unreachable_Destination_Sequence_Numbers

Table 5. Route Reply Acknowledgment (Route_Rep_Ack)

Type
MUST_be_sent_in_response_to_a_Route_Rep_message

4.2.2 Dynamic Source Routing

As the name itself specified, DSR is source routing based on demand routing protocol. It is very scalable and highly dynamic protocol because every node posses a source cache that updates all entries about new route. Similar to AODV, DSR also works on two phases i.e. route discovery and route maintenance. When a source node wishes to communicate with the destination node, before broadcasting the RREQ message it first looks to its route cache. If there is unexpired route of that particular destination node then it will send data via this route, if the route is expired then it will broadcast RREQ message to its neighbor. When the neighbor node receives the RREQ message, it also looks out its route cache. If there is expired route to the destination node, then it will add its address in route record of the packet. This technique avoids the duplicate message to be sent in the network because, if a node sees its own address in the route record it discards that particular packet. In this way, only limited numbers of packets are propagated in the network that also reduce the complexity of the network. This process is continued until the destination node will not receive the RREQ message. When the destination node receives the RREQ message it saves the route in its cache and send the RREP message viva the same route. Every node stores the route in its cache. If a destination node receives RREQ message from multiple routes, then it will select the shortest path based on the address mention in the route record of the packet.

Figure 9 shows the working of DSR routing. Sender node A wish to communicate with Node F so it broadcast RREQ message to its neighbor nodes C and B. C will again broadcast the RREQ packet but append its address and then send to E, Node E will do the same and broadcast to node D. Node D will again broadcast the message to C, but this time C will discard the packet because it will see its own address in the route record. When Node F will receive the RREQ message from Node D, it will send RREP message viva the route mentioned in the route record. The route maintenance is done with the of ERROR message. If a node senses a transmission problem that cannot be recovered, it sends an ERROR message to the Source Node A. The sender will discard that particular route in which error is produced (Johnson, 2001).

- **DSR Route Discovery:** In DSR route discovery, when a node wants to send a packet to another node normally it stored routing information in the header of the packet. The routing information in DSR is in the form of "sequence of nodes". This routing information had been learned by every node. If no routing information is available, source node uses the "route discovery" mechanism to find the appropriate route to reach destination. The whole process to find the route is called route discovery.

The route discovery mechanism can be better defined with the given figure 10. In this figure, node "A" starts the discovery process to find the route to node "E". So node "A" called initiator and node "E" is called target. When route discovery starts, an initiator sends "*discovery request*" to nodes that are

Figure 9. Route discovery in DSR

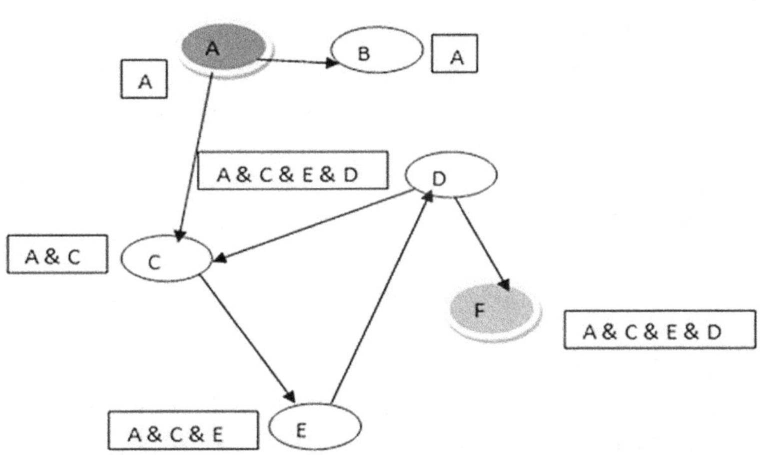

Figure 10. DSR route discovery

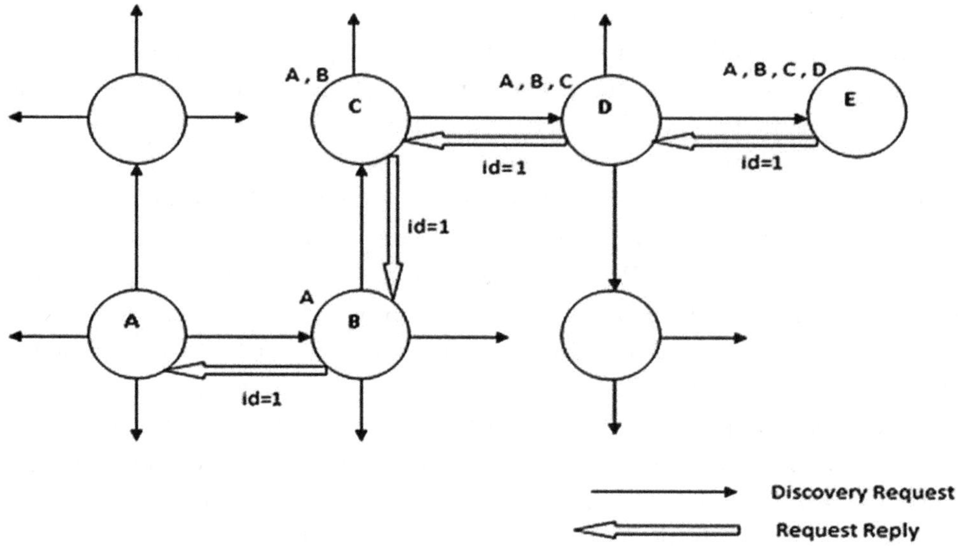

within its wireless range. The discovery request contains initiators, target´s and route information. In the start, route record is set to empty by the initiator. When any node receives the discovery request, it checks the target information. If a receiver is not targets itself, it adds its information to the route record and forwards the discovery request to all nodes in the wireless range. But when the target node receives the discovery requests it send the request reply with final route information containing the complete intermediate path. The target node can send the request reply using its route cache or by reversing the order or discovery request. Thus, in this way DSR discover the route from source to destination.

- **DSR Route Maintenance:** In DSR when any node sends a packet using source route, it also has the responsibility for its delivery confirmation. If consider the above scenario again, then the route maintenance mechanism as follows:

If node A transmit a packet to node E, through node B, C and D. Node A will retransmit the packer until it received delivery confirmation from packet C. Same like A, B will retransmit until confirmation from C, C will retransmit until confirmation from D and D will retransmit until it received confirmation from E. This retransmission process continues for a limited number called "maximum number of attempts".

In case, if a node retransmits a packet for the maximum number of attempts and there is no acknowledgement received yet the node will send error message to the source node.

Then the source node can use another route from a route cache to send packet or it can start route discovery again. This simple mechanism DSR used in route maintenance.

4.2.3 Temporally Ordered Routing Algorithm

TORA (Park, 2000) is a highly adaptive, loop-free, distributed routing algorithm. TORA routing belongs to a family of link reversal routing algorithms where a directed acyclic graph (DAG) toward the destination is built based on the height of the tree rooted at the source. The directed acyclic graph directs the flow of packets and ensures reachability to all nodes. When a node has a packet to send, it broadcasts the packet. Its neighbor only broadcasts the packet if it is the sending node's downward link based on the DAG.

TORA is proposed to operate in a highly dynamic mobile networking environment. It is source-initiated and provides multiple routes for any desired source/destination pair. The key design concept of TORA is the localization of control messages to a very small set of nodes near the occurrence of a topological change. To accomplish this, nodes need to maintain routing information about adjacent (one-hop) nodes.

TORA uses a metric referred to as the "height" of the node to assign a direction to links for forwarding packets to a given destination. The node heights can be totally ordered lexicographically, and thus define a directed acyclic graph rooted at the destination.

The TORA protocol performs three basic functions:

- Route Creation,
- Route Maintenance, and
- Route Erasure.

The route creation algorithm generates a directed acyclic graph from source to destination based on a propagation parameter, called "height". The first step toward the Route creation is the Route Discovery. In Figure 11 sources broadcast route request to the destination through all possible paths.

TORA can suffer from unbounded worst-case convergence time for very stressful scenarios. TORA has a unique feature of maintaining multiple routes to the destination so that topological changes do not require any reaction at all.

The protocol reacts only when all routes to the destination are lost. In the event of network partitions the protocol is able to detect the partition and erase all invalid routes. The algorithm starts by assigning the height of the destination to 0 (base) and all other node's height undefined (NULL), which is shown in Figure 12.

Each node along the way to the source increases its height by one and rebroadcasts the Route Reply Packet with its updated heights, which is shown in Figure 13.

Figure 11. TORA route discovery

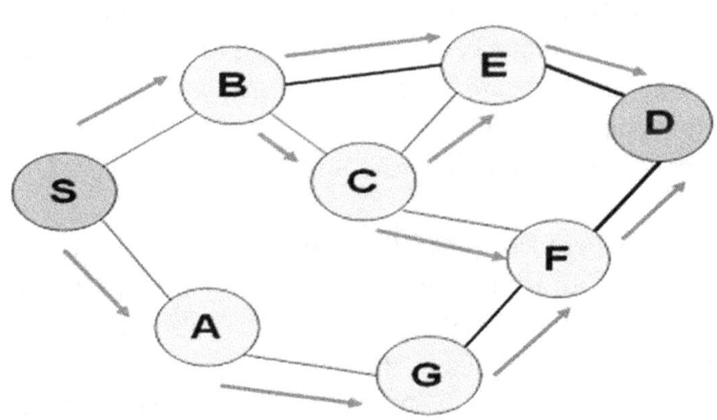

Figure 12. TORA: Initial height

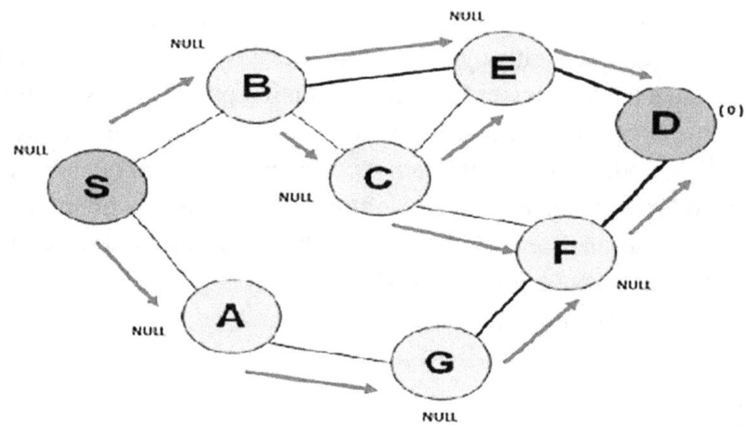

Figure 13. TORA: Height calculations

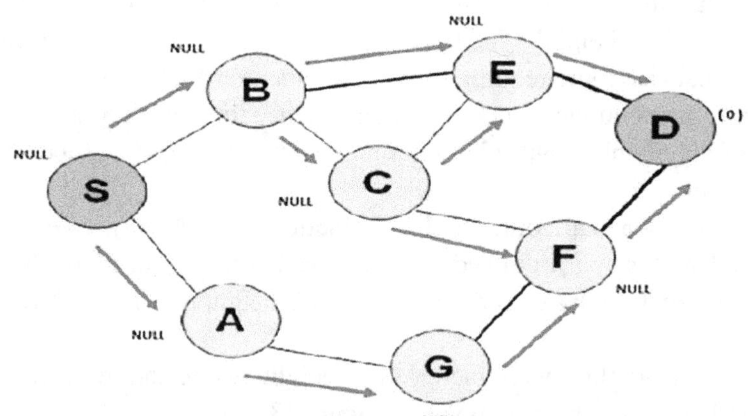

4.3 Hybrid Routing Protocols

The major challenges faced by MANET are asymmetric links and rapidly changing nature of network. Reactive and proactive both routing protocols individually are not much effective under the same conditions, so concept of Hybrid Routing Protocol is introduced. Hybrid routing combines characteristics of both reactive and proactive routing protocols to make routing more scalable and efficient (Abolhasan, 2004). Mostly hybrid routing protocols are zone based; it means the number of nodes is divided into different zones to make route discovery and maintenance more reliable for MANET.

Hybrid Routing Protocols have been mainly categorized into two types. They are as listed below:

- **Hybrid Ad-hoc Routing Protocol (HARP) (Abolhasan, 2004; Nikaein, 2001):** As the name indicates, this routing protocol is proactive and reactive routing protocol. HARP is designed for improving the delay caused by path failure. It is offers a stable path between source and destination with minimum delay by anticipating the path failures. A hierarchical topology is used by HARP to reduce delay which is given by DDR. The mechanism of HARP is based on two levels:

- **Intra Zone Level:** Intra Zone Level is a proactive based approach that is responsible for forwarding the packets in a pre-decided path. This feature is derived from DDR.
- **Inter Zone Level:** Inter Zone Level is a reactive based approach that is responsible for route discovery and route maintenance with the help of Request PREP and Reply PREP.

- **Zone Routing Protocol (ZRP) (Haas, Pearlman & Samar, 2002; Beijar, 2002):** It has been noticed, from various studies that, reactive routing protocol has disadvantages of request delay and flooding, whereas, proactive routing protocols consumes too much bandwidth for maintain routing information. ZRP is the alternative solution for these problems. As the name indicates, ZRP is based on routing zones, separately for each node, and these zones overlap each other of neighboring node. A zone is not in terms of geographical area but it can be seen as group of neighboring nodes, and every zone has a central node. A zone with radius *R* will have collections of all nodes whose at most distance from central node is *T*. ZRP protocols have two components namely IntrA-Zone Routing Protocol (IARP) and IntEr- Zone Routing Protocol (IERP). IARP is based on proactive routing protocol, that is responsible for maintain the nodes in a particular zone. IERP is based on reactive routing protocol responsible for route discovery and maintenance. ZRP use *Bordercasting* instead of broadcasting the packet. *Bordercasting Routing Protocol* is responsible for direct the request to border of zone using information given y IARP. Nevertheless, it is very effective technique as compare with any single proactive and reactive routing protocol but it is too complicated to implement and geographical based routing protocols can perform better than ZRP.

5. CONCLUSION

AODV and DSR are both on demand routing protocol. Both protocols are loop free and guarantee the shortest path. AODV and DSR routing is source routing based where AODV is based on both source routing and distance vector. In DSR route maintenance is done by route cache where AODV maintains the route table. HELLO packets and Sequence numbers are not available in DSR routing but they are in AODV routing.

The study of vehicular cloud computing helps the young scientist and scholars to find the new opportunities and challenges in vehicular networks. VCC provides an efficient way to manage the three key resources i.e. computing, storage and internet for communication and decision making. The study of their architecture in respect to on demand routing protocols helps the authors to perform their simulation in an efficient manner. It is the new paradigm which is going to enhance the intelligent transport system at real time.

REFERENCES

Abolhasan, M., Wysocki, T., & Dutkiewicz, E. (2004). A review of routing protocols for mobile ad hoc networks. *Ad Hoc Networks*, *2*(1), 1–22. doi:10.1016/S1570-8705(03)00043-X

Alam, K. M., Saini, M., & El Saddik, A. (2015). Toward social internet of vehicles: Concept, architecture, and applications. *IEEE Access: Practical Innovations, Open Solutions*, *3*, 343–357. doi:10.1109/ACCESS.2015.2416657

Aloqaily, M., Kantarci, B., & Mouftah, H. T. (2015, November). Vehicular clouds: State of the art, challenges and future directions. In *Applied Electrical Engineering and Computing Technologies (AEECT), 2015 IEEE Jordan Conference on* (pp. 1-6). IEEE.

Azouqa, M., Jamoos, A., & Bali, S. (2013). *Investigation of vehicle-to-vehicle communication in a road intersection. Department of Electronic Engineering Al-Quds University.*

Balon, N. (2006). *Introduction to vehicular ad hoc networks and the broadcast storm problem*. Academic Press.

Beijar, N. (2002). Zone routing protocol (ZRP). Networking Laboratory, Helsinki University of Technology.

Belding-Royer, E., & Das, S. (2003). *Ad hoc on-demand distance vector (AODV) routing. draft-ietf-manet-aodv-13. txt.* INTERNET DRAFT.

Bitam, S., Mellouk, A., & Zeadally, S. (2015). VANET-cloud: A generic cloud computing model for vehicular Ad Hoc networks. *IEEE Wireless Communications*, *22*(1), 96–102. doi:10.1109/MWC.2015.7054724

Chakeres, I. (2008). Dynamic MANET on-demand (DYMO) routing. *Internet Draft (draft-ietf-manet-dymo-14).*

Chakeres, I. D., & Belding-Royer, E. M. (2004, March). AODV routing protocol implementation design. In *Distributed Computing Systems Workshops, 2004. Proceedings. 24th International Conference on* (pp. 698-703). IEEE. 10.1109/ICDCSW.2004.1284108

Clausen, T., & Jacquet, P. (2003). *Optimized link state routing protocol (OLSR)* (No. RFC 3626).

Darus, M. Y., & Bakar, K. A. (2011). A review of congestion control algorithm for event-driven safety messages in vehicular networks. *Channels, 10*, 13.

Darus, M. Y. B., & Bakar, K. A. (2011). Congestion control framework for disseminating safety messages in Vehicular Ad-Hoc Networks (VANETs). *International Journal of Digital Content Technology and Its Applications, 5*(2), 173-180.

De Fuentes, J. M., González-Manzano, L., González-Tablas, A. I., & Blasco, J. (2014). Security models in vehicular ad-hoc networks: A survey. *IETE Technical Review*, *31*(1), 47–64. doi:10.1080/02564602.2014.890844

De Oliveira Barros, M. T. A., de Morais Gomes, R. C., & da Costa, A. F. B. F. (2013). A Top-down Multi-layer Routing Architecture for Vehicular Ad-Hoc Networks. *IEEE Latin America Transactions*, *11*(6), 1344–1352. doi:10.1109/TLA.2013.6710382

Frodigh, M., Johansson, P., & Larsson, P. (2000). Wireless ad hoc networking: the art of networking without a network. *Ericsson Review, 4*(4), 249.

Fujise, M., Kato, A., Sato, K., & Harada, H. (2002). Intelligent transport systems. In *Wireless Communication Technologies: New Multimedia Systems* (pp. 171–200). Boston, MA: Springer. doi:10.1007/0-306-47326-7_8

Gerla, M. (2012, June). Vehicular cloud computing. In *Ad Hoc Networking Workshop (Med-Hoc-Net), 2012 The 11th Annual Mediterranean* (pp. 152-155). IEEE. 10.1109/MedHocNet.2012.6257116

Gerla, M., Hong, X., & Pei, G. (2002). *Fisheye state routing protocol (FSR) for ad hod networks*. draft-ietf-manet-fsr-03. txt.

Gerla, M., Pei, G., & Lee, S. J. (1999). Wireless, mobile ad-hoc network routing. *Proceedings of IEEE/ ACM FOCUS'99*, 1112-1124.

Haas, Z. J., Pearlman, M. R., & Samar, P. (2002). *The zone routing protocol (ZRP) for ad hoc networks*. Academic Press.

Hassanein, H. S., Abdelhamid, S., & Elgazzar, K. (2015, October). A framework for vehicular cloud computing. In *Connected Vehicles and Expo (ICCVE), 2015 International Conference on* (pp. 238-239). IEEE. 10.1109/ICCVE.2015.51

Jacquet, P., Muhlethaler, P., Clausen, T., Laouiti, A., Qayyum, A., & Viennot, L. (2001). Optimized link state routing protocol for ad hoc networks. In *Multi Topic Conference, 2001. IEEE INMIC 2001. Technology for the 21st Century. Proceedings. IEEE International* (pp. 62-68). IEEE. 10.1109/INMIC.2001.995315

Jiang, H., & Garcia-Luna-Aceves, J. J. (2001). Performance comparison of three routing protocols for ad hoc networks. In *Computer Communications and Networks, 2001. Proceedings. Tenth International Conference on* (pp. 547-554). IEEE. 10.1109/ICCCN.2001.956320

Johnson, D., Hu, Y. C., & Maltz, D. (2007). *The dynamic source routing protocol (DSR) for mobile ad hoc networks for IPv4* (No. RFC 4728).

Johnson, D. B., Maltz, D. A., & Broch, J. (2001). DSR: The dynamic source routing protocol for multi-hop wireless ad hoc networks. *Ad Hoc Networks*, *5*, 139–172.

Kumar, A., & Singla, A. K. (2011). Performance evaluation of MANET routing protocols on the basis of TCP traffic pattern. *International Journal of Information Technology Convergence and Services*, *1*(5), 41.

Lee, E., Lee, E. K., Gerla, M., & Oh, S. Y. (2014). Vehicular cloud networking: Architecture and design principles. *IEEE Communications Magazine*, *52*(2), 148–155. doi:10.1109/MCOM.2014.6736756

Lee, S. J., Gerla, M., & Toh, C. K. (1999). A simulation study of table-driven and on-demand routing protocols for mobile ad hoc networks. *IEEE Network, 13*(4), 48-54.

Li, B., Mirhashemi, M. S., Laurent, X., & Gao, J. (2010). Wireless access for vehicular environments. *Telematics Communication Technologies and Vehicular Networks: Wireless Architectures and Applications*, 90-107.

Li, F., & Wang, Y. (2007). Routing in vehicular ad hoc networks: A survey. *IEEE Vehicular Technology Magazine*, *2*(2), 12–22. doi:10.1109/MVT.2007.912927

Li, M. W., Wu, T. H., Lin, W. Y., Lan, K. C., Chou, C. M., & Hsu, C. H. (2011). On the feasibility of using 802.11 p for communication of electronic toll collection systems. *ISRN Communications and Networking*.

Loo, J., Mauri, J. L., & Ortiz, J. H. (Eds.). (2016). *Mobile ad hoc networks: current status and future trends*. CRC Press.

Morgan, Y. L. (2010). Managing DSRC and WAVE standards operations in a V2V scenario. *International Journal of Vehicular Technology*.

Murthy, S., & Garcia-Luna-Aceves, J. J. (1996). An efficient routing protocol for wireless networks. *Mobile Networks and Applications*, *1*(2), 183–197. doi:10.1007/BF01193336

Nagaraj, U., & Dhamal, P. (2012). Broadcasting routing protocols in VANET. *Network and Complex Systems*, *1*(2), 13–19.

Park, V. (2000). *Temporally-ordered routing algorithm (TORA) version 1 functional specification*. draft-ietf-manet-tora-spec-03. txt.

Perkins, C., Belding-Royer, E., & Das, S. (2003). *Ad hoc on-demand distance vector (AODV) routing* (No. RFC 3561).

Perkins, C. E., & Bhagwat, P. (1994, October). Highly dynamic destination-sequenced distance-vector routing (DSDV) for mobile computers. *Computer Communication Review*, *24*(4), 234–244. doi:10.1145/190809.190336

Perkins, C. E., Royer, E. M., Das, S. R., & Marina, M. K. (2001). Performance comparison of two on-demand routing protocols for ad hoc networks. *IEEE Personal communications, 8*(1), 16-28.

Pesel, R., & Maslouh, O. (2011). *Vehicular Ad Hoc Networks (VANET) applied to Intelligent Transportation Systems*. Universite de Limoges, France: ITS.

Racehorse, V. J., Zhang, G., Sussman, A., Jian, A., & Parker, T. (2014). Bus rapid transit system deployment for high quality and cost-effective transit service: A comprehensive review and comparative analysis. *IET Intelligent Transport Systems*, *9*(2), 175–183. doi:10.1049/iet-its.2013.0176

Royer, E. M., & Perkins, C. E. (2000). An implementation study of the AODV routing protocol. In Wireless Communications and Networking Confernce, 2000. WCNC. 2000 IEEE (Vol. 3, pp. 1003-1008). IEEE. doi:10.1109/WCNC.2000.904764

Sallam, G., & Mahmoud, A. (2015, April). Performance Evaluation of OLSR and AODV in VANET Cloud Computing Using Fading Model with SUMO and NS3. In *Cloud Computing (ICCC), 2015 International Conference on* (pp. 1-5). IEEE. 10.1109/CLOUDCOMP.2015.7149649

Sanzgiri, K., Dahill, B., Levine, B. N., Shields, C., & Belding-Royer, E. M. (2002, November). A secure routing protocol for ad hoc networks. In *Network Protocols, 2002. Proceedings. 10th IEEE International Conference on* (pp. 78-87). IEEE. 10.1109/ICNP.2002.1181388

Sharma, M. K., & Kaur, A. (2015, September). A survey on vehicular cloud computing and its security. In *Next Generation Computing Technologies (NGCT), 2015 1st International Conference on* (pp. 67-71). IEEE.

Shastri, A., Dadhich, R., & Poonia, R. C. (2011). Performance analysis of on-demand Routing protocols for vehicular ad-hoc Networks. *International Journal of Wireless & Mobile Networks*, *3*(4), 103–111. doi:10.5121/ijwmn.2011.3407

Singh, R., Singh, D. K., & Kumar, L. (2011). Performance Evaluation of DSR and DSDV Routing Protocols for Wireless Ad Hoc Networks. *Int. J. Advanced Networking and Applications*, *2*(04), 732–737.

Steenstrup, M. (1995). Routing in communications networks. Prentice Hall International (UK) Ltd.

Tsirigos, A., & Haas, Z. J. (2001). Multipath routing in the presence of frequent topological changes. *IEEE Communications Magazine*, *39*(11), 132–138. doi:10.1109/35.965371

Tyagi, A. K., & Niladhuri, S. (2016, August). Providing trust enabled services in vehicular cloud computing. In *Proceedings of the International Conference on Informatics and Analytics* (p. 3). ACM.

Vidhale, B., & Dorle, S. S. (2011, August). Performance analysis of routing protocols in realistic environment for vehicular Ad Hoc networks. In *Systems Engineering (ICSEng), 2011 21st International Conference on* (pp. 267-272). IEEE. 10.1109/ICSEng.2011.55

Wang, S. Y., & Lin, C. C. (2008, September). NCTUns 5.0: A network simulator for IEEE 802.11 (p) and 1609 wireless vehicular network researches. In *Vehicular Technology Conference, 2008. VTC 2008-Fall. IEEE 68th* (pp. 1-2). IEEE. 10.1109/VETECF.2008.464

Whaiduzzaman, M., Sookhak, M., Gani, A., & Buyya, R. (2014). A survey on vehicular cloud computing. *Journal of Network and Computer Applications*, *40*, 325–344. doi:10.1016/j.jnca.2013.08.004

Yu, R., Zhang, Y., Gjessing, S., Xia, W., & Yang, K. (2013). Toward cloud-based vehicular networks with efficient resource management. *IEEE Network*, *27*(5), 48–55. doi:10.1109/MNET.2013.6616115

Chapter 8
Smart Vehicles for Traffic Management and Systems Using Cloud Computing

Ashish Thomas
Rajasthan University Jaipur, India

Gaurav Singal
Bennett University, India

Riti Kushwaha
MNIT Jaipur, India

ABSTRACT

A vehicular ad hoc network (VANET) is the network of mobile devices as well as stationary objects that can communicate with each other. This technology comprises of both vehicle-to-vehicle (V2V) and vehicle-to-infrastructure (V2I) data transmission links. At present, vehicular mobility framework has a lot of limitations, which includes lack of real-time operations, frequent disconnects due to dynamic-restricted topology, tracking vehicle that break rules, lane-changes, exceed speed-limit, etc. These characteristics call for a new type of high class of protocol. This chapter presents a survey report on "smart traffic control" system that incorporates traffic-related parameters to further augment the control and management of vehicular movements on the roadways. This can support efficient management of traffic in the city, and cops can communicate with each other, get real-time, accurate, status update of the traffic, track the vehicular movements, etc. In addition, jam control mechanism can be placed on heavy traffic days to optimize the routes. This system introduces artificial intelligence (AI) that can optimize deployed of cops and find alternate routes for the driver to reach the destination address without much fuel consumption.

DOI: 10.4018/978-1-5225-3981-0.ch008

1. INTRODUCTION

VANET is a sort of Mobile Ad-hoc NETwork (MANET). The chief characteristic of VANET comprises of movable nodes or vehicles that send and receive message or information from one movable node to another by following a straight constraint path called roads or lanes. As described in the above figure 1.0, they can communicate with mobile nodes on the road, a.k.a. Vehicle-to-Vehicle communication) as well as stationary nodes available on the roadside which are not movable, a.k.a. Vehicle-to-Vehicle communication. Such, mobile gadgets kept on the side of the highways that are used to communicate with vehicles are called Road Side Units (RSU). In the future vehicles would be computer operated and would automate a lot of manual tasks done presently (Lind, 1999). However, in the present circumstance, traffic security, traffic congestion management, traffic rule adherence mechanism, etc. have of paramount importance.

Electronic sensor and device panel placed inside a vehicle enable the driver to navigate appropriately on roadways. Driver has to adhere to traffic laws or instructions, such as speed limits, possible danger and other policies assure traffic is orderly and safe. Traffic security, traffic congestion management, traffic rule adherence mechanism, etc. are also required to properly manage traffic on Roads. A typical VANET network is formed by vehicles moving on the roadways. These are highly dynamic networks. These networks are similar to MANET and change network topology rapidly, connection lasts for very short span, communication may occur just once. Both MANETs and VANET are self-organization of nodes that are independent and dynamic, VANET has additional attributes, which includes, network dependence on driver's way to navigate a node in a straight path for a very short duration with higher speed, dynamics, and higher mobility amongst nodes in comparison to MANETs. In addition, to that VANET has frequent fragmentation due to swift and fairly predictable topological change. Changing the transmission range of the vehicle with respect to distance increases the network throughput. When

Figure 1. Vehicle to vehicle and vehicle to infrastructure

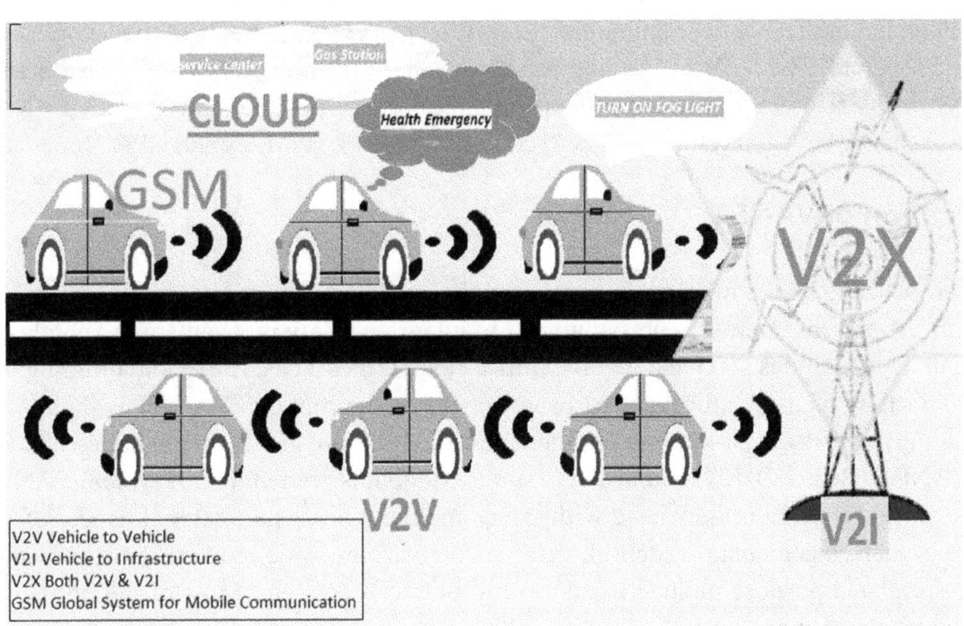

the distance is small transmission range can be lowered and it can be increased when the vehicle moves far apart (Artimy, & Maen, 2005). VANETs is recognized as the eminent technology that can advance competence and security in transportation structure (Blum, 2006). For instance, some of the application may include neighboring vehicles gossip with each other and suggest alternative route to other vehicles, and transmute traffic congestion related information to nearby vehicles early in order to reduce traffic jams in and around the affected areas. UN report on road safety 2015 states that approximately 1.25 million people get killed every year on road accidents worldwide. It further forecasts that the roadside mishaps is on the rise and if this pattern continues and no measures are taken by the governments, then, it would eventually become the 7th major cause of death in the world by the year 2030. Hence, providing road safety for all users is a major challenge in recent times. Application of "SMART TRAFFIC CONTROL" mechanism using VANET can enhance the existing road traffic system and look at transportation system from an entirely different perspective. These vehicles can connect to the Internet to obtain real-time data, updated path information reports, location-specific details without actually visiting those nearby location and congestion, weather and route related estimations based on available information through V2V or V2I interactions. V2V together with V2I is termed as V2X communication.

1.1 VANET

The Vehicular Ad-Hoc Network comprise of nodes that are moving vehicles or nodes that form a network of mobile nodes. VANETis similar to that of MANET, only difference between them is that VANET has the additional capability of complex devices and resources with better processing ability than MANET. But these devices can only run in a predefined straight roadway. These technologies also give way to an opportunity to produce the vehicles that are robot-like and do not need a driver to operate. In such network every moving node becomes a wireless router that perms vehicles that are close to each other to communicate. Un-interrupted connectivity in VANET is more challenging task (Kim, 2009) because the vehicles move in random path. Nowadays almost all vehicles are equipped with On Board Units (OBU). The OBU would contain all the necessary devices, such as, sensors, information exchange units, to perform inter and intravehicular communication with internal components of the vehicle and close by wireless infrastructure. The V2V connection has direct and multi-hop communication, between nearby vehicles. A typical V2I interaction involves data exchange between vehicles and RSUs. V2V is short range information exchange between vehicles and V2I consists of Wi-Fi, WiMAX and GPRS based application.

1.2 Application of VANET

VANET is proposed to be the most prominent technology to enhance the existing legacy road transportation system. At present, there is a pressing need to integrate safety mechanism in vehicles to prevent road accident. This includes send and receive traffic-related messages, traffic data mining, congestion avoidance and safety measures through roadways traffic prediction, etc. VANET is highly mobile and rapidly changing network of fast-moving vehicles and quickly fading interaction of nodes. The topological change of VANET (Bento, 2012) is rapid and to some extend it is predictable. A typical VANET has two types of relevance. First one is associated with safety and the other is non-safety linked. Safety is linked with functions such as accidental watchful, disaster forewarning, side road circumstance cautions etc. Non-safety associated purpose include infotainment, online access etc. Most of the online application require uninterrupted connectivity (Hellbruck, 2009). The problem with flawless connectivity grow to be even tougher when vehicles travel across beyond varied wireless environments. In such case frequent disconnects lead to degraded network performance.

1.3 V2V and V2I Communications

Vehicular movement and RSU are constantly engaged in traffic activity that makes these units available and records any traffic events happening in different parts of the world. These units are capable of gathering and transmits most up-to-date data traffic-related information to the neighboring nodes. Thus, gathers data related to accident spots, real-time roadways update and keeps track of optimal routes to reach a destination by maintaining updated information on the current traffic density and mobility on various routes. Weather forecast and data exchange between vehicles can aid in understanding the routes condition keep the driver abreast of changing traffic density, environment conditions, and safety wakefulness. It can also play a vital role in increased entertainment application to passengers. Risk and alarm wakefulness can also be handled by using Smart traffic management. Sensors are placed inside the vehicles to sense issues or failures in a particular component of the vehicles. This V2V and V2I communication develop an ecosystem for VANET communication. Here appropriate quality and standardization must be placed so that the communicating vehicles are able to interpret the message and comprehend the messages exchanged with each other. As node movement speed in VANET is higher in comparison to MANET, opportunistic connectivity is used to connect traffic moving in opposite direction this is also known as bridging technique. In short, V2V communications has the following benefit over other communication:

- It allows small to intermediate communication span.
- It aids in lower cost of deployment.
- It supports minuscule message delivery.
- Diminishes in the communication links.
- Similarly, V2I communications has the following benefit over other communication.
- Since it represents stationary point in the network infrastructure, information can easily be broadcasted with the use of VANETs superior antennas.
- V2I is interoperable in VANET as well as Cellular network with ease.
- Integration of WiMAX and Wi-Fi is possible, and it is cheaper in vehicular networks.
- V2V communications present the following shortcomings that can be solved with the integration with V2I, such as:
    - Due to highly mobile node, there are repeated topology detachments.
    - There are issues in higher range communication.
    - Legacy and protocols used in MANET cannot be used here.
    - In High-density conditions, there is a chance of broadcast storm issue.

V2X Communication and Challenges: Connected V2X (Vehicle-to-Vehicle or Vehicle-to-Infrastructure) infrastructure can generate and transmit data that are user-centric. Circulating environmental data such as foggy weather can alert the entire network V2X network and suggested counteractive actions to be taken such as, turn on the fog lights or speed reduction, activate parking lights, etc. Such action gets triggered automatically that can be figured out by the rest of the network to prevent road-side accidents etc. However, Integration of these eminent technology, such as ITS, IoT and Cloud platform is challenging.

Cloud technology makes use of RESTful web services and IP technology for iterations (Datta, 2016). Latency is a major challenge and quality of service is minuscule and may not be meet the actual data transmission rate required to be operational in a dynamic application. Analyzing and processing data at the source of origin has to be evaluated using Edge computing mechanism by autonomous vehicles. Comprehending the quality of data transmitted from different sensors is of paramount importance as well as necessary to probe and predict outcome depending on such patterns. However, the vast amount of data collected from heterogeneous sensors and controllers belong to different domains which make it challenging to apply standardization to data communication and processing. IP based communication as not designed to consider mobility. Hence, the Information-Centric Network has to use for this communication. Data-centric approach to promote interoperability within heterogeneous data capturing devices for Seamless integration of vehicular network.

1.4 Vertical and Horizontal Handover

The method of toggle from serving network to the target network is known as Vertical handover (VHO). This can be motivated by the automobiles or by the RSUs and is carried out according to as well-defined assessment norms. Handover can also be termed as handoffs. This event occurs when a mobile node exits a wireless cell and enters into another cell. The main intent of this movement is to handover services in such a manner that this is no loss of connectivity and interruption of services can be avoided. This handover is termed as Horizontal Handover (HHO) when it happens within the same wireless access technology. On the other hand, when it happens amid varied wireless access technology it is termed as Vertical handover (Blum, 2006). This chapter is structured as follows. Section1 VANET. Section3 Cloud Platform Section3 SMART CARS

1.4.1 Vanets Technologies

A high honored serial data communication bus for dynamic applications was developed by Bosh Controller Area Network (CAN) (Hellbruck, 2009), this is one of the most effective automotive control networks. There can be more than one CAN in a vehicle; it works at varied speed and data rate. Depending on its usage, for instance, if comforts in a vehicle are considered such has window movement, adjusting the seat position, etc. The required compatibly needs of the CAN for this type operating is of the order of few hundred Kbps speed, it can also be kept in a dormant state when that particular functionality is left unused, thereby, conserve energy for other important vehicular usages. For dynamic operations that are time critical may require high speed CAN (Fang, 2008), that is, frequent and real-time data exchange, such as, collision avoidance, management of engine, monitoring the real-time status of all the components of the vehicle.

1.4.2 FlexRay

FlexRay communication system is another system for automobile application which is a time-activated way to network communication. Such communication system is essential for a system that needs to be safeguarded on a real-time basis. FlexRay (Okrouhly, 2011) has high data rate; its chief goal is deterministic approach and two-channel topology. Due to the necessity of high bandwidth, Media Oriented

Systems Transport is developed for information and entertainment. Networking in the Vehicular industry (Kim, 2008).

1.4.3 VLC (Visible Lighting Communications)

Usage of Visible Lighting Communications provides a valid technology for VANETs. Visible light spectrum has a density that exceeds femtocells in wireless access. Such technology provides high data rate and visible bright light while communication occurs. Packets collisions are radically reduced due to Line Of Sight (LOS) (Salian, 2013). VLC can be used improve the way we live, which include the directing traffic signals, safety communication, situations where high density and interference of WIFI and noise is a factor. Vehicles can communicate directionally using LOS links as light follow a straight path. Vehicles can be encumbered with short-range transceivers that can communicate with similar vehicles. They can provide vital information to adjust speed according to the speed of the other vehicles or possibly avoiding a collision scenario (Pathak, 2015). Smart Transportation System in VANET involves many technologies such as GSM UMTS Wi-Max and WAVE (Wireless Access in Vehicular Environment) IEEE 802.11/p. A vehicle must be equipped with the network card. It suggests novel Dedicated Short Range Communication protocol, it aids low latency in V2V and V2I interaction using IEEE 802.11p and WAVE standard. The first band is reserved for V2V communication, the 2nd and 3rd is reserved for private channels 4th is kept for control channel 5th and 6th for public safety services for short range and the last channel, 7th channel, for public safety intersections.

1.4.4 CALM

Continuous Air-Interface, Long and Medium Range (CALM) consists of a networking terminal that joins vehicles and roadside systems flawlessly to stop frequent disconnections in network communication. It uses, DSRC/IR, Mobile Terminal, WLAN network (Fernandes, 2007; Dobashi, 2005).

1.4.5 Lane Change Assistance (LCA)

System constantly servile the rear side region of the vehicle. It checks the movement of the driver. It tracks changes in the roadway vehicular movements, for instance, when the vehicle changes lane on the highway, intersection and crossing roads etc. There are different modes of operation of such application, the first on is active mode, wherein information is exchanged between vehicles that are in close proximity. Here, interactions occur between vehicles and it updates the current status of each vehicles and any action that needs to be taken gets examined/explained. In the passive mode, this application lousily monitors and records detection and ranging procedure. Traffic light scheduling mechanism can help drivers to drive easily through different intersection easily. Which eventually reduces travel time of the vehicle and increases the throughput (Alonso, 2008; Liu, 2007). For instance, a token based interaction among vehicles can reduce traffic delays and wait time. Traffic congestion in highways a greatly reduce by efficiently managing and optimizing distance travelled by the vehicles. Eventually, fuel consumption and environment pollution get reduced comprehensively due to optimize movement and waiting time of vehicles on roadways.

1.4.6 Congestion Road Notification (CRN)

Congestion Road Notification can be used to trace out and advise about road congestion, such that route-trips can be planned efficiently and subsequently reduce the travel time for all drivers (Nassar, 2016). GPS system also suggests details and routes to the user and searches out alternative ways to travel in ways that have no congestion. However, the information present is limited and not as detailed for taking appropriate decision. In the next generation of congestion monitoring system Radio Data System Traffic Message Channel (RDS-TMC) (Geetanjali, 2014) collects detailed information about congestion in the routes and streets. This message is very comprehensive. It further augments the present information fetched using GPS technology, below are the extra details that gets available to the user in the message content of this system.

- **Identification:** Root cause of the congestion is detailed out to the Driver. Its level of alertness, estimated time of clearance of the traffic.
- **Direction:** Roadways that would get affected by the current road block or congestion is explained.
- **Stretch:** Extend of this problem in different direction possible prediction of the occurrence of the problem in various routes.
- **Duration:** Traffic congestion based on the severity of this issue can predict the approximate time till the congestion my last.
- **Diversion:** Alternate routes are suggested to avert route congestion.

This message service is updated in every 30 seconds. These reports are broadcasted to the entire network via FM Radio Data system (RDS). It can be customized to be compatible with V2V or V2I type configurations. As these information's are huge and could add additional overhead of processing in vehicles. In such a short travel period for a vehicle. It is however, easer for land infrastructure RSU present in V2I configuration to process the pile/chunk of data and extract meaningful information to be transited to the vehicular traffic that is useful for navigation of automobiles. Such RSU can process the data and divert traffic to appropriate direction when fewer congestion routes are present. Since data processing is done on road side unit as it has more time and efficient resources to process and predict reliable data that can be exchanged with Vehicles traveling in that direction.

1.4.7 CCW (Cooperative Collision Warning) System

This system work hand in hand with VANET components and technologies, such as GPS, Sensors on Board, RSU, to correct spot and flood information in the network and alerting other vehicles regarding a stopped vehicle or slowing down vehicle, the vehicle in downhill, etc. (Zardosht, 2013). That enters or exits a network. The main rationale behind this system is to keep the network aware of the attitude and mindset of the other drivers in the network. This information has to be floated in the network to keep drivers informed about other driver's immediate situation who is in their proximity and sent a possible alert if any. This information exchange has two phases, in the first phase, with the use of V2V information related to a particular vehicle is floated to a delocalized group of vehicles. Then, later on, in the next phase, this information is updated every 30 seconds by the RSU using V2I communication (Cho, 2014). This system also enables the exchange of information regarding the local restaurants, hotels and other emanates while vehicles pass through the different lane, streets, highways based on context-aware applications.

1.4.8 MOST (Media Oriented Systems Transport) Protocol

As we know that the bandwidth of the CAN-Bus is just enough to be able to move control signals, and audio and video signals are out of scope, such protocol is no longer sufficient to meet these demands now a day. Hence, a pressing need to meet these challenges. MOST communication system that has innovative and flexible design meets these demands and used by many different manufacturers. It can transmit audio signals synchronously, like, telecommunication; it is readily used in the multimedia system in cars. MOST technology products are generated by the teamwork of car manufacturers and dealers. The MOST committee comprises of founder members that are also the running/functioning body of this committee. The requirement document of MOST specifies programming of the audio and video signals and the structure designers make the blueprint of multi-media structure on a higher abstraction level. MOST (Media Oriented Systems Transport) presents high bandwidth transportation mechanism for video application and Ethernet channel for IP-based packet data that occur at the same time. It supports widespread video applications, as well as an Ethernet channel for efficient transport of IP-based packet data. MOST gives physical layer for Ethernet that can be used in Smart Cars. In the like manner, MOST unwrap itself to a diverse variety of IP based platforms. It allows alteration to the bandwidth of the usual streaming connection and the IP communication. For instance, the late version of the MOST protocol has more bandwidth to send and receive audio and video signals efficiency and without any overhead in addressing, collision detection/recovery or broadcasting. HD video streams surround sound with a superior class of service can be transmitted, in parallel with moving high loads of packet data around the network. Also, MOST infotainment support has grown recently. However, the interaction of these devices connected to MOST must be orchestrated is such a manner that they are participating in a gig or concert, so that, other devices can only be controlled via a central operating surface to ensure drivers are not unnecessarily distracted.

1.4.9 Features That Should Be Present in Vehicular Communication

1. V2X arrangement should enable reliable data exchanged in real time between infrastructure and vehicles;
2. Monitor the environmental condition of the vehicle by using high-quality sensors;
3. Processing information generated by the vehicles and manage the data;
4. Integration of communication networks with other devices that are faultless;
5. Develop a framework to enable inter-vehicular components and surrounding environment to connect with each other. To accomplish that, there must be additional ingredients including –
    a. Information that is collected from different domains has to be synthesized.
    b. Process the data and transmit the data to the destined sensors in the vehicle.
    c. Proper illustration and storage of data.
    d. Choice of correct protocol that consumes less power and a complementary network.

Temperature and humidity sensors can also be attached to collect data from a particular position and values of prevailing climatic variables of the longitude/latitude can be transmitted to cloud for processing.

This aids in measuring, air quality and pollution level of the passing by street. These equipment's can be deployed on front/rare side of the car to gather accurate information. Short-range wireless technologies like Wi-Fi, Bluetooth, ZigBee technologies can be used. Raspberry-Pi and these sensors can collect information happening in the surrounding. Capture data on a real-time basis and pass it to the sensor. These sensors log this information as it is on a continuous basis. Hence, generate huge collection of data that can be further analysis to find patterns. These data can be analyzed to perform accurate prediction. Information from the environment is fetched by the web server and display it/email to the owner. This enables the owner to access the data from any place to perform data analytics. The sensors collect data and dispatch to the controller and actuator as provide as service to different client layers. Client can access these base station data remote and gather data related to temperature pressure, moisture, light and pollution. For instance, Delhi has only six major air quality detectors in the entire city to calculate cities pollution but with the employment of such strategy it would be possible to send and receive such information from different sources there by accurate data received. Also, due to its cheap setup cost and numerous people using Smart Cars accurate information of weather and prediction can be exchanged.

1.4.10 IVHS-Eliminating Roads

With the efficient use of control, co-ordination, communication and computing the road traffic can be better managed, and alternate routes can be mapped out in case of heavy traffic scenarios in cities. Thus, optimal traffic flow can be achieved in roadways without constructing new roads (Varaiya, 1993). The decisions can be classified into the following categories before-journey, during-journey, and after-journey. Intelligent Vehicle/Highway System (IVHS) enlists necessary goals for each judgment. The sub tasks are attached to each goal and the outcomes are specified. Involvement policies recommend proper decision to be taken, taking control or even giving away remunerations to the vehicles that follow the rules ardently are laid out by the system.

1.4.11 Intelligent Transportation System (ITS)

Vehicles are the essential resource for IOT framework that makes available end user centric functionality to join with vehicle's domain. Some smart vehicles have On Board Units (OBU) which collect data from different access points from highways infrastructure. These applications run on cloud platform to store, send and receive data for high performance computing. A Smart Car application (running in a cloud) contains sensors to gauge air and noise pollutions levels on a particular road/street that it passes by and transmit the data to other communication media for further processing. Such dynamic interaction of car with roadside infrastructure saves installation cost of setting up dedicated resources to measure environmental conditions. These data can be utilized to navigate through different regions that have lower pollution level. The chief ITS components include: 1.) Security; 2.) Environmental hazard computing; 3.) Seamless Integration; 4.) Efficient connectivity; 5.) Mobility; 6.) Deterministic associations. Security (Devi, 2016): Encryption and encoding of information are necessary to be transmitted in the entire network. To make the connection more secure intrusion detection and prevention mechanism must be in place to fully secure communication. This requires the tough firewall to stop intruders. The deployment and management should be done in a centralized manner. This has to be managed and secured by using suitable authorization and authentication policies.

- **Environmental Hazard Computing:** Edge computing network enables the processing of information. It is the local cloud information gatherer and processor. Other techniques are implemented using cloud computing that includes Virtualization, Information as service, Platform and Service, Software as service and Robotics as Service. Problems faced by such network include. Loss of data when information is transmitted between vehicles that are connected with each other by some giant intruder or skyscrapers on the roadside. Due to large density of flowing vehicles, there is a change of Doppler Effect, path fading and shadowing. Fast moving vehicles cause network topology too abruptly. The Data link between vehicles gets disrupted because of these obstacles. Challenges include amalgamation of the heterogeneous rudiments of IoT architectures in linking vehicles. Smart devices join together to form an intelligent transportation. Need to synchronize information gathered from heterogeneous sensors in a uniform manner. Cloud platform has the ability to hold, process and take appropriate action of these intelligent interactions.

2. CLOUD COMPUTING

Cloud computing initiated the consciousness that hardware infrastructure that required to run big business can be leased out instead out owning them by individual business. High-performance hardware could be rented-out to run their software to process data. Cloud computing enables high-speed data processing with scalable resources availability. Developers that use cloud systems do not require huge capital and hardware to build their service-oriented applications. This usefulness of cloud computing can be leveraged in Vehicular Cloud V-Cloud that uses the best of two worlds, that is, mobile technology and highly efficient computing using Cloud system. Nowadays, vehicles are that use internet has humongous processing, storage, and sensing capacities on-board. These abilities can be considered as gigantic ranch of workstations that are on the move always. While moving on the highway most of these services are unused and are in dormant state. However, one of the very practical application to leverage the full potential of these idle machine on the move is to use them in a pay-when-you-move manner. In that way these vehicles become closely associated with high-end computing devices that provide storage and computing services on a pay-as-you-move manner, this term, can also be described as cloud computing in vehicle in the form of Edge processing. Here processing at the end of smart device sensors. These devices are not idle be are constantly in touch with other infrastructure on the network whether moving or stationary. This process involves active communication with various other nodes in the network that cooperate with many authorities for problem-solving, resource sharing and participating in different events. While moving on the highways, few vehicles may not be connected to the internet while other maybe connected to the network indirectly. An interesting thing to note here is that with the use of MANET those vehicles that are not connected can send out message to nearby vehicles that they want to take part in sharing. It is possible that many drivers on roadways will have persistent connectivity to the internet through cellular networks and other fixed access points on the road while driving. In the meantime, vehicles that have internet connectivity and are ready to distribute their reserve and it will send out notification to those nodes that are willing to participate. In this interaction or information exchange processing occurs with devices that are near to them on roadways/highways. As a result, this information could be passed from one hop to another hop. These vehicles communicate with each other and exchange information regarding their presence to vehicles that are willing to participate in sharing the internet through mobile ad hoc network (MANET/VANET). In this manner, each vehicle present in a particular region can be incorporated in the cloud architecture.

2.1 Cloud in VANET

The integration of VANET with Cloud computing can aid in the development of Network as a Service (NaaS) or Storage as a Service (SaaS). The moving vehicles can act as internet access point. These vehicles can share internet connection to nearby vehicles to access internet using these access points. This access can by granted upon request by nodes/vehicles that don't have internet access. Similarly, with SaaS the vehicles with abundant storage faculty can share storage nearby vehicles which require storage capacity for short-term purpose.

2.2 Smart Traffic Using Cloud Computation

The vehicles are loaded with wireless devices that act as the base of the network. These nodes are free to move in the entire network or roadways/highways. The nodes include both movable and stationary nodes on the roadways. Due to the inclusion of several type and categories of nodes the network, so formed, has different hierarchy and layer of information exchange, also known as multi-tier network. Packets get exchanged in this process and form personal mobile node within the adhoc network. These packets are routed within the ad-hoc network; they also include set of fixed base stations. This network is stable and the network formed by hop nodes and base station becomes part of a Adhoc Cloud network. Absence of a fixed station in the nearby vicinity is not a problem due to the ad hoc network structure, all the nodes form a network by hopping the signal ultimately to the adjacent base stations. VANET aids in communication with nearby vehicles through V2V network. This also help in the management to traffic in a controlled fashion. Congestion avoidance strategy are placed when these nodes /vehicles come very close to each other in a certain area. With the use of V2V these vehicles may transmit alert signals to others present in the same route to generate enough space for mobility. The traffic signals can be managed according to allow vehicle to pass without stopping them as this may lead to huge traffic jam.

2.3 Operation Phases of Smart Cars and Its Benefits

It encourages a data-driven approach free of the deployed infrastructure. The structure chiefly develops an edge computation stage that aid in storage and processing of data using machine to machine interaction and higher level of machine intelligence. Here, cloud platform is considered as a repository of information gathered in the form of datasets, queries and domain-specific classification for web integration into the edge server. Use web technologies to develop a standard way of interfacing vehicle and standardize interaction with vehicles, smart device sensors, controllers and actuators. The organization of the framework and its newness.

- **Detection Phase:** IoT advocates ecology of resources that maneuver without much human participation, detection of resources is important. This phase permit probing of smart utilities and related equipment, such as, sensors and controllers. To assist in the detection of these assets, their ability and ways to contact and the arrangement of the assets have to be illustrated. However, the ability to communicate and describe these heterogeneous sensors and actuators in standard way is mammoth task. In order to utilize their web-based data synthesis they need to be semantically described. Subsequently, to detect this equipment these vehicles must be registered and linked into an edge computing framework. The controllers, sensors and actuators are illustrated in expressions of its events, attributes and actions. Their illustrations can be fashioned at the OBU or a vehicular entry point. Resource detection component tell its type and its corresponding domain of working. Also, fetches extra data pertaining to its use. Hence, a greater detail is attached on the whole to IoT platform. The OBU generates device description that is circulated to the edge server in the network. The API used in the setup process pull out the resource description and stores it in its local repository. In the detection phase, the setup API searches the local storage space to fetch the device. This interaction generates list details that can be gathered. Sensor Markup Language can be used to gather heterogeneous data that aids in encoding various attributes in the form of metadata. This phase overcomes the challenge pertaining to standard data collection and identification/naming/describing of resources. Permits device sensor that are integrated in Smart Car as well as surrounding vehicles included in detection process.

- **Preparation/Planning Phase:** In this phase edge servers are prepared for information synthesis and data analytics. As we know that in the detection phase sensors are searched to gather crude metadata. However, in this phase the classification of device's type, form and its domain of its work is examined, this is fetched form SenML metadata and descriptions. This preparatory information is transmitted to cloud computing platform that comprise of Machine-to-Machine Measurement (M3) framework (home of this semantics). The framework has all the essential development related models consisting of 'domain-specific classification', datasets, policy for semantic reasoning, queries etc. for data synthesis and analytics. Only things that are require for this scenario and for the application requirement are chosen to be placed into the edge computing framework.
- **Data Synthesis, Analytics, and Storage Phase:** This phase figures out the research issues related to transforming the crude information gathered from devices and sensors into intelligent information. This intelligence may be known to travelers seated in vehicles or other autonomous vehicles moving around. Based on the inputs and processed data, an intelligent decision can be taken. Subsequent, goals or command can be generated to be sent to different controllers of Smart-Cars. This acquired intelligence can be sent to actuators and controllers of the Smart Car to allow the linked vehicles to respond to its surrounding in an intelligent way. Semantic web technologies for data synthesis has these advantages: (i) consistent handling of SenML metadata into superior level of intelligence enables interoperability in IoT; (ii) process is autonomous and is not dependent on any communication network. The downloaded model combines sensor metadata from different domains. The expected metadata on the edge serve has to be transformed into Semantic information just prior semantic rules are submitted in order to find the new domain perception. Then after classifications is done in accordance to domain. Afterwards, domain dataset is submitted. Last and final step is reasoning which finishes the data synthesis and analytic phase, now, the main outcome of his activity is "superior intelligence". Consequently, Data is stored locally and index in accordance to Named Data Networking (NDN). As a result, thin edition of the M3 framework is generated to run on an edge computing platform. Information broadcasting phase: In this phase mobility is considered while the resultant intelligence is distributed. As explain earlier, IP technologies do not hold mobility. This issue is mitigated by the use of NDN that broadcasts information. With NDN there is no need of host name resolution and it puts forward scalability, security. NDN chiefly deals with two forms of data packets. First one is called the interest packet

and the second one is data packet. The interest packets align itself to the need of the travelers and the Smart Car. For instance, if air sensor of wheel detects that the air in one of the wheel is lesser than the minimum capacity, then the air sensor metadata sends alerts indicating that the air level is "low", the Smart car can call for nearest air filling stations as an interest. These interests are mapped to a specific URI and passed to NDN router which intern passes the request to the specified node that contains the data, for instance, GPS location of the closest air filling station are searched. Distribution and broadcasting of intelligence is performed across NDN. Actuation phase: In this phase, IoT devices of passengers and vehicles participate to take decision and send and receive commands to Smart cars and controller to acclimatize to the changing environmental parameter and conditions. When the Smart car receives signal that it is raining it automatically switches on the wipers switch for the driver.

2.4 Innovative Solution

In today's competitive world vehicle manufacturer are seeking innovation in technologies to make vehicular journey more reliable, safe and provide infotainment to its customers. Smart cars have to be fuel efficient, it should cause less carbon emission. For this, advance embedded technology has to be considered that make use of supreme electronics architecture. As the devices integrated in cars are operated electronically there are chances of failure. The cause maybe random but in the event of a major electronic failure the consequences may be life threating.

- **Diagnostics of Smart Cars:** Electronic system must have high quality sensor deployed to detect and correct faults in the system. Fault in a system are recorded on a continuous basis and automatically. It starts executing remedial step to recover the system from the issue or otherwise, acquaints the driver about this fault by sending alerts.

2.5 IOT and Cloud

The key aspect of communication now a days includes machine to machine communication. This is performed by introducing intelligent sensor for smart communication. As smart gadgets are on the rise this also calls for its integration with vehicular management system, that is, they should be made intelligent to interact with these systems efficiently. A car system should be secured by embedding intelligence in it. This can be achieved with high end computing, access to internet, GPS and Cloud platform. Devices integrated in cars must be capable to comprehend application related domains, for instance, healthcare, ecological system, agriculture, transportation, security, and energy/power. Deploying application of specific microcontroller, sensors and actuators possess these functionality, such as, signal taming, signal exception detection, interaction, smart failure detection and remediation management system. The conglomeration of these modules co-operate with each other that possess similar capacities, thus, form an ecosystem. Likewise, in the case of a theft these cars can be tracked and could be controlled remotely by using smart techniques, for instance, to lock the car or to activate video, audio signals back to the owner and protect the car from thieves. Smart car generate data related to speed/movement of car and current condition of the car to its owner. It automatically updates/notifies the owner by sending detailed reports to owner via a text message etc. There could be alerts that can be installed to capture data of car if its speed exceeds a certain limit.

2.6 Connected Car and IOT Automotive Cloud Services

IoT and cloud services realization to connect Smart Cars is causing a paradigm shift in car manufacturing industry. The Internet of Things is laying new foundation on the traditional concept of a car production process and breaking fresh ground by introducing entirely new stratum not observed before. This advance concept to connect with vehicles that behave as intelligent receptors and transmitters of information enables the owners to stay in touch with things happening around the world and augmenting the level of understanding even the minute changes in our surrounding by the use of smart sensor and actuators. This change, puts forward a diverse infotainment provision and car appliances for car owners, this opens the scope for automotive manufacturers to overtake other industry and grow to be an IoT leader among other business and act as a stimulant to other industries to walk on their footsteps. Big players in the automotive sector, entertainment and maintenance service industry should mandate IoT components that streamline development of connected cars/Smart Cars. This would eventually ease handling numerous Smart vehicles exchange information isochronously and robotically balance out height loads in cloud services. Integration of various automotive hardware such as, domain specific sensors, actuators, control unit, lighting and steering unit, engine and other parts into a cohesive component by introduce numerous smart features that connect with each other support the fundamental design of a Smart Car. The sky's the limit for IoT and cloud integration in the case of Smart Car, there are innumerable ways and measures that can be taken to give Smart Cars its wings to fly.

2.7 Cloud Robotics

Cloud platforms provide services and resources to users by contracting out resources. It comprises of networks, storage and operational utilities. It was originally used to provide software as service but in the recent times it has stretched out to other devices such as vehicles like cars, smart phones and UAVs. With the introduction of low cost mobile robots in the monitoring of climatic condition is possible. Tiny bots a.k.a. Boebots collect images that are processed by a clouds to gather real data. Low cost devices send, receive and store real data to process, that is, built using Raspberry Pi and Bluetooth which is cost efficient alternative. Such equipment's monitor environmental temperature, pressure etc. Sensors in this device gather data and store the same in cloud server. The data collected on a real-time basis can be stored for big data analysis or displayed on website (real time interface). For instance, a Raspberry Pi microcontroller can be used to send data to the cloud server, email the environment condition report to the end user. Connecting vehicles with the cloud platform enhances computational and data processing power of the data coming from Car and other transport machinery: (a)Consistent reporting and data gathering mechanisms from vehicular actuators and sensors; (b) amalgamating smart utilities into transportation schemes; (c) standardized method of information synthesis and analytics; (d) assimilate heterogeneous rudiments into a typical IoT architecture of allied vehicles

3. SMART VEHICLES

The main aim of Smart vehicles is to incorporate mechanism, namely, information exchange, precise spotting, manage, organize, decision making and cognitive learning through acquiring information about the environment. There are different requirements for VANET network application, for instance, alarms/

alert for collision detection have to be made very precise, of the order of few milliseconds and it has to be very reliable (Prasad, 2016). However, monitoring traffic congestion and traffic information collection are wait tolerant, this information can be gathered with ease and no latency constraint is placed.

Quality of service in VANET is a major concern in VANET due to high speed vehicular movement and topological changes. Road can be classified based on speed limits, traffic density (high density or sparse), type of road particular roads (highways, rural roads, street lane), types of vehicles (car, bikes, bicycles, trucks) (Lozano, 2011). Hence, customization in protocols have to be made to cater to different traffic/mobility scenarios. Next generation automobiles will be equipped with both inter and intra sensor devices. Due to technological advancements a range of sensors would be used to control and communicate within and outside world. These devices are used, namely, Universal Mobile Telecommunications System, Bluetooth, WiMax and Radar, Ladar (Wang, 2002) to efficiently manage transport and the prime focus point is optimizing flows of vehicles by dropping the travel time and keep the vehicle away from any traffic congestion. This can be applied to further sense traffic congestion on highways and according the speed of the vehicles can be altered. Using Artificial Intelligence (AI), the probability of an accidents can be avoided and various sensor placed on different parts of the car can anticipate any crash based on prior data point collection in similar traffic scenario with the use of Artificial Intelligence. Such data can be circulated to the entire VANET nodes to appraise fellow travelers etc. Future vehicles would be very different from the present once as these would be on-board with many sensor/wireless devices that can sense lot of information happening in its environment. Apart from safety specific application, there could be gossip based application that talk with other vehicles on a regular basis and gather data for the future travel and augmenting it database of information. There are different layers of functions in SMART vehicles, for instance, intra vehicles communication has the braking system, crash detection sensors, data recording, control the central unit of engine, car stability, steering control, information and entertainment in cart, lighting system, belt and air bag sensors, tire pressure sensors etc. Emergency situation drills and stoppage of vehicles in case of failures. Vehicles would be on-board with cameras to recode and communicate to the driver and take remedial steps. Smart vehicles are loaded with devices such as Central processing unit, standardized protocols are implemented for communication and applica-

Figure 2. Modern vehicle architecture

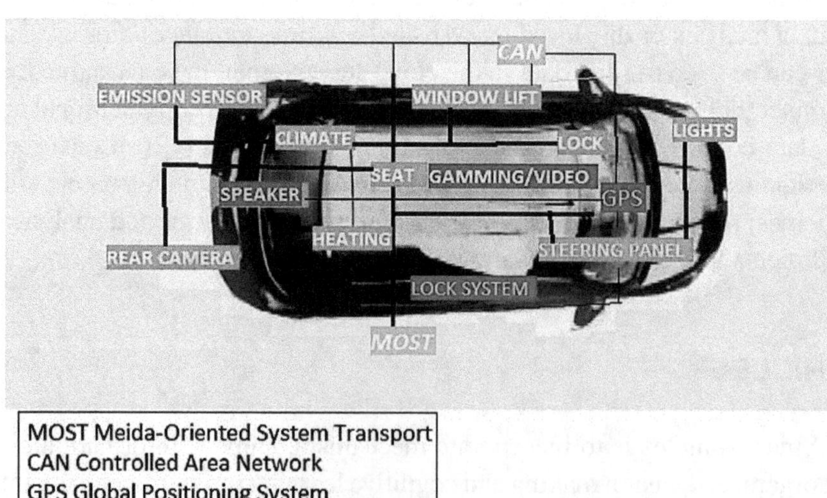

tion purposes, Wireless receiver and sender of information to V2V and V2I to Road Side Units, Global Positioning Service is in place for navigation and positioning purpose, Vehicles are equipped with user interface to interact with humans, sensors related to speed, collision avoidance, acceleration are also fully loaded here. To address safety issue, monitor, interact and processing mechanisms are deployed. Smart vehicles and be used to alter driver's decisions and prevent wrong driving manners, this utility is added to smart vehicles electronic devices. This smartness can be used to optimize fuel consumption, path optimization and detect any anomalous behavior in vehicles.

3.1 Integration of IOT into Smart Cars

The Smart Car is a highly mechanized vehicle featuring all-pervading computing, discerning human-computer interface and open application ecosystems. Like smart-phones, drivers can modify their Smart Car by installing customized application for their cars with application specific IoT interfacing. As a general rule, the smart gadgets are illustrated by three major attributes: Ubiquitous computing: evaluating data or interactive and autonomous accessing data send by heterogeneous sensor over any medium using various wireless protocols.

- **Human Computer Interaction (HCI):** (Huang, 2016) It is main interface between humans and computers as it provides way to connect with computers. Applications platform: It permits user to install vendor application to customize smart devices according to end users' choice.
- **Smart Car Platform:** Overall architecture of Smart Car exhibition platform is divided into three major parts the communication unit, with a transparent windshield monitor to show information rapidly instead of a conventional windshield. It also comprises of majority of intra- vehicular sensors that send and receive information to end users.
- **The Computing and Communication Unit:** It gives entry and which implementation of both outside and interior services that has access the internet/cloud.
- **The Application Unit:** It aids in installing vendor application. Smart cars have two classification of input sensors to talk about information regarding the environment outside. The vehicle and the owners propose actions, the other class is output utility that is used to propagate information that illustrate the opinion of the computing and communication unit.
- **Key-In Sensors:** Vehicle enclose sensors: These are six image sensors along with GPS sensor just around the Smart Car exhibition framework to gather road landscapes and position the car co-ordinates. Image sensors combine aid the driver to get full view of the vehicle. The behavior of the user is sensed by behavioral sensors. Smart Car display framework comprise of three kinds of motion sensors. Gesture sensor, voice sensor, eye tracking sensor that are placed on the car dashboard (Sehgal, 2016). It constantly monitors driver's behavior during a trip. This sensor transmits and receives command.
- **Output Gadgets – Transparent Windshield Display:** The Smart Car exhibits platform has a transparent display built into its windshield. Hence, transparent windshield monitors show information rapidly and displays on the fly data which contains augmented reality with the use of visible interface. This on-demand interface fetch information to the driver about the status of different IoT devices placed in Smart Car.

- **Skype Call:** Video calls inside the Smart Car are possible through live video stream using Skype. The user is able to make skype call by calling keywords like "skype" to activate or trigger events like auto login in to owner's skype id and dialing to the specified destination address. This feature can be used as spy camera that helps user to identify the driver and record voice for reference in the near future. This data can be transmitted to web API and cloud for any future reference.
- **GPS Tracking:** This sensor starts recording the position of the Smart car, its latitude and longitude values and sends the data to the cloud framework that intern connects to high definition Maps. For instance, user can contact to the nearby hospital, police station to report any accident information etc.
- **Turn Off the Car:** The owner of the Smart Car can control the car's engine by turning it on or off from any location anytime. These request signal from the owner could be "turn off the car" or "turn on the car". Smart Car can safeguard itself from any external threats when the owner has parked the car and the owner is away from the car or at night when the owner when is sleeping. It can also be useful at times when guests arrive they could be greeted. The Smart car would be able to recognize their face using face recognition system. Safety signal or text message, such as, "Guest in your Car", "Theft" would be transmitted.

3.2 Health Emergencies: A Pain Point

The health care system is changing, with the use of modern equipment patients can work together with to medical staff on daily bases using video conferencing, etc. for treatment and diagnose of any health mishap. Here, revolution in electronic gadgets using Internet of Things (IoT) would play a critical role in connecting gadgets the smart way. Data related to health could be uploaded on cloud that sends out detailed report about patient's health condition using various applications via communication channels. In the case of Smart Car, various smart devices would be installed for any roadside emergency. Various sensor can be installed for doctors to treat patients. As these sensors would send real time data about patient's health. Doctors can treat patient using high definition video calls and send instruction to treat and monitor the patient's condition till Doctor's crew arrive. Health IoT applications would make the care process more in-timed and proficient then every before. These applications could be operated, processed, transmitted and exchanged all at the same time to smart gadgets from different locations. Cloud system would then process this data and intelligent information could be exchanged between patients and the medical staff/surgeon /doctor/specialist etc.

3.3 Road Accidents and Mishaps

There is a need to drastically decrease traffic mishaps occurring daily. This usually occurs due to human error. Also, to keep the environment safe from carbon and other pollutant emission from the exhaust of the vehicle needs to be addressed and controlled to prevent air pollution. Car manufacturers, such as Audi has started selling its "stage 3" autonomous car, likewise, Ford its autonomous version and Tesla's, on the other hand has researched that there is a forty percent decline of roadside calamity by using its Autopilot mode used in this vehicle. These are some of the instances of Smart Car becoming a reality in these troublesome times when actual pain point is getting solved by leading Car manufactures. However, there are many problems yet to be solved. For example, how will these autonomous vehicles make ethical decisions, for instance, "trolley problem", killing one person or killing four is more ethical, how can pain be termed as a comparable entity? How will cities that are non-smart currently change? These questions may not be answered just by innovation but the need of the hour is a major change in technology. The big players in the industry must have the motivation to invest in these technologies and design an ecosystem to make things work.

3.4 Smart Parking Solution

The sales of car in India have crossed over 2 million marks. Also, there is an increasing trend of purchase of new cars, that is, more that 7% annually. The increase in car purchase implies increase in need of parking space. At present, people search for parking space in the parking bays and are unaware of the actual available space available in the parking vicinity. Due to lack of information either people need to search each parking space or wait a lot until space is available or even travel long distance (Ravishankar, 2017). Consequent, lot of fuel energy and time get wasted in this process. Connecting Smart car will local parking IoT device can help in tracking information of available parking space. Apache Cordova Tools is used. Angular JavaScript is used to ease of this application that can be ported to Android devices, Windows and IOS platforms. In addition, many smart features can be installed that help user not to look at mobile phones while driving these smart assists can prompt the users various intelligent decisions that can be taken and user can choose one among them. Also, it can call out the number of empty parking bays for the driver to park his car. The application reports any faults in the sensors. This application can be used to fetch this information from the sensors. This data can be accessed by the Smart Car using interaction device to fetch vacant space and optimal distance etc. This turns out to be the most inexpensive and effective way to solve parking bay problem. In addition to that application related insights can be used to get analytical data such as pattern of filling the parking space and preference pertaining to zones and levels etc. Raspberry Pi can be used to collect this information using Infrared sensor. It has the following collections: a power source, an Ethernet cable and 12 sensors. A voltage divider circuit is used to connect the sensors to the GPIO of the Raspberry Pi. It generates output of 5Volts. Raspberry pi takes inputs of 3.3Volts. The gadget gets register by Web API which is considered part of the network.

3.5 Future of Transport

There is a debate in the industry that the future cars would be driver-less or users will prefer only public transportation or companies such as, Mazda, which foresees consumers would always love to drive. Hence, they build and sell to customer who loves driving. No matter what becomes tomorrow's reality, Cars have to become smarter. These viewpoints are different fragments of the future bazaar of car industry. Processing generated the data: Vehicles of the future will both consume and produce enormous data. Smart Car requires dynamic data to educate its networks. Map and pilot in such a manner to avoid road mishaps, control and regulate velocity, understand rule of parking. Customize the data related to driver to calculate efficient use of limited resources for the owner's benefit. Such computation of trip rides and expenses need to be pre-calculated. These vehicles are bound to generate humongous data that could be of the order of terabytes in each ride or travel these include, data capture by cameras, temperature and humidity sensor, Radar, Lidar, Sonar, GPS and other sensors. This data processing requires new IoT devices and infrastructure including software components newer industrial models for dispensation, distribution and handling of information. Numerous businesses firms are joining together to obtain access to construct new set of data around high-definition maps to gain entry to design layouts for their autonomous vehicles. Feedback loops are created and huge amount of data is generated for trial and testing. However, the issues ownership of data remains a decisive question for industries and managers in the next several years.

3.6 Smart Car's Value Chain

Operating system gives a standardized way to interact with the machine hardware and aids the developers to develop and work with the hardware kernel easily, such is the case with existing mobile and desktop application. User friendly and easy-to-use interfacing and techniques enable larger audience to purchase the product. Smart Cars should capable to understand how the user feels. Smart vehicle should be able to detect drivers' current emotion in a journey. Emotion of the driver can be read using sensors. Also, data related to car maneuvering details can be captured, like, sudden drift, sudden acceleration burst and abrupt starts and stops to predict the current state of mind of the driver can be determined by these patterns. These data can be uploaded and processed to cloud to help evaluate the possible routes to fix any issue in vehicular movement. Smart Car integration should be more human like, so that the information exchanged between these devices are intelligent, timely; They also need to be efficient enough to provide few options to the user to choose from the suggested option.

4. CONCLUSION

We have explained the different types of technologies used in smart vehicles. There are different reliability needs of these vehicles, for instance, alarms and alerts for collision detection is very precise, it is of the order of few milliseconds. On the other hand, monitoring traffic congestion and traffic information collection can be wait tolerant. Due to the variant needs of this system, these technologies have to be integrated together in the best possible manner to leverage the maximum benefits of this system. For Inter vehicular communication, these technologies work with V2V and V2I network to retrieve most up-to-date and relevant information for smart communication. And various sensors and other monitoring technology are used to handle intra vehicle communications. The moving vehicles can act as an internet access point. These vehicles can share internet connection to nearby vehicles to access internet using these access points. This access can by granted upon request by nodes/vehicles that done have internet access. Similarly, with SaaS/RaaS the vehicles with abundant storage and processing faculty can share storage nearby vehicles which require storage capacity for short-term purpose and also share intelligent information to take appropriate action in response to some environmental condition etc. Improvement in Quality of Service in VANET would eventually increase the ability of SMART vehicles to take intelligent decisions like humans in the near future.

REFERENCES

Alonso, J. D., Vidal, E. R., Rotter, A., & Mühlenberg, M. (2008). Lane-Change Decision Aid System Based on Motion-Driven Vehicle Tracking. *IEEE Transactions on Vehicular Technology*, *57*(5), •••.

Artimy, M. M., Robertson, W., & Phillips, W. J. (2005), Assignment of dynamic transmission range based on estimation of vehicle density. *Proceedings of the Second International Workshop on Vehicular Ad Hoc Networks*. 10.1145/1080754.1080761

Bento, L. C., Parafita, R., & Nunes, U. (2012). Intelligent traffic management at intersections supported by V2V and V2I communications. *Intelligent Transportation Systems (ITSC)15th International IEEE Conference on*.

Blum, J. J., Eskandarian, A., & Hoffman, L. J. (2004, December). Challenges of inter vehicle ad hoc networks. *IEEE Transactions on Intelligent Transportation Systems*, *5*(4), 347–351. doi:10.1109/TITS.2004.838218

Chen, H., & Tian, J. (2009). Research on the Controller Area Network. *Networking and Digital Society, 2009. ICNDS '09. International Conference on, 2,* 251-254. 10.1109/ICNDS.2009.142

Cho, H., & Kim, B. (2014). Cooperative Intersection Collision-Warning System Based on Vehicle-to-Vehicle Communication. *Contemporary Engineering Sciences, 7*(22), 1147-1154.

Datta, S. K. & Costa, D. C., & Härri, J. (2016). *Christian Bonnet Communication Systems Department, France Integrating Connected Vehicles in Internet of Things Ecosystems: Challenges and Solutions.* French research project and the HIGHTS project funded by the European Commission (636537-H2020).

Devi, Y. U., & Rukmini, M. S. S. (2016). IoT in Connected Vehicles: Challenges and Issues- A Review. *International conference on Signal Processing,Communication, Power and Embedded System.*

Dobashi, K. (2005). Adaptive MAC Protocol for high-load Inter-Vehicle Communication. *Wireless and Mobile Computing, Networking and Communications, 2005. (WiMob'2005), IEEE International Conference on, 3.* 10.1109/WIMOB.2005.1512896

Fernandes, P., & Nunes, U. (2007). *Vehicle communications: A short survey*. Institute of Systems and Robotics Department of Electrical and Computer Engineering, University of Coimbra.

Hartenstein, H., & Laberteaux, K. (2010). *VANET Vehicular Applications and Inter-Networking Technologies*. Academic Press.

Hellbruck, H., Cao, J., & Zuo, T. (2009). Fast prototyping for VANET applications with PDAs Wireless Communication, Vehicular Technology, Information Theory and Aerospace & Electronics Systems Technology, 2009. *Wireless VITAE 2009.1st International Conference on.*

Huang, S. C., Chen, B. H., Chou, S. K., Hwang, J. N., & Lee, K. H. (2016). Smart Car. *IEEE Computational Intelligence Magazine*.

Kim, Seo & Kim. (2008). A gateway system for an automotive system: LIN, CAN, and FlexRay Industrial Informatics. *6th IEEE International Conference*.

Kim, J., Jo, M. J., & Jung, Y. (2009), Efficient Routing Scheme Using Circular Node for VANET Ubiquitous Information Technologies & Applications, 2009. *ICUT '09. Proceedings of the 4th International Conference on.*

Li, F., & Wang, L. & Can. (2008). Bus Communication System Based on Matlab/Simulink Wireless Communications, Networking and Mobile Computing. *WiCOM '08. 4th International Conference on.*

Lind, R. (1999). The Network Vehicle-a glimpse into the future of mobile multi-media. *IEEE Aerospace and Electronic Systems Magazine, 14*(9), 27-32.

Liu, W., Wen, X. Z., Duan, B., Yuan, H., & Wang, N. (2007). Rear Vehicle Detection and Tracking for Lane Change Assist, Member. *IEEE Proceedings of the 2007 IEEE Intelligent Vehicles Symposium.*

Lozano, J. G., Montero, M. I., & Martínez, M. A. (2011). Power Electrical and Electronic Systems (PE&ES). *School of Industrial Engineering (University of Extremadura) Compatibility and Power Electronics (CPE), 7th International Conference-Workshop.*

Nassar, L., & Karray, F. (2016) Fuzzy Logic in VANET Context Aware Congested Road & Automatic Crash Notification Electrical & Computer Engineering University of Waterloo Waterloo, ON, Canada. *IEEE International Conference on Fuzzy Systems (FUZZ).*

Okrouhly, M., & Waraus, D. (2011). Anti-Lock Braking System Based on FlexRay Protocol, Czech Technical University in Prague. *The 6th IEEE International Conference on Intelligent Data Acquisition and Advanced Computing Systems: Technology and Applications.*

Pathak, P. H., Feng, X., Hu, P., & Mohapatra, P. (2015). Visible Light Communication, Networking and Sensing: A Survey, Potential and Challenges. Computer Science Department, Electrical and Computer Engineering Department, University of California, Davis, CA.

Prasad, K. V., Hodges, S., & Noble. (2016). Vehicle Spaces. *IEEE Pervasive Computing, 15*(1).

Ravishankar, S., & Theetharappan, N. (2017). Cloud Connected Smart Car Park. *International conference on I-SMAC- IoT in Social, Mobile, Analytics and Cloud.*

Salian, P. P., Prabhu, S., Amin, P., Naik, S.K., & Parashuram, M. K. (2013). Visible Light Communication. *Institute of Technology, Nitte, Karkala, Udupi Texas Instruments India Educators' Conference.*

Sehgal, V. K., Mehrotra, S., & Marwah, H. (2016). Car Security using Internet of Things. *1st IEEE International Conference on Power Electronics, Intelligent Control and Energy Systems.*

Singh, G., Chakrabarty, N., & Gupta, K. (2014). Traffic Congestion Detection and Management Using Vehicular Ad-Hoc Networks (VANETs) In India, Traffic Engineering and Safety Division,CSIR-CRRI. *International Journal of Advanced Computer Technology.*

Varaiya, P. (1993). Smart Cars on Smart Roads: Problems of Control. IEEE Transactions on Automatic Control, 38(2).

Wang, W. (2002). A digital-driving system for smart vehicles. *IEEE Intelligent Systems, 17*(5).

Zardosht, B., Beauchemin, S., & Bauer, M. A. (2013). A Decision Making Module for Cooperative Collision Warning Systems Using Vehicular Ad-Hoc Networks. *Proceedings of the 16th International IEEE Annual Conference on Intelligent Transportation Systems.* 10.1109/ITSC.2013.6728481

KEY TERMS AND DEFINITIONS

Controller Area Network (CAN): A high honored serial data communication bus for dynamic applications, was developed by Bosh.

Lane Change Assistance (LCA): System constantly servile the rear side region of the vehicle. It checks the movement of the driver. It tracks changes in the roadway vehicular movements.

Media-Oriented Systems Transport (MOST): Media-oriented systems transport is a communication system with a new and flexible architecture. It transmits audio signals synchronously. MOST has high bandwidth transportation mechanism for video application and Ethernet channel for IP-based packet data.

Named Data Networking (NDN): It is a future internet architecture that aims to expand a new Internet architecture which can takes advantage of strengths and addresses weaknesses of the Internet's current host-based, point-to-point communication.

Sensor Markup Language (SenML): Simple information model for sensor measurements and device.

V2X (Vehicle-to-Vehicle or Vehicle-to-Infrastructure): V2X arrangement enables reliable data exchanged in real time between infrastructure and vehicles that are on the move.

Vehicular Ad Hoc Network (VANET): A vehicular ad hoc network (VANET) is network of mobile devices as well as stationary objects that can communicate with each other. This technology comprises of both vehicle-to-vehicle (V2V) and vehicle-to-infrastructure (V2I) data transmission links.

Chapter 9
Vehicular Fog Computing Paradigm:
Scenarios and Applications

Jyoti Grover
Manipal University Jaipur, India

ABSTRACT

The main objective of vehicular ad hoc networks (VANETs) is to improve driver safety and traffic efficiency. Most VANET applications are based on periodic exchange of safety messages between nearby vehicles and between vehicles and nearby road side communication units (e.g., traffic lights, road-side lights, etc.). This periodic communication generates huge amount of data that have typical storage, computation, and communication resources needs. In recent years, there has been huge developments in automotive industry, computing, and communication technologies. This has led to vehicular cloud computing (VCC) as a solution to satisfy the requirements of VANETs such as computing, storage, and networking resources. Vehicular fog computing (VFC) is a standard that comprehends cloud computing and related services to the proximity of a network. Since VANET applications have special mobility, low latency, and location awareness requirements, fog computing plays a significant role in VANET applications and services. In urban cities, vehicles parked at shopping malls, offices and similar other places are under-utilized. These can offer great opportunity and value to implement applications of VFC by utilizing vehicles as an infrastructure. In this chapter, we present real time scenarios and applications of VANET that can be implemented using VFC. VANET applications and quality of service can be enhanced by aggregating the resources of these vehicles. We discuss different types of scenarios of moving and parked vehicles as computational, communication, storage and network infrastructures. We have also discussed the challenges and open problems to implement VFC system. This chapter provides the thorough understanding of novel research paradigm and about vehicular communication infrastructures.

DOI: 10.4018/978-1-5225-3981-0.ch009

1. INTRODUCTION

Driving has become integral part of our everyday lives. Globally, average driving time of people is increasing up to 84 minutes per day (Ding, D. & Gebel, K. & Phongsavan, P. & Bauman, A. E. & Merom, D., 2014). Some of the major reasons of traffic accidents are poor road conditions, traffic congestion and long driving time. With the large scale improvement in computation and wireless communication technologies, vehicles can communicate with other vehicles and communication units on roadside. Such communication creates new opportunities for enhancing road safety. VANET has emerged as a solution to many road safety problems by providing safety information to drivers on time (Bai, F. & Krishnan, H. & Sadekar, V. & Holl, G. & Elbatt, T., 2006). VANET is a special class of Mobile Ad Hoc Network (MANET) (Yousefi, S. & Fathy, M., 2006). Main characteristics of VANET (Biswas, S. & Tatchikou, R. & Dion, F., 2006) that distinguish it from MANET are high mobility of nodes, scalability and frequent topology changes. VANETs use short range radios (Jiang, D. & Taliwal, V. & Meier, A. & Holfelder, W. & Herrtwich., R., 2006) in each vehicle, which allows various vehicles to communicate with other vehicles and roadside infrastructure. Safety and traffic management applications require real time information and can play significant role in life and death decisions. VANET is considered as a most important component of future intelligent transportation systems (Khabazian, M. & Aissa, S. & Mehmet, M., 2011) and support various mobile services such as content sharing applications, emergency information of any disaster or terrorist attack etc.

In VANET, Vehicle to Vehicle (V2V) and Vehicle to Roadside (V2R) communication is used to propagate safety information to nearby vehicles on time. Apart from the communication unit, each vehicle is also equipped with high computation and storage unit. But most of the time, these units are under-utilized. Vehicular cloud computing (VCC) is the paradigm (Whaiduzzaman, M. & Sookhak, M. & Gani, M., 2014) emerged to utilize VANET resources efficiently by taking advantages of cloud computing. It serves the drivers of VANET with a pay as you go model. In VCC, group of vehicles cooperate with each other to dynamically share computing, sensing and communication resources for decision making on the road in order to improve traffic management and road safety. Some examples of VCC applications are:

1. Local traffic condition can be collected from nearby vehicles for route planning.
2. Current transportation system can be improved by big data processing of traffic information by local traffic authorities.
3. Collaborative image of critical events can be reconstructed such as car accident, congestion on the road etc.

VCC is a very promising solution to share the computation and storage resources among the vehicles and roadside units in order to implement these applications. But VCC is not sufficient for many VANET applications due to mobility of vehicles and the latency sensitive requirements imposed by these. It is difficult to meet the Quality of Service (QoS) requirements using VCC. So, a new approach is designed that comprehend cloud computing with VANET applications named as Fog computing. Fog computing leverage computation infrastructure that is closer to the network edge to compliment cloud computing in providing latency sensitive applications and services. Fog computing is similar to cloud computing but the only difference is that time sensitive applications can be implemented at the network edge rather than sending huge amount of data to remote cloud. The idea of Fog computing (Stojmenovic, I.

& Wen, S., 2014) is to place cloud-like facility at the proximity of application users. It provides storage, computation and application services to the edge of network (nodes within proximity), thereby reduces burden on the cloud and facilitates real time VANET applications implementation (Kai, K. & Cong, W. & Tao, L., 2016). In VANET, any node having communication, computation and storage resources can become fog nodes. Fog computing is very useful to implement VANET applications as these are time sensitive and real time which is the prime agenda of Intelligent Transportation system (ITS) (Jiang, D. & Taliwal, V. & Meier, A. & Holfelder, W. & Herrtwich, R., 2006).

Vehicular network and its associated applications have been developed dramatically due to emergence of advanced technologies such as cellular networks and cloud computing. This has led a serious issue i.e. dynamic demand of data communication and computation capability which is required by advanced applications such as augmented reality (AR) techniques (Tachi, S. & Inami, M. & Uema, Y. ,2014), self-driving etc. To meet these ever-expanding requirements, vehicular clouds can be employed at company parking lot, shopping malls etc. to collaborate with each other to handle communication and computation requirements.

Existing solutions to solve the communication and computation problem such as 3G/4G cellular networks and mobile computing (Huang, D. & Xing, T. & Fang, S., 2013) are not appropriate because of the following reasons: (a) Cellular network covers the mobile nodes by augmenting the communication. But these are controlled by network operators which is not suitable from application perspective. Mobile Cloud Computing brings the heavy computational resources within the reach of mobile users, but it's not suitable for real time applications because of its cost and longtime requirements. This is due to the fact that mobile information needs to be uploaded on cloud periodically for performing high computation operations. It also requires Internet connection of high quality to make connection with the remote cloud (Tuli, A. & Hasteer, N. & Sharma,. M. & Bansal, A., 2013).

This chapter discusses various applications of VANET which can be implemented using fog computing. The rest of the paper is organized as follows. Section 2 provides overview of cloud computing and its emergence with VANET. It also discusses the concept of VCC. Fog computing and its related work in relation with VANET is discussed in Section 3. Some of the VANET real time applications which can be implemented using fog computing are discussed in Section 4. Finally, concluding remarks with future work are covered in Section 5.

2. VEHICULAR CLOUD COMPUTING

Cloud computing is an Internet-based computing technique that provides shared computer processing resources and data to computers and other devices on demand. It enables the users with various capabilities to store and process their data in privately owned or third party data centers that may be located far from user (in different city or country). Cloud computing (Eltoweissy, Olariu, & Younis, 2010) provides high computation power, cheap cost of service, high performance, scalability, accessibility and availability etc.

Inspired by the success of cloud computing integrated with mobile communications, authors (Olariu, Hristov, & Yan, 2013) have introduced the concept of vehicular cloud computing (VCC). Figure 1 describes vehicular cloud environment in VANET.

Huge advancements in automotive industry, computation and communication technologies has led the dream of smart cars true. Ford has announced the manufacturing of four millions vehicles with SYNC systems (having integrated in-vehicle communication system, GPS and Radar etc.). Recently, Google

Figure 1. Vehicular cloud environment for vehicular ad hoc network

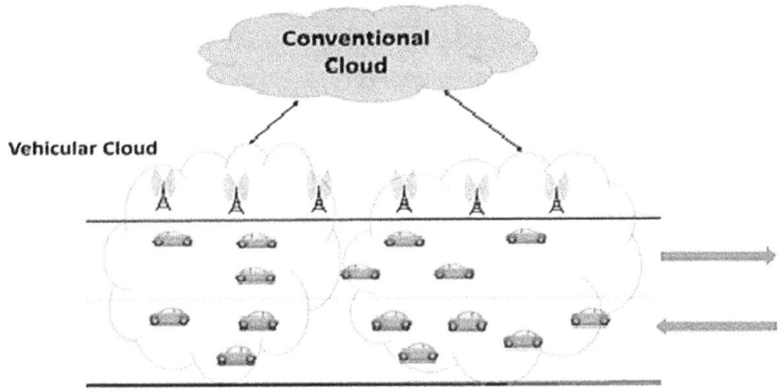

has also tested a self-driving car which is equipped with computation and communication device, radar sensors, GPS and Video camera etc.

Nowadays, vehicles are just more than transportation machines because these have high computation and storage capability. People can get different types of services (such as navigation, weather and entertainment etc.) when they are driving. Most of the time (especially when vehicles are parked in offices, shopping malls etc.), computing and storage resources of these vehicles are untapped. These vehicles could be the potential resources for vehicular cloud (VC). The main issue in implementing VC is due to dynamic behavior of vehicles. VC services can be switched to traditional cloud placed nearby in the case of vehicles move in/out a specific area.

3. VEHICULAR FOG COMPUTING

Fog computing is defined as a platform where large number of heterogeneous, ubiquitous and distributed devices communicate and cooperate by forming a network in order to perform storage and processing functions without the intervention of third parties. These functions can be providing support to new applications that can run in sandboxed environment. Most of VANET applications are area specific and needs to be handled locally such as navigation, infotainment and safety information etc. In these cases, fog computing just fits in because it allows the contents and application services as close as possible to drivers on the road in particular road segment. Fog computing addresses the limitation of location awareness of cloud computing. For example, a person visiting a new city would like to seek information on the places of interest, news and weather conditions of this city rather than interested in other city information. Fog computing is well suited for these types of applications. However, fog computing paradigm needs extensive research before its practical deployment (Firdhous, Ghazali, & Hassan, 2014).

Cloud computing is the central portal of information, does not have location awareness. Fog computing overcome this issue by providing localized services to specific deployment sites. Basic architecture of storage, computing and networking for cloud and fog computing is presented in (Bonomi, 2011). Fog computing is a most suitable approach that can control critical resources like energy, traffic and healthcare etc. (Vauero & Merino, 2014) presented a border and integrative view of fog computing. Different authors presented their different perception of fog computing, (Yi, Li, & Li, 2015) strongly differentiated the fog computing from related technologies. It is shown in (Sarkar, & Misra, 2016) that for a scenario where 25% of applications demand real-time low latency services, mean energy expenditure in fog computing is 40% less than conventional cloud computing model. This analysis motivates the research and academia to explore fog computing more extensively.

As can be seen from Figure 2, fog computing is overlaid onto cloud computing in VANET scenario. It creates a distributed computing platform for VANET components to tackle with data processing and storage services. The role of fog computing is like near-end computing proxies between the front-end devices and the far-end servers. There are three types of connections made by fog server in VANETs:

- **RSU to Fog Server:** In VANET, V2V and V2R communication is used to propagate safety/non-safety information. RSUs are also able to communicate with each other. Hence RSUs act as a backbone of fog. There is wired/wireless connection between RSU and fog server. In the Figure 2, RSUC is the fog server.
- **Fog Server to Fog Server:** Fog servers at different locations manage a pool of resources for localized area. There can be direct wired/wireless communication between peer fog servers, or these can be connected via. vehicular control center. It allows the collaborative service provision and content delivery among peered fog servers, thereby improve the performance of entire system.
- **Fog Server to Cloud:** The cloud is the central controller of fog servers deployed at different locations. Each fog server provides services to users at specific locations. Cloud aggregates the information received from fog servers and perform centralized computation and Fog servers convey the information received from cloud to application users.

3.1 Features of Vehicular Fog Computing

In VFC, all the services are performed at the edge of network in order to reduce service latency, improve Quality of Service (QoS). Some of the characteristics of VFC are listed as follows:

Figure 2. Architecture of vehicular fog computing

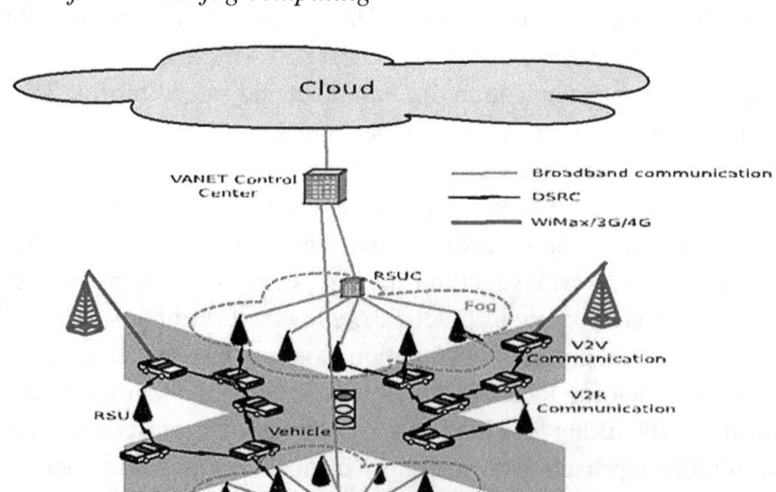

1. **Geographical Distribution:** VFC contains large number of distributed vehicles. As the applications of VFC are localized, geographically nearby vehicles can become part of it and their resources collectively contribute towards accomplishment of applications. Fog can play an important role in dynamic traffic light optimization, propagation of on-road safety messages, mitigation of recurring congestion etc.
2. **Location Awareness and Low Latency:** As VFC supports endpoints with localized services, thereby it reduces the data movement across network remarkably. Fog nodes can gather data from different nearby vehicular nodes and intelligently use the perceived data in making appropriate decisions.
3. **Heterogeneity:** VFC supports multi-tiered architecture. VFC nodes can be heterogeneous such as vehicles, nearby access points, traffic lights, roadside lights etc.
4. **Interoperability:** Due to heterogeneous nature of VFC, it must be able to interoperate vast range of services which require the cooperation of different service providers.
5. **Real Time Interactions:** Most of the VFC applications are real time, so it requires real time interaction among its components for fast service with low latency.

3.2 Difference between VCC and VFC

The differences between fog and cloud computing is presented in (Luan, Gao, & Xiang, 2015). Some of the main features of fog computing compared to cloud computing are:

1. It simplifies the development model for huge number of decentralized heterogeneous devices.
2. Fog computing enhances service quality with increased data rate and reduced service latency and response time.
3. It avoids the back-and-forth traffic between cloud and users. This saves the network bandwidth and reduces the energy consumption.
4. Fog users can use the resources of fog or of cloud reactively based on their demands.

Some of the differences between Vehicular Cloud Computing and Vehicular Fog Computing are shown in Table 1.

Table 1. Differences between vehicular cloud computing and vehicular fog computing

Features	Vehicular Cloud Computing	Vehicular Fog Computing
Bandwidth Constraint	Yes	No
Communication Delay	More	Less
Expensive Deployment	Yes	No
Geo-distribution	No	Yes
Supervisory	Remote	Local
Deployment expense	High	Low
Robustness in Emergency Situations	No	Yes

3.3 Different Scenarios of Vehicular Fog Computing

Vehicular Fog Computing (VFC) can be formed by employing moving and parked vehicles in public places, hotels, colleges, malls, stadium etc. Slow moving vehicles on the road can also form a vehicular cloudlet if these are able to communicate with each other through Vehicle-to-Vehicle communication.

1. **Parked Vehicles:** In metropolitan cities, there are huge number of parked vehicles (e.g. shopping malls, universities, corporate offices etc.). These vehicles are parked for fixed time duration and their position is also fixed. The storage, computation and communication resources of these vehicles are under-utilized. These vehicles form a small data center and deal with various complex tasks which require large computation power. These vehicles can serve as static backbones and service infrastructure for VFC. Parked vehicles on roadside can be used to relay packets from vehicles on road to other communication units nearby such as for collecting the traffic congestion information on the road segment of sending vehicle. These vehicles can serve as static backbones and service infrastructure for VFC as illustrated in Figure 3.
2. **Moving Vehicles Forming VFC:** In vehicular networks, two communication devices are used: Vehicles and Road Side Units (RSUs), which are deployed in the form of roadside and traffic lights. VANET supports two types of communication: Vehicle to Vehicle (V2V) and Vehicle to RSU (V2R). V2V communication is formed when two vehicles moving on the road are within each other's communication range. VFC uses the characteristics of V2V communication to relay the information from one place to another place. Moving vehicles can form new connections with passing-by vehicles and relay the information in particular direction. Parked vehicles can form a communication hub and can connect the passing by vehicles with mobile access points. In this way fog is formed and instead of sending information to remote cloud server, complex tasks can be computed by using the communication and computation resources locally. Moving vehicles can form new connections with passing-by vehicles and relay the information in particular direction as shown in Figure 4.

4. VFC APPLICATIONS

Applications of fog computing in vehicular scenario are not as much as for VCC. Authors (Giang, Victor, & Rodger, 2016) argued that traditional cloud computing is not sufficient for many VANET applications due to mobility of vehicles and their real-time applications. VFC leverage computation infrastructure that is closer to the network edge, thereby supports applications with low latency requirements. Some of the applications are listed below:

4.1 Smart Traffic Lights

A Dynamic Traffic Light System (DTLS) is an important application of VANET that optimize traffic at traffic light junctions efficiently by using the concept of fog computing. As this application is distributed (location proximity specific and time sensitive), centralized solution such as VCC cannot be used. Different fogs can be used at traffic light junctions to compute the duration of each signal based on the traffic around the junction. The current system of traffic lights allots a dedicated time towards each direction,

Figure 3. Vehicles parked in parking lot as infrastructure

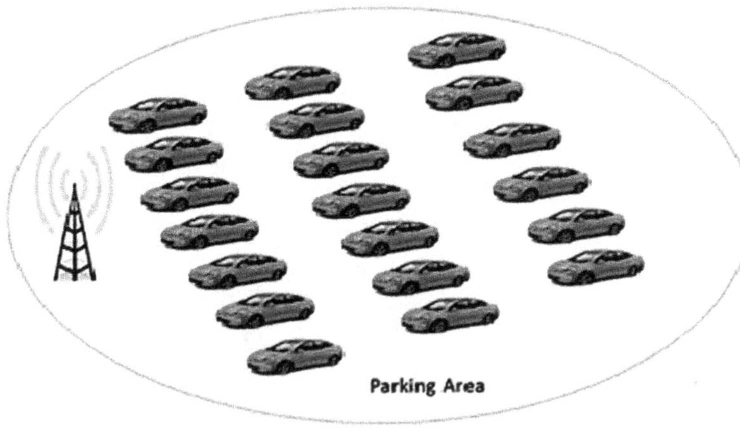

Figure 4. Vehicles parked on roadside are used to relay the information

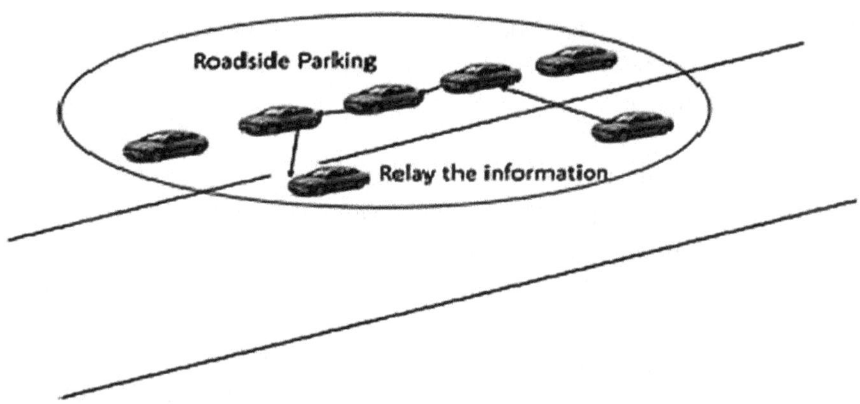

thereby making the whole system static. The limitation of this system is that traffic from one side has to wait for the green light even if there is no vehicle crossing the road in the current direction of green light. A dynamic DTLS allots the time of each signal based on the traffic present around the junction. A sample smart traffic light system is shown in Figure 5.

Traffic lights can serve as fog devices to send warning messages to approaching vehicles in particular direction. The interaction between vehicle and RSUs are based on Dedicated Short Range Communication (DSRC). RSUs can communicate with each other to form a fog and send the aggregated perceived safety information to traffic signal. Traffic lights can be switched from one mode to another as number or type of vehicles approach towards them.

DTLS is one of the main application of fog computing. Communication range of each RSU is generally 500–1000 meters and it is a usual norm that each vehicle sends beacon packets periodically (Jiang, & Delgrossi, 2008). Hence, RSUs are able to estimate the traffic on particular road segment. Based upon the computed information, RSUs can propagate the information to fog servers and thereby duration of traffic light can be decided accordingly. Now, we discuss different scenarios of DTLS.

Figure 5. Smart traffic light system using fog computing

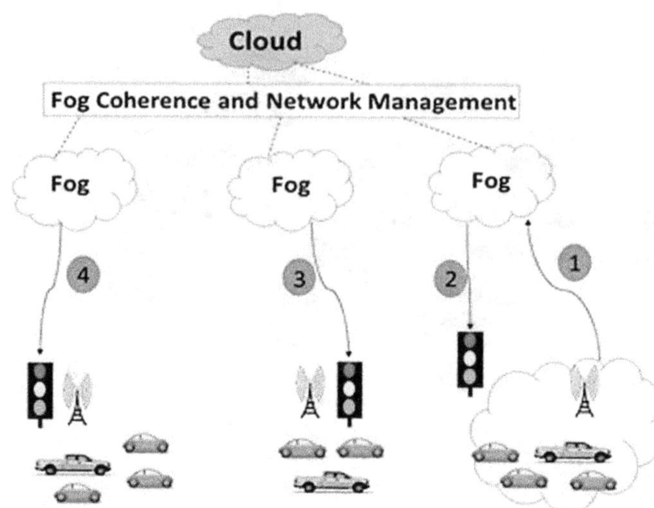

1. If there is heavy traffic on one side of road segment and other side of the road is free, traffic signal can be turned green for the heavy traffic s ide and red for the free road. This can be implemented using fog servers collecting information from RSUs and timely making decision by performing computations.
2. In a scenario, where emergency vehicle such as an ambulance is approaching to-wards road junction. Nearby RSUs can identify ambulance by information sent by it. Video cameras deployed in RSUs are also able to recognize flashing lights. RSU observing the event spread this information in the fog. Neighboring smart lights serving as fog can coordinate and create green light signal in addition to sending warning signals to approaching vehicles.
3. Roadside lights can also detect the presence of vehicles and measures the speed and distance of nearby vehicles. Sensors deployed on fog servers can facilitate in turning on these lights on identification of vehicles and vic e-versa. Smart lights along the road side serve as fog devices can collaboratively make efficient use of resources.

4.2 Parking System

Finding an empty parking space in rush hour is really a big problem in urban areas. Authors (Kim, Dang Tri, Nguyen, Tran, & Hong, 2015) have presented their work on solving parking problems to relieve the traffic congestion, reduce air pollution and enhance driving experience. They have shown how fog computing and roadside clouds can be used to find a vacant s pot. They assume that unused public and private parking slot (e.g. in hotels, parks, universities etc.) can be shared for the vehicles. Matching theory is used to solve the parking problem by bringing profit to the owners of these places.

4.3 Formation of Datacenter at Airport, Parking Area, and Malls

Datacenters can be formed at airports parking lots where cars are parked for days while their owners are on travel. The travelers can allow their vehicles to participate in VFC by sharing their travel plan in order to make proper utilization of these resources. These pool of computational, storage resources can be used for airport routines which otherwise needs to offload their work to remote cloud. In return of letting the vehicles participate in VFC, travelers can enjoy free parking along with other advantages also.

Similarly, in a parking area of offices, universities and public places, the resources of parked vehicles are idle for fixed period on daily basis. These resources can form cluster of computational resources with huge storage distribution facility, which no office can afford to waste. The offices can provide incentives to employees who rent the resources of their vehicles on a per-hour, per-day or per-month basis. The resources of cars can be used in universities with hundreds or even thousands of cars parked in its various parking lots. The resources in those cars are likely to sit idle for many hours each day while their owners are at work or taking classes on campus. Several solutions are proposed to manage parking availability (Yan G. & Rawat, D. B. & Olariu, S. 2011). However, these solutions are centralized and reports from individual parking lots are aggregated at a central place and depending upon the location of parking requester, specific parking lot in nearby parking area is suggested. The problem in this solution is that information that reaches the public is often outdated.

In a shopping mall, hundreds of vehicles are parked for couple of hours every day. Mall management can use these underutilized resources by "pay as you go" for customers over the Internet. The mall management may provide various incentives to make it attractive for the shoppers to share the resources in their parked vehicles. These may include discounts at various stores at the mall, free parking, and many more of similar types. Finding an empty parking space in rush hour is really a big problem in urban areas. Authors (Kim, O. T. T. & Dang Tri, N. & Nguyen, V. D. & Tran, N. H. & Hong, C. S., 2015) have presented their work on solving parking problems to relieve the traffic congestion, reduce air pollution and enhance driving experience. They have shown how fog computing and roadside clouds can be used to find a vacant space. They assume that unused public and private parking slot (e.g. in hotels, parks, universities etc.) can be shared for the vehicles. Matching theory is used to solve the parking problem by bringing profit to the owners of these places.

4.4 Content Distribution

Fog computing can be used for distribution of any type of information (safety on non-safety). For example, if a particular section of a road is blocked due to some accident or natural hazard. This information can be conveyed by the fog servers to vehicles approaching towards this site. In developed countries, cameras are fixed at every road intersections and along the roadside. These cameras track the lines on the road and help the driver stays in lane. In VANET, vehicles on the road are able to detect the traffic conditions based upon the received information from nearby vehicles in its vicinity. Contemporary VANET design cannot pull together the required solution and foster the level of coordination needed for providing these safety.

4.5 Decision Support System

Safety of drivers can be improved by using intelligent decision support system (DSS) using fog computing. Different fog servers can exchange their localized conceived in-formation with each other and form DSS in order to monitor traffic violation and safety system for drivers.

VCC has vast number of applications as compared to fog computing. But fog computing fulfils the service requests on localized information, thereby reduces the service latency and response time.

5. CHALLENGES AND OPEN ISSUES IN VFC

Most of the services supported by vehicular networks cannot be practically implemented unless the communication and computation capacity is not sufficiently powerful. Vehicular fog computing uses vehicles as infrastructures and employs nearby vehicular resources (such as parked vehicles, traffic lights, roadside lights etc.). There are numerous challenges required to turn the VFC concept into reality. In this section, we present and discuss the real challenges and open issues existing in VFC.

5.1 VANET Mobility Model

In VANET vehicles move from one place to other with very high speed on road segments. So the connections established between these vehicles is only for short duration of time. A suitable mobility model is required that can predict and evaluate the movement pattern of vehicles on the road and nearby.

Some work has been done on mobility models as discussed in (Naboulsi & Fiore, 2013; Viriyasitavat, Tonguz, & Bai, 2009). However, these models are inaccurate and insufficient for VFC scenario. An effective mobility model is required that can simulate accurate behavior of vehicles such as their speed and availability in time and area domain both. It is the need of VFC to study and design mobility models for various VANET environments for their practical implacability. It can further enable us to design the applications associated with VFC.

5.2 System Implementation

In order to be a part of VFC, all the vehicles need to be equipped with GPS device, sensing devices, digital maps, and communication and computation devices. These devices facilitate drivers to acquire real-time traffic information on road such as congestion, road accident, traffic diversion etc. Some car companies have already started to manufacture cars equipped with aforementioned devices. However, improvement is still required (Arif, 2012) in terms of better hardware and software to support VFC. There is no concern regarding the power supply in VANET as battery of vehicles automatically get charged when it moves. Rechargeable vehicular batteries can be used in order to perform the functionality of VFC when engines of these vehicles are switched off. However, it is essential to develop more energy efficient wireless communication and onboard computation framework to support VFC applications. Wireless communication and computation equipment within vehicles are also supposed to support connections between cloud control center and vehicles. It requires integration of individual parts of VANET system in order to efficient design of VFC.

5.3 Applications

It is not too far when VFC applications will be operational in our daily lives such as driver safety warning, entertainment services and emergency networks etc. VFC plays a crucial role for applications which require rich computational processing power and have time constraints. For example, if some catastrophic event (natural or man-made) happens on the road, and road traffic diversion is required. This requires rescheduling of all the traffic lights on connected roads in the city. Therefore it requires high computation power and time sensitive also. Using VFC, computation power of connected vehicular resources can be used to control the traffic lights accordingly. It is required to categorize various applications by identifying these and then implement using VFC. As different applications have different purposes and computation and communication requirement, interoperability issue also needs to be handled so that all types of applications can be supported by VFC simultaneously.

5.4 Resource Management

IT is one of the major challenge to implement VFC in an efficient and optimized way. The main objective is to take the advantage of available vehicular computation and storage resources comprehensively. Some of the methods to encourage the sharing of resources between nodes in network are presented such as reward-based approach (Zhong, Chen, & Yang, 2003), penalty-based mechanisms (Balakrishnan, Deng, & Varshney, 2005). These approaches require some guidelines and legal agreements in order to promote the resource utilization of VFC. In order to efficiently utilize the available resources of vehicle, these need to join the VFC and authorize to exchange the data with each other. Some sort of security protocols (Kleberger, Olovsson, & Jonsson, 2011) are required to implement these efficiently. VFC can provide people better communication service and many localized applications in reward of the use of their vehicular resources so that more people will be attracted towards joining VFC.

5.5 Security and Privacy

Security and privacy of VFC is very important because of diversity of VANET applications (Djenouri, Khelladi, & Badache, 2005). Security can be breached because of sharing of contents and accessing each other's data. Some of the security problems are: Weak authentication, vulnerabilities in protocol, insufficient protection. Privacy is also important as mobile users detail such as name, e-mail address may be obtained by social networking sites or third parties, thereby resulting in privacy leakage (Lin, Sun, Ho, & Shen, 2007). Developing a secure VFC is of paramount importance. It is also very important to decide where exactly these security and privacy algorithms will be run (either at cloud or fog). It is one of the hottest area to be explored for VFC.

5.6 Quality of Service (QoS)

It is one of the important parameter for the services which are provided by VFC. There are four main aspects of QoS: Connectivity, reliability, capacity and delay. It is very difficult to maintain a VFC system connected due to dynamic nature of vehicular nodes and 3G/4G/WiFi. But to some extent, it is required to employ some algorithms to ensure the system connected by using the embedded system in fog nodes.

It is also essential to identify and repair fog nodes as rescheduling and check pointing may not be suitable in VFC. Provisioning QoS is also the key area of research in VFC.

6. CONCLUSION

In this chapter, we have presented the concept of Vehicular Fog Computing and discussed how fog computing can be used in implementing real-time VANET applications We have considered the different scenarios of moving and parked vehicles as a fog that can be used as communication and computation infrastructure. We have also presented the architecture of VCC and fog computing in VANET environment. VFC has the potential to handle the data generated by VANET components and take appropriate actions in order to implement desired VANET applications. The features of VFC like mobility, proximity to end users, low latency, heterogeneity, location awareness, make it best suited for implementation of some applications of VANET. We have justified that communication and computation capacity can be enhanced by the use of VFC. We also have discussed different real time applications of VANET that can be implemented using VFC. Most vehicular applications using mobile cloud computing can be implemented by improving the communication and computation capacity of vehicular nodes. In our future work, we would like to design the layout and implementation of these applications using VFC.

REFERENCES

Arif, S., Olariu, S., Wang, J., Yan, G., Yang, W., & Khalil, I. (2012). Data Center at the Airport: Reasoning about time dependent parking lot occupancy. *IEEE Transactions on Parallel and Distributed Systems*, *23*(11), 2067–2080. doi:10.1109/TPDS.2012.47

Bai, F., Krishnan, H., Sadekar, V., Holl, G., & Elbatt, T. (2006). Towards characterizing and classifying communication-based automotive applications from a wire- less networking perspective. *IEEE Workshop on Automotive Networking and Applications (AutoNet)*. doi:10.1371/journal.pone.0094602

Balakrishnan, K., Deng, J., & Varshney, P. K. (2005). TWOACK: Preventing Selfishness in Mobile Ad Hoc Networks. *Proceedings WCNC*, 2137-2142.

Biswas, S., Tatchikou, R., & Dion, F. (2006). Vehicle-to-Vehicle wireless communication pro-tocols for enhancing highway traffic safety. *IEEE Communications Magazine*, *44*(1), 74–82. doi:10.1109/MCOM.2006.1580935

Bonomi, F. (2011). Connected Vehicles, Internet of things and fog computing *Proceedings of 8th ACM international Workshop on Vehicular Inter-Networking*.

Bonomi, F. Milito, R. & Zhu, Z. and Sateesh Addepalli. (2012). Fog computing and its role in the internet of things. In Proceedings of the first edition of the MCC workshop on Mobile cloud computing (MCC '12). ACM, New York, NY, USA, 13-16.

Ding, D., Gebel, K., Phongsavan, P., Bauman, A. E., & Merom, D. (2014). Driving: A Road to Unhealthy Lifestyles and Poor Health Outcomes. *PLoS One*, *9*(6).

Djenouri, D. & Khelladi, L. & Badache, A. N. (2005). A survey of Security Issues in mobile Ad Hoc Networks. *IEEE Communication Surveys Tutorials, 7*(4), 2-28.

Eltoweissy, M., Olariu, S., & Younis, M. (2010). *Towards autonomous Vehicular clouds Ad Hoc Networks*. Berlin, Germany: Springer; doi:10.15439/2014F503

Firdhous, M., Ghazali, O., & Hassan, S. (2014). Fog computing: will it be the future of cloud computing? *Proceedings of the 3rd International Conference on Informatics and Applica-tions (ICIA'14)*, 8–15.

Giang, N. K., Victor, C. M., & Rodger, L. (2016). On Developing Smart Transportation Appli-cations in Fog Computing Paradigm. *Proceedings of the 6th ACM Symposium on develop-ment and Analysis of Intellegent Vehicular Networks and Applications (DIVANet'16)*, 91–98.

Hu, X., Wang, L., & Sheng, Z. & TalebiFard, P. & Zhou, L. & Liu, J. & r C.M. Leung, V. (2014). Towards a service centric contextualized vehicular cloud. In Proceedings of the fourth ACM international symposium on Development and analysis of intelligent vehicular networks and applications (DIVANet '14). ACM, New York, NY, USA, 73-80.

Huang, D., Xing, T., & Fang, S. (2013). Mobile Cloud Computing Service Models: A User Centric Approach. *IEEE Network, 27*(5), 6–11.

Jiang, D., & Delgrossi, L. (2008). IEEE 802.11p: Towards an International Standard for Wireless Access in Vehicular Environments. *Vehicular Technology Conference, VTC Spring 2008, IEEE*, 2036–2040.

Jiang, D., Taliwal, V., Meier, A., Holfelder, W., & Herrtwich, R. (2006). Design of 5.9 GHz DSRC-based vehicular safety communication. *IEEE Wireless Communications, 13*(5), 36–43. doi:10.1109/WC-M.2006.250356

Kai, K., Cong, W., & Tao, L. (2016). Fog computing for Vehicular Ad hoc Networks: Paradigms, Scenarios and Issues. *Journal of China Universities of Posts and Telecommunications, 23*(2), 56–96.

Khabazian, M., Aissa, S., & Mehmet, M. (2011). Performance Modeling of message dissemination in vehicular ad hoc networks with priority. *IEEE Journal on Selected Areas in Communications, 29*(1), 61–71.

Kim, O. T. T., Dang Tri, N., Nguyen, V. D., Tran, N. H., & Hong, C. S. (2015). A Shared Parking Model in Vehicular Network Using Fog and Cloud environment. *7th Asia Pacific Network Operations and Management Symposium (APNOMS)*, 321–326. 10.1109/APNOMS.2015.7275447

Kleberger, P., Olovsson, T., & Jonsson, E. (2011). Security Aspects of the in vehicle network in the connected car. *Proceedings of IEEE Intelligent Vehicle Symposium*, 528-533. 10.1109/IVS.2011.5940525

Lin, X., Sun, X., Ho, P. H., & Shen, X. (2007). GSIS: A Secure and Privacy preserving protocol for vehicular communications. *IEEE Transactions on Vehicular Technology, 56*(6), 3442–3456. doi:10.1109/TVT.2007.906878

Liu, N., Liu, M., Lou, W., Chen, G., & Cao, G. (2011). *PVA in VANETs: Stopped cars are not silent* (pp. 431–435). Shanghai: IEEE INFOCOM.

Luan, T. H., Gao, L. X., & Xiang, Y. (2015). *Fog computing: focusing on mobile users at the edge.* CoRR, arXiv:1502.0181v1

Malandrino, F., Casetti, C., & Chiasserini, C. F & Sommer. & C. Dressler, F. (2014). The Role of Parked Cars in Content Downloading for Vehicular Networks. *IEEE Transactions on Vehicular Technology, 63*(9), 4606–4617.

Naboulsi, D., & Fiore, M. (2013). On the instantaneous topology of a large scale urban vehicular network: The Cologne case. *Proceedings of ACM MobiHoc*, 167-176. 10.1145/2491288.2491312

Olariu, S., Hristov, T., & Yan, G. (2013). The next paradigm shift: from vehicular networks to vehicular clouds. In S. Basagni, M. Conti, S. Giordano, & I. Stojmenovic (Eds.), *Mobile Ad Hoc Networking: Cutting Edge Directions* (2nd ed.). Hoboken, NJ: John Wiley & Sons, Inc. doi:10.1002/9781118511305.ch19

Sarkar, S., & Misra, S. (2016). Theoretical modelling of fog computing: A Green Computing to support IoT Applications. *IET Networks, 5*(2), 23–29. doi:10.1049/iet-net.2015.0034

Stojmenovic, I., & Wen, S. (2014). The Fog Computing Paradigm: Scenario and security Issues. *Proceedings of the 2014 Federated Conference on Computer Science and Information Systems (FedCSIS'14)*, 8–16.

Tachi, S., Inami, M., & Uema, Y. (2014). Augmented Reality helps drivers see around blind spots. *IEEE Spectrum*, 31.

Tuli, A., Hasteer, N., Sharma, M., & Bansal, A. (2013). Exploring Challenges in Mobile Cloud Computing: An Overview. *Proceedings of 4th International Conference on Next Generation Information Technical Summit*, Noida, India, 496-501.

Vauero, L. M., & Rodero Merino, L. (2014). Finding your Way in the Fog: Towards a Comprehensive Definition of Fog computing. *Computer Communication Review, 44*(5), 27–32. doi:10.1145/2677046.2677052

Viriyasitavat, W., Tonguz, O. K., & Bai, F. (2009). Network Connectivity of VANET in Urban Areas. *Proceedings IEEE SECON*, 1-9. 10.1109/SAHCN.2009.5168949

Whaiduzzaman, M., Sookhak, M., Gani, M., & Buyya, R. (2014). A survey on Vehicular cloud computing. *Journal of Network and Computer Applications, 40*(1), 325–344. doi:10.1016/j.jnca.2013.08.004

Yan, G., Rawat, D. B., & Olariu, S. (2011). Smart Parking: A secure and intelligent parking system. In IEEE Intelligent Transportation Systems Magazine 3(1):18–30.

Yi, S. H., Li, C., & Li, Q. (2015). A Survey of fog computing, concepts, applications and Issues. *Proceedings of the 2015 workshop on mobile big data (MobiData'15)*, 37–42. 10.1145/2757384.2757397

Yousefi, S., & Fathy, M. (2006). Vehicular Ad-hoc NETworks (VAN ETs): Challenges and Per-spectives. *6th International Conference on ITS Telecommunications, IEEE*, 761–766.

Zhong, S., Chen, J., & Yang, Y. R. (2003). Sprite: A Simple, Cheat Proof, Credit based System for Mobile Ad Hoc Networks. *Proceedings INFOCOM*, 1987-1997. 10.1109/INFCOM.2003.1209220

KEY TERMS AND DEFINITIONS

Cloud: A network of servers.

Cloud App: The phrase used to describe a software application that is never installed on a local computer. Instead, it is accessed via the internet.

Cloud Provider: A service provider who offers customers storage or software solutions available via the internet.

Cloudlet: A cloudlet is a new architectural element that emerges from the convergence of mobile computing and cloud computing. It represents the middle tier of a 3-tier hierarchy: mobile device–cloudlet–cloud.

Computing: The process of utilizing computer technology to complete a task.

ITS: An application which provides innovative services related to different modes of transport and traffic management and enable users with safety and coordinated features by smart use of transport networks.

Utility: The measurement of "useful-ness" that a user obtains from any good. Air, water, food, cloth, etc. satisfy people's wants and possess utility.

Related References

To continue our tradition of advancing information science and technology research, we have compiled a list of recommended IGI Global readings. These references will provide additional information and guidance to further enrich your knowledge and assist you with your own research and future publications.

Aalmink, J., von der Dovenmühle, T., & Gómez, J. M. (2013). Enterprise tomography: Maintenance and root-cause-analysis of federated erp in enterprise clouds. In P. Ordóñez de Pablos, H. Nigro, R. Tennyson, S. Gonzalez Cisaro, & W. Karwowski (Eds.), *Advancing information management through semantic web concepts and ontologies* (pp. 133–153). Hershey, PA: IGI Global. doi:10.4018/978-1-4666-2494-8.ch007

Abu, S. T., & Tsuji, M. (2011). The development of ICT for envisioning cloud computing and innovation in South Asia. *International Journal of Innovation in the Digital Economy*, *2*(1), 61–72. doi:10.4018/jide.2011010105

Abu, S. T., & Tsuji, M. (2012). The development of ICT for envisioning cloud computing and innovation in South Asia. In *Grid and cloud computing: Concepts, methodologies, tools and applications* (pp. 453–465). Hershey, PA: IGI Global. doi:10.4018/978-1-4666-0879-5.ch207

Abu, S. T., & Tsuji, M. (2013). The development of ICT for envisioning cloud computing and innovation in South Asia. In I. Oncioiu (Ed.), *Business innovation, development, and advancement in the digital economy* (pp. 35–47). Hershey, PA: IGI Global. doi:10.4018/978-1-4666-2934-9.ch003

Adams, R. (2013). The emergence of cloud storage and the need for a new digital forensic process model. In K. Ruan (Ed.), *Cybercrime and cloud forensics: Applications for investigation processes* (pp. 79–104). Hershey, PA: IGI Global. doi:10.4018/978-1-4666-2662-1.ch004

Adeyeye, M. (2013). Provisioning converged applications and services via the cloud. In D. Kanellopoulos (Ed.), *Intelligent multimedia technologies for networking applications: Techniques and tools* (pp. 248–269). Hershey, PA: IGI Global. doi:10.4018/978-1-4666-2833-5.ch010

Aggarwal, A. (2013). A systems approach to cloud computing services. In A. Bento & A. Aggarwal (Eds.), *Cloud computing service and deployment models: Layers and management* (pp. 124–136). Hershey, PA: IGI Global. doi:10.4018/978-1-4666-2187-9.ch006

Ahmed, K., Hussain, A., & Gregory, M. A. (2013). An efficient, robust, and secure SSO architecture for cloud computing implemented in a service oriented architecture. In X. Yang & L. Liu (Eds.), *Principles, methodologies, and service-oriented approaches for cloud computing* (pp. 259–282). Hershey, PA: IGI Global. doi:10.4018/978-1-4666-2854-0.ch011

Ahuja, S. P., & Mani, S. (2013). Empirical performance analysis of HPC benchmarks across variations in cloud computing. *International Journal of Cloud Applications and Computing*, *3*(1), 13–26. doi:10.4018/ijcac.2013010102

Ahuja, S. P., & Rolli, A. C. (2011). Survey of the state-of-the-art of cloud computing. *International Journal of Cloud Applications and Computing*, *1*(4), 34–43. doi:10.4018/ijcac.2011100103

Ahuja, S. P., & Rolli, A. C. (2013). Survey of the state-of-the-art of cloud computing. In S. Aljawarneh (Ed.), *Cloud computing advancements in design, implementation, and technologies* (pp. 252–262). Hershey, PA: IGI Global. doi:10.4018/978-1-4666-1879-4.ch018

Ahuja, S. P., & Sridharan, S. (2012). Performance evaluation of hypervisors for cloud computing. *International Journal of Cloud Applications and Computing*, *2*(3), 26–67. doi:10.4018/ijcac.2012070102

Akyuz, G. A., & Rehan, M. (2013). A generic, cloud-based representation for supply chains (SC's). *International Journal of Cloud Applications and Computing*, *3*(2), 12–20. doi:10.4018/ijcac.2013040102

Al-Aqrabi, H., & Liu, L. (2013). IT security and governance compliant service oriented computing in cloud computing environments. In X. Yang & L. Liu (Eds.), *Principles, methodologies, and service-oriented approaches for cloud computing* (pp. 143–163). Hershey, PA: IGI Global. doi:10.4018/978-1-4666-2854-0.ch006

Al-Zoube, M., & Wyne, M. F. (2012). Building integrated e-learning environment using cloud services and social networking sites. In Q. Jin (Ed.), *Intelligent learning systems and advancements in computer-aided instruction: Emerging studies* (pp. 214–233). Hershey, PA: IGI Global. doi:10.4018/978-1-61350-483-3.ch013

Alam, N., & Karmakar, R. (2014). Cloud computing and its application to information centre. In S. Dhamdhere (Ed.), *Cloud computing and virtualization technologies in libraries* (pp. 63–76). Hershey, PA: IGI Global. doi:10.4018/978-1-4666-4631-5.ch004

Alhaj, A., Aljawarneh, S., Masadeh, S., & Abu-Taieh, E. (2013). A secure data transmission mechanism for cloud outsourced data. *International Journal of Cloud Applications and Computing*, *3*(1), 34–43. doi:10.4018/ijcac.2013010104

Alharbi, S. T. (2012). Users' acceptance of cloud computing in Saudi Arabia: An extension of technology acceptance model. *International Journal of Cloud Applications and Computing*, *2*(2), 1–11. doi:10.4018/ijcac.2012040101

Ali, S. S., & Khan, M. N. (2013). ICT infrastructure framework for microfinance institutions and banks in Pakistan: An optimized approach. *International Journal of Online Marketing*, *3*(2), 75–86. doi:10.4018/ijom.2013040105

Aljawarneh, S. (2011). Cloud security engineering: Avoiding security threats the right way. *International Journal of Cloud Applications and Computing*, *1*(2), 64–70. doi:10.4018/ijcac.2011040105

Aljawarneh, S. (2013). Cloud security engineering: Avoiding security threats the right way. In S. Aljawarneh (Ed.), *Cloud computing advancements in design, implementation, and technologies* (pp. 147–153). Hershey, PA: IGI Global. doi:10.4018/978-1-4666-1879-4.ch010

Alshattnawi, S. (2013). Utilizing cloud computing in developing a mobile location-aware tourist guide system. *International Journal of Advanced Pervasive and Ubiquitous Computing*, *5*(2), 9–18. doi:10.4018/japuc.2013040102

Alsmadi, I. (2013). Software development methodologies for cloud computing. In K. Buragga & N. Zaman (Eds.), *Software development techniques for constructive information systems design* (pp. 110–117). Hershey, PA: IGI Global. doi:10.4018/978-1-4666-3679-8.ch006

Anand, V. (2013). Survivable mapping of virtual networks onto a shared substrate network. In X. Yang & L. Liu (Eds.), *Principles, methodologies, and service-oriented approaches for cloud computing* (pp. 325–343). Hershey, PA: IGI Global. doi:10.4018/978-1-4666-2854-0.ch014

Antonova, A. (2013). Green, sustainable, or clean: What type of IT/IS technologies will we need in the future? In P. Ordóñez de Pablos (Ed.), *Green technologies and business practices: An IT approach* (pp. 151–162). Hershey, PA: IGI Global. doi:10.4018/978-1-4666-1972-2.ch008

Ardissono, L., Bosio, G., Goy, A., Petrone, G., Segnan, M., & Torretta, F. (2011). Collaboration support for activity management in a personal cloud environment. *International Journal of Distributed Systems and Technologies*, *2*(4), 30–43. doi:10.4018/jdst.2011100103

Ardissono, L., Bosio, G., Goy, A., Petrone, G., Segnan, M., & Torretta, F. (2013). Collaboration support for activity management in a personal cloud environment. In N. Bessis (Ed.), *Development of distributed systems from design to application and maintenance* (pp. 199–212). Hershey, PA: IGI Global. doi:10.4018/978-1-4666-2647-8.ch012

Argiolas, M., Atzori, M., Dessì, N., & Pes, B. (2012). Dataspaces enhancing decision support systems in clouds. *International Journal of Web Portals*, *4*(2), 35–55. doi:10.4018/jwp.2012040103

Arinze, B., & Anandarajan, M. (2012). Factors that determine the adoption of cloud computing: A global perspective. In M. Tavana (Ed.), *Enterprise Information Systems and Advancing Business Solutions: Emerging Models* (pp. 210–223). Hershey, PA: IGI Global. doi:10.4018/978-1-4666-1761-2.ch012

Arinze, B., & Sylla, C. (2012). Conducting research in the cloud. In L. Chao (Ed.), *Cloud computing for teaching and learning: Strategies for design and implementation* (pp. 50–63). Hershey, PA: IGI Global. doi:10.4018/978-1-4666-0957-0.ch004

Arshad, J., Townend, P., & Xu, J. (2011). An abstract model for integrated intrusion detection and severity analysis for clouds. *International Journal of Cloud Applications and Computing*, *1*(1), 1–16. doi:10.4018/ijcac.2011010101

Arshad, J., Townend, P., & Xu, J. (2013). An abstract model for integrated intrusion detection and severity analysis for clouds. In S. Aljawarneh (Ed.), *Cloud computing advancements in design, implementation, and technologies* (pp. 1–17). Hershey, PA: IGI Global. doi:10.4018/978-1-4666-1879-4.ch001

Arshad, J., Townend, P., Xu, J., & Jie, W. (2012). Cloud computing security: Opportunities and pitfalls. *International Journal of Grid and High Performance Computing*, *4*(1), 52–66. doi:10.4018/jghpc.2012010104

Baars, T., & Spruit, M. (2012). Designing a secure cloud architecture: The SeCA model. *International Journal of Information Security and Privacy*, *6*(1), 14–32. doi:10.4018/jisp.2012010102

Bai, X., Gao, J. Z., & Tsai, W. (2013). Cloud scalability measurement and testing. In S. Tilley & T. Parveen (Eds.), *Software testing in the cloud: Perspectives on an emerging discipline* (pp. 356–381). Hershey, PA: IGI Global. doi:10.4018/978-1-4666-2536-5.ch017

Baldini, G., & Stirparo, P. (2014). A cognitive access framework for security and privacy protection in mobile cloud computing. In J. Rodrigues, K. Lin, & J. Lloret (Eds.), *Mobile networks and cloud computing convergence for progressive services and applications* (pp. 92–117). Hershey, PA: IGI Global. doi:10.4018/978-1-4666-4781-7.ch006

Balduf, S., Balke, T., & Eymann, T. (2012). Cultural differences in managing cloud computing service level agreements. In *Grid and cloud computing: Concepts, methodologies, tools and applications* (pp. 1237–1263). Hershey, PA: IGI Global. doi:10.4018/978-1-4666-0879-5.ch512

Banerjee, S., Sing, T. Y., Chowdhury, A. R., & Anwar, H. (2013). Motivations to adopt green ICT: A tale of two organizations. *International Journal of Green Computing*, *4*(2), 1–11. doi:10.4018/jgc.2013070101

Barreto, J., Di Sanzo, P., Palmieri, R., & Romano, P. (2013). Cloud-TM: An elastic, self-tuning transactional store for the cloud. In D. Kyriazis, A. Voulodimos, S. Gogouvitis, & T. Varvarigou (Eds.), *Data intensive storage services for cloud environments* (pp. 192–224). Hershey, PA: IGI Global. doi:10.4018/978-1-4666-3934-8.ch013

Belalem, G., & Limam, S. (2011). Fault tolerant architecture to cloud computing using adaptive checkpoint. *International Journal of Cloud Applications and Computing*, *1*(4), 60–69. doi:10.4018/ijcac.2011100105

Belalem, G., & Limam, S. (2013). Fault tolerant architecture to cloud computing using adaptive checkpoint. In S. Aljawarneh (Ed.), *Cloud computing advancements in design, implementation, and technologies* (pp. 280–289). Hershey, PA: IGI Global. doi:10.4018/978-1-4666-1879-4.ch020

Ben Belgacem, M., Abdennadher, N., & Niinimaki, M. (2012). Virtual EZ grid: A volunteer computing infrastructure for scientific medical applications. *International Journal of Handheld Computing Research*, *3*(1), 74–85. doi:10.4018/jhcr.2012010105

Bhatt, S., Chaudhary, S., & Bhise, M. (2013). Migration of data between cloud and non-cloud datastores. In A. Ionita, M. Litoiu, & G. Lewis (Eds.), *Migrating legacy applications: Challenges in service oriented architecture and cloud computing environments* (pp. 206–225). Hershey, PA: IGI Global. doi:10.4018/978-1-4666-2488-7.ch009

Biancofiore, G., & Leone, S. (2014). Google apps as a cloud computing solution in Italian municipalities: Technological features and implications. In S. Leone (Ed.), *Synergic integration of formal and informal e-learning environments for adult lifelong learners* (pp. 244–274). Hershey, PA: IGI Global. doi:10.4018/978-1-4666-4655-1.ch012

Bibi, S., Katsaros, D., & Bozanis, P. (2012). How to choose the right cloud. In *Grid and cloud computing: Concepts, methodologies, tools and applications* (pp. 1530–1552). Hershey, PA: IGI Global. doi:10.4018/978-1-4666-0879-5.ch701

Bibi, S., Katsaros, D., & Bozanis, P. (2012). How to choose the right cloud. In X. Liu & Y. Li (Eds.), *Advanced design approaches to emerging software systems: Principles, methodologies and tools* (pp. 219–240). Hershey, PA: IGI Global. doi:10.4018/978-1-60960-735-7.ch010

Bitam, S., Batouche, M., & Talbi, E. (2012). A bees life algorithm for cloud computing services selection. In S. Ali, N. Abbadeni, & M. Batouche (Eds.), *Multidisciplinary computational intelligence techniques: Applications in business, engineering, and medicine* (pp. 31–46). Hershey, PA: IGI Global. doi:10.4018/978-1-4666-1830-5.ch003

Bittencourt, L. F., Madeira, E. R., & da Fonseca, N. L. (2014). Communication aspects of resource management in hybrid clouds. In H. Mouftah & B. Kantarci (Eds.), *Communication infrastructures for cloud computing* (pp. 409–433). Hershey, PA: IGI Global. doi:10.4018/978-1-4666-4522-6.ch018

Bonelli, L., Giudicianni, L., Immediata, A., & Luzzi, A. (2013). Compliance in the cloud. In D. Kyriazis, A. Voulodimos, S. Gogouvitis, & T. Varvarigou (Eds.), *Data intensive storage services for cloud environments* (pp. 109–131). Hershey, PA: IGI Global. doi:10.4018/978-1-4666-3934-8.ch008

Boniface, M., Nasser, B., Surridge, M., & Oliveros, E. (2012). Securing real-time interactive applications in federated clouds. In *Grid and cloud computing: Concepts, methodologies, tools and applications* (pp. 1822–1835). Hershey, PA: IGI Global. doi:10.4018/978-1-4666-0879-5.ch806

Boukhobza, J. (2013). Flashing in the cloud: Shedding some light on NAND flash memory storage systems. In D. Kyriazis, A. Voulodimos, S. Gogouvitis, & T. Varvarigou (Eds.), *Data intensive storage services for cloud environments* (pp. 241–266). Hershey, PA: IGI Global. doi:10.4018/978-1-4666-3934-8.ch015

Bracci, F., Corradi, A., & Foschini, L. (2014). Cloud standards: Security and interoperability issues. In H. Mouftah & B. Kantarci (Eds.), *Communication infrastructures for cloud computing* (pp. 465–495). Hershey, PA: IGI Global. doi:10.4018/978-1-4666-4522-6.ch020

Brown, A. W. (2013). Experiences with cloud technology to realize software testing factories. In S. Tilley & T. Parveen (Eds.), *Software testing in the cloud: Perspectives on an emerging discipline* (pp. 1–27). Hershey, PA: IGI Global. doi:10.4018/978-1-4666-2536-5.ch001

Calcavecchia, N. M., Celesti, A., & Di Nitto, E. (2012). Understanding decentralized and dynamic brokerage in federated cloud environments. In M. Villari, I. Brandic, & F. Tusa (Eds.), *Achieving federated and self-manageable cloud infrastructures: Theory and practice* (pp. 36–56). Hershey, PA: IGI Global. doi:10.4018/978-1-4666-1631-8.ch003

Calero, J. M., König, B., & Kirschnick, J. (2012). Cross-layer monitoring in cloud computing. In H. Rashvand & Y. Kavian (Eds.), *Using cross-layer techniques for communication systems* (pp. 328–348). Hershey, PA: IGI Global. doi:10.4018/978-1-4666-0960-0.ch014

Cardellini, V., Casalicchio, E., & Silvestri, L. (2012). Service level provisioning for cloud-based applications service level provisioning for cloud-based applications. In A. Pathan, M. Pathan, & H. Lee (Eds.), *Advancements in distributed computing and internet technologies: Trends and issues* (pp. 363–385). Hershey, PA: Information Science Publishing. doi:10.4018/978-1-61350-110-8.ch017

Cardellini, V., Casalicchio, E., & Silvestri, L. (2012). Service level provisioning for cloud-based applications service level provisioning for cloud-based applications. In *Grid and cloud computing: Concepts, methodologies, tools and applications* (pp. 1479–1500). Hershey, PA: IGI Global. doi:10.4018/978-1-4666-0879-5.ch611

Carlin, S., & Curran, K. (2013). Cloud computing security. In K. Curran (Ed.), *Pervasive and ubiquitous technology innovations for ambient intelligence environments* (pp. 12–17). Hershey, PA: IGI Global. doi:10.4018/978-1-4666-2041-4.ch002

Carlton, G. H., & Zhou, H. (2011). A survey of cloud computing challenges from a digital forensics perspective. *International Journal of Interdisciplinary Telecommunications and Networking*, *3*(4), 1–16. doi:10.4018/jitn.2011100101

Carlton, G. H., & Zhou, H. (2012). A survey of cloud computing challenges from a digital forensics perspective. In *Grid and cloud computing: Concepts, methodologies, tools and applications* (pp. 1221–1236). Hershey, PA: IGI Global. doi:10.4018/978-1-4666-0879-5.ch511

Carlton, G. H., & Zhou, H. (2013). A survey of cloud computing challenges from a digital forensics perspective. In M. Bartolacci & S. Powell (Eds.), *Advancements and innovations in wireless communications and network technologies* (pp. 213–228). Hershey, PA: IGI Global. doi:10.4018/978-1-4666-2154-1.ch016

Carpen-Amarie, A., Costan, A., Leordeanu, C., Basescu, C., & Antoniu, G. (2012). Towards a generic security framework for cloud data management environments. *International Journal of Distributed Systems and Technologies*, *3*(1), 17–34. doi:10.4018/jdst.2012010102

Casola, V., Cuomo, A., Villano, U., & Rak, M. (2012). Access control in federated clouds: The cloud-grid case study. In M. Villari, I. Brandic, & F. Tusa (Eds.), *Achieving Federated and Self-Manageable Cloud Infrastructures: Theory and Practice* (pp. 395–417). Hershey, PA: IGI Global. doi:10.4018/978-1-4666-1631-8.ch020

Casola, V., Cuomo, A., Villano, U., & Rak, M. (2013). Access control in federated clouds: The cloudgrid case study. In *IT policy and ethics: Concepts, methodologies, tools, and applications* (pp. 148–169). Hershey, PA: IGI Global. doi:10.4018/978-1-4666-2919-6.ch008

Celesti, A., Tusa, F., & Villari, M. (2012). Toward cloud federation: Concepts and challenges. In M. Villari, I. Brandic, & F. Tusa (Eds.), *Achieving federated and self-manageable cloud infrastructures: Theory and practice* (pp. 1–17). Hershey, PA: IGI Global. doi:10.4018/978-1-4666-1631-8.ch001

Chaka, C. (2013). Virtualization and cloud computing: Business models in the virtual cloud. In A. Loo (Ed.), *Distributed computing innovations for business, engineering, and science* (pp. 176–190). Hershey, PA: IGI Global. doi:10.4018/978-1-4666-2533-4.ch009

Chang, J. (2011). A framework for analysing the impact of cloud computing on local government in the UK. *International Journal of Cloud Applications and Computing, 1*(4), 25–33. doi:10.4018/ijcac.2011100102

Chang, J. (2013). A framework for analysing the impact of cloud computing on local government in the UK. In S. Aljawarneh (Ed.), *Cloud computing advancements in design, implementation, and technologies* (pp. 243–251). Hershey, PA: IGI Global. doi:10.4018/978-1-4666-1879-4.ch017

Chang, J., & Johnston, M. (2012). Cloud computing in local government: From the perspective of four London borough councils. *International Journal of Cloud Applications and Computing, 2*(4), 1–15. doi:10.4018/ijcac.2012100101

Chang, K., & Wang, K. (2012). Efficient support of streaming videos through patching proxies in the cloud. *International Journal of Grid and High Performance Computing, 4*(4), 22–36. doi:10.4018/jghpc.2012100102

Chang, R., Liao, C., & Liu, C. (2013). Choosing clouds for an enterprise: Modeling and evaluation. *International Journal of E-Entrepreneurship and Innovation, 4*(2), 38–53. doi:10.4018/ijeei.2013040103

Chang, V., De Roure, D., Wills, G., & Walters, R. J. (2011). Case studies and organisational sustainability modelling presented by cloud computing business framework. *International Journal of Web Services Research, 8*(3), 26–53. doi:10.4018/JWSR.2011070102

Chang, V., Li, C., De Roure, D., Wills, G., Walters, R. J., & Chee, C. (2011). The financial clouds review. *International Journal of Cloud Applications and Computing, 1*(2), 41–63. doi:10.4018/ijcac.2011040104

Chang, V., Li, C., De Roure, D., Wills, G., Walters, R. J., & Chee, C. (2013). The financial clouds review. In S. Aljawarneh (Ed.), *Cloud computing advancements in design, implementation, and technologies* (pp. 125–146). Hershey, PA: IGI Global. doi:10.4018/978-1-4666-1879-4.ch009

Chang, V., Walters, R. J., & Wills, G. (2012). Business integration as a service. *International Journal of Cloud Applications and Computing, 2*(1), 16–40. doi:10.4018/ijcac.2012010102

Chang, V., & Wills, G. (2013). A University of Greenwich case study of cloud computing: Education as a service. In D. Graham, I. Manikas, & D. Folinas (Eds.), *E-logistics and e-supply chain management: Applications for evolving business* (pp. 232–253). Hershey, PA: IGI Global. doi:10.4018/978-1-4666-3914-0.ch013

Chang, V., Wills, G., Walters, R. J., & Currie, W. (2012). Towards a structured cloud ROI: The University of Southampton cost-saving and user satisfaction case studies. In W. Hu & N. Kaabouch (Eds.), *Sustainable ICTs and management systems for green computing* (pp. 179–200). Hershey, PA: IGI Global. doi:10.4018/978-1-4666-1839-8.ch008

Chang, Y., Lee, Y., Juang, T., & Yen, J. (2013). Cost evaluation on building and operating cloud platform. *International Journal of Grid and High Performance Computing, 5*(2), 43–53. doi:10.4018/jghpc.2013040103

Chao, L. (2012). Cloud computing solution for internet based teaching and learning. In L. Chao (Ed.), *Cloud computing for teaching and learning: Strategies for design and implementation* (pp. 210–235). Hershey, PA: IGI Global. doi:10.4018/978-1-4666-0957-0.ch015

Chao, L. (2012). Overview of cloud computing and its application in e-learning. In L. Chao (Ed.), *Cloud computing for teaching and learning: Strategies for design and implementation* (pp. 1–16). Hershey, PA: IGI Global. doi:10.4018/978-1-4666-0957-0.ch001

Chauhan, S., Raman, A., & Singh, N. (2013). A comparative cost analysis of on premises IT infrastructure and cloud-based email services in an Indian business school. *International Journal of Cloud Applications and Computing*, *3*(2), 21–34. doi:10.4018/ijcac.2013040103

Chen, C., Chao, H., Wu, T., Fan, C., Chen, J., Chen, Y., & Hsu, J. (2011). IoT-IMS communication platform for future internet. *International Journal of Adaptive, Resilient and Autonomic Systems*, *2*(4), 74–94. doi:10.4018/jaras.2011100105

Chen, C., Chao, H., Wu, T., Fan, C., Chen, J., Chen, Y., & Hsu, J. (2013). IoT-IMS communication platform for future internet. In V. De Florio (Ed.), *Innovations and approaches for resilient and adaptive systems* (pp. 68–86). Hershey, PA: IGI Global. doi:10.4018/978-1-4666-2056-8.ch004

Chen, C. C. (2013). Cloud computing in case-based pedagogy: An information systems success perspective. *International Journal of Dependable and Trustworthy Information Systems*, *2*(3), 1–16. doi:10.4018/jdtis.2011070101

Cheney, A. W., Riedl, R. E., Sanders, R., & Tashner, J. H. (2012). The new company water cooler: Use of 3D virtual immersive worlds to promote networking and professional learning in organizations. In Organizational learning and knowledge: Concepts, methodologies, tools and applications (pp. 2848-2861). Hershey, PA: IGI Global. doi:10.4018/978-1-60960-783-8.ch801

Chiang, C., & Yu, S. (2013). Cloud-enabled software testing based on program understanding. In S. Tilley & T. Parveen (Eds.), *Software testing in the cloud: Perspectives on an emerging discipline* (pp. 54–67). Hershey, PA: IGI Global. doi:10.4018/978-1-4666-2536-5.ch003

Chou, Y., & Oetting, J. (2011). Risk assessment for cloud-based IT systems. *International Journal of Grid and High Performance Computing*, *3*(2), 1–13. doi:10.4018/jghpc.2011040101

Chou, Y., & Oetting, J. (2012). Risk assessment for cloud-based IT systems. In *Grid and cloud computing: Concepts, methodologies, tools and applications* (pp. 272–285). Hershey, PA: IGI Global. doi:10.4018/978-1-4666-0879-5.ch113

Chou, Y., & Oetting, J. (2013). Risk assessment for cloud-based IT systems. In E. Udoh (Ed.), *Applications and developments in grid, cloud, and high performance computing* (pp. 1–14). Hershey, PA: IGI Global. doi:10.4018/978-1-4666-2065-0.ch001

Cohen, F. (2013). Challenges to digital forensic evidence in the cloud. In K. Ruan (Ed.), *Cybercrime and cloud forensics: Applications for investigation processes* (pp. 59–78). Hershey, PA: IGI Global. doi:10.4018/978-1-4666-2662-1.ch003

Cossu, R., Di Giulio, C., Brito, F., & Petcu, D. (2013). Cloud computing for earth observation. In D. Kyriazis, A. Voulodimos, S. Gogouvitis, & T. Varvarigou (Eds.), *Data intensive storage services for cloud environments* (pp. 166–191). Hershey, PA: IGI Global. doi:10.4018/978-1-4666-3934-8.ch012

Costa, J. E., & Rodrigues, J. J. (2014). Mobile cloud computing: Technologies, services, and applications. In J. Rodrigues, K. Lin, & J. Lloret (Eds.), *Mobile networks and cloud computing convergence for progressive services and applications* (pp. 1–17). Hershey, PA: IGI Global. doi:10.4018/978-1-4666-4781-7.ch001

Creaner, G., & Pahl, C. (2013). Flexible coordination techniques for dynamic cloud service collaboration. In G. Ortiz & J. Cubo (Eds.), *Adaptive web services for modular and reusable software development: Tactics and solutions* (pp. 239–252). Hershey, PA: IGI Global. doi:10.4018/978-1-4666-2089-6.ch009

Crosbie, M. (2013). Hack the cloud: Ethical hacking and cloud forensics. In K. Ruan (Ed.), *Cybercrime and cloud forensics: Applications for investigation processes* (pp. 42–58). Hershey, PA: IGI Global. doi:10.4018/978-1-4666-2662-1.ch002

Curran, K., Carlin, S., & Adams, M. (2012). Security issues in cloud computing. In L. Chao (Ed.), *Cloud computing for teaching and learning: Strategies for design and implementation* (pp. 200–208). Hershey, PA: IGI Global. doi:10.4018/978-1-4666-0957-0.ch014

Dahbur, K., & Mohammad, B. (2011). Toward understanding the challenges and countermeasures in computer anti-forensics. *International Journal of Cloud Applications and Computing*, *1*(3), 22–35. doi:10.4018/ijcac.2011070103

Dahbur, K., Mohammad, B., & Tarakji, A. B. (2011). Security issues in cloud computing: A survey of risks, threats and vulnerabilities. *International Journal of Cloud Applications and Computing*, *1*(3), 1–11. doi:10.4018/ijcac.2011070101

Dahbur, K., Mohammad, B., & Tarakji, A. B. (2012). Security issues in cloud computing: A survey of risks, threats and vulnerabilities. In *Grid and cloud computing: Concepts, methodologies, tools and applications* (pp. 1644–1655). Hershey, PA: IGI Global. doi:10.4018/978-1-4666-0879-5.ch707

Dahbur, K., Mohammad, B., & Tarakji, A. B. (2013). Security issues in cloud computing: A survey of risks, threats and vulnerabilities. In S. Aljawarneh (Ed.), *Cloud computing advancements in design, implementation, and technologies* (pp. 154–165). Hershey, PA: IGI Global. doi:10.4018/978-1-4666-1879-4.ch011

Daim, T., Britton, M., Subramanian, G., Brenden, R., & Intarode, N. (2012). Adopting and integrating cloud computing. In E. Eyob & E. Tetteh (Eds.), *Customer-oriented global supply chains: Concepts for effective management* (pp. 175–197). Hershey, PA: IGI Global. doi:10.4018/978-1-4666-0246-5.ch011

Davis, M., & Sedsman, A. (2012). Grey areas: The legal dimensions of cloud computing. In C. Li & A. Ho (Eds.), *Crime prevention technologies and applications for advancing criminal investigation* (pp. 263–273). Hershey, PA: IGI Global. doi:10.4018/978-1-4666-1758-2.ch017

De Coster, R., & Albesher, A. (2013). The development of mobile service applications for consumers and intelligent networks. In I. Lee (Ed.), *Mobile services industries, technologies, and applications in the global economy* (pp. 273–289). Hershey, PA: IGI Global. doi:10.4018/978-1-4666-1981-4.ch017

De Filippi, P. (2014). Ubiquitous computing in the cloud: User empowerment vs. user obsequity. In J. Pelet & P. Papadopoulou (Eds.), *User behavior in ubiquitous online environments* (pp. 44–63). Hershey, PA: IGI Global. doi:10.4018/978-1-4666-4566-0.ch003

De Silva, S. (2013). Key legal issues with cloud computing: A UK law perspective. In A. Bento & A. Aggarwal (Eds.), *Cloud computing service and deployment models: Layers and management* (pp. 242–256). Hershey, PA: IGI Global. doi:10.4018/978-1-4666-2187-9.ch013

Deed, C., & Cragg, P. (2013). Business impacts of cloud computing. In A. Bento & A. Aggarwal (Eds.), *Cloud computing service and deployment models: Layers and management* (pp. 274–288). Hershey, PA: IGI Global. doi:10.4018/978-1-4666-2187-9.ch015

Deng, M., Petkovic, M., Nalin, M., & Baroni, I. (2013). Home healthcare in cloud computing. In M. Cruz-Cunha, I. Miranda, & P. Gonçalves (Eds.), *Handbook of research on ICTs and management systems for improving efficiency in healthcare and social care* (pp. 614–634). Hershey, PA: IGI Global. doi:10.4018/978-1-4666-3990-4.ch032

Desai, A. M., & Mock, K. (2013). Security in cloud computing. In A. Bento & A. Aggarwal (Eds.), *Cloud computing service and deployment models: Layers and management* (pp. 208–221). Hershey, PA: IGI Global. doi:10.4018/978-1-4666-2187-9.ch011

Deshpande, R. M., Patle, B. V., & Bhoskar, R. D. (2014). Planning and implementation of cloud computing in NIT's in India: Special reference to VNIT. In S. Dhamdhere (Ed.), *Cloud computing and virtualization technologies in libraries* (pp. 90–106). Hershey, PA: IGI Global. doi:10.4018/978-1-4666-4631-5.ch006

Dhamdhere, S. N., & Lihitkar, R. (2014). The university cloud library model and the role of the cloud librarian. In S. Dhamdhere (Ed.), *Cloud computing and virtualization technologies in libraries* (pp. 150–161). Hershey, PA: IGI Global. doi:10.4018/978-1-4666-4631-5.ch009

Di Martino, S., Ferrucci, F., Maggio, V., & Sarro, F. (2013). Towards migrating genetic algorithms for test data generation to the cloud. In S. Tilley & T. Parveen (Eds.), *Software testing in the cloud: Perspectives on an emerging discipline* (pp. 113–135). Hershey, PA: IGI Global. doi:10.4018/978-1-4666-2536-5.ch006

Di Sano, M., Di Stefano, A., Morana, G., & Zito, D. (2013). FSaaS: Configuring policies for managing shared files among cooperating, distributed applications. *International Journal of Web Portals, 5*(1), 1–14. doi:10.4018/jwp.2013010101

Dippl, S., Jaeger, M. C., Luhn, A., Shulman-Peleg, A., & Vernik, G. (2013). Towards federation and interoperability of cloud storage systems. In D. Kyriazis, A. Voulodimos, S. Gogouvitis, & T. Varvarigou (Eds.), *Data intensive storage services for cloud environments* (pp. 60–71). Hershey, PA: IGI Global. doi:10.4018/978-1-4666-3934-8.ch005

Distefano, S., & Puliafito, A. (2012). The cloud@home volunteer and interoperable cloud through the future internet. In M. Villari, I. Brandic, & F. Tusa (Eds.), *Achieving federated and self-manageable cloud infrastructures: Theory and practice* (pp. 79–96). Hershey, PA: IGI Global. doi:10.4018/978-1-4666-1631-8.ch005

Djoleto, W. (2013). Cloud computing and ecommerce or ebusiness: "The now it way" – An overview. In *Electronic commerce and organizational leadership: perspectives and methodologies* (pp. 239–254). Hershey, PA: IGI Global. doi:10.4018/978-1-4666-2982-0.ch010

Dollmann, T. J., Loos, P., Fellmann, M., Thomas, O., Hoheisel, A., Katranuschkov, P., & Scherer, R. (2011). Design and usage of a process-centric collaboration methodology for virtual organizations in hybrid environments. *International Journal of Intelligent Information Technologies*, *7*(1), 45–64. doi:10.4018/jiit.2011010104

Dollmann, T. J., Loos, P., Fellmann, M., Thomas, O., Hoheisel, A., Katranuschkov, P., & Scherer, R. (2013). Design and usage of a process-centric collaboration methodology for virtual organizations in hybrid environments. In V. Sugumaran (Ed.), *Organizational efficiency through intelligent information technologies* (pp. 45–64). Hershey, PA: IGI Global. doi:10.4018/978-1-4666-2047-6.ch004

Dreher, P., & Vouk, M. (2012). Utilizing open source cloud computing environments to provide cost effective support for university education and research. In L. Chao (Ed.), *Cloud computing for teaching and learning: Strategies for design and implementation* (pp. 32–49). Hershey, PA: IGI Global. doi:10.4018/978-1-4666-0957-0.ch003

Drum, D., Becker, D., & Fish, M. (2013). Technology adoption in troubled times: A cloud computing case study. *Journal of Cases on Information Technology*, *15*(2), 57–71. doi:10.4018/jcit.2013040104

Dunaway, D. M. (2013). Creating virtual collaborative learning experiences for aspiring teachers. In R. Hartshorne, T. Heafner, & T. Petty (Eds.), *Teacher education programs and online learning tools: Innovations in teacher preparation* (pp. 167–180). Hershey, PA: IGI Global. doi:10.4018/978-1-4666-1906-7.ch009

Dykstra, J. (2013). Seizing electronic evidence from cloud computing environments. In K. Ruan (Ed.), *Cybercrime and cloud forensics: Applications for investigation processes* (pp. 156–185). Hershey, PA: IGI Global. doi:10.4018/978-1-4666-2662-1.ch007

El-Refaey, M., & Rimal, B. P. (2012). Grid, SOA and cloud computing: On-demand computing models. In *Grid and cloud computing: Concepts, methodologies, tools and applications* (pp. 12–51). Hershey, PA: IGI Global. doi:10.4018/978-1-4666-0879-5.ch102

El-Refaey, M., & Rimal, B. P. (2012). Grid, SOA and cloud computing: On-demand computing models. In N. Preve (Ed.), *Computational and data grids: Principles, applications and design* (pp. 45–85). Hershey, PA: IGI Global. doi:10.4018/978-1-61350-113-9.ch003

Elnaffar, S., Maamar, Z., & Sheng, Q. Z. (2013). When clouds start socializing: The sky model. *International Journal of E-Business Research*, *9*(2), 1–7. doi:10.4018/jebr.2013040101

Elwood, S., & Keengwe, J. (2012). Microbursts: A design format for mobile cloud computing. *International Journal of Information and Communication Technology Education*, *8*(2), 102–110. doi:10.4018/jicte.2012040109

Emeakaroha, V. C., Netto, M. A., Calheiros, R. N., & De Rose, C. A. (2012). Achieving flexible SLA and resource management in clouds. In M. Villari, I. Brandic, & F. Tusa (Eds.), *Achieving federated and self-manageable cloud infrastructures: Theory and practice* (pp. 266–287). Hershey, PA: IGI Global. doi:10.4018/978-1-4666-1631-8.ch014

Etro, F. (2013). The economics of cloud computing. In A. Bento & A. Aggarwal (Eds.), *Cloud computing service and deployment models: Layers and management* (pp. 296–309). Hershey, PA: IGI Global. doi:10.4018/978-1-4666-2187-9.ch017

Ezugwu, A. E., Buhari, S. M., & Junaidu, S. B. (2013). Virtual machine allocation in cloud computing environment. *International Journal of Cloud Applications and Computing*, *3*(2), 47–60. doi:10.4018/ijcac.2013040105

Fauzi, A. H., & Taylor, H. (2013). Secure community trust stores for peer-to-peer e-commerce applications using cloud services. *International Journal of E-Entrepreneurship and Innovation*, *4*(1), 1–15. doi:10.4018/jeei.2013010101

Ferguson-Boucher, K., & Endicott-Popovsky, B. (2013). Forensic readiness in the cloud (FRC): Integrating records management and digital forensics. In K. Ruan (Ed.), *Cybercrime and cloud forensics: Applications for investigation processes* (pp. 105–128). Hershey, PA: IGI Global. doi:10.4018/978-1-4666-2662-1.ch005

Ferraro de Souza, R., Westphall, C. B., dos Santos, D. R., & Westphall, C. M. (2013). A review of PACS on cloud for archiving secure medical images. *International Journal of Privacy and Health Information Management*, *1*(1), 53–62. doi:10.4018/ijphim.2013010104

Firdhous, M., Hassan, S., & Ghazali, O. (2013). Statistically enhanced multi-dimensional trust computing mechanism for cloud computing. *International Journal of Mobile Computing and Multimedia Communications*, *5*(2), 1–17. doi:10.4018/jmcmc.2013040101

Formisano, C., Bonelli, L., Balraj, K. R., & Shulman-Peleg, A. (2013). Cloud access control mechanisms. In D. Kyriazis, A. Voulodimos, S. Gogouvitis, & T. Varvarigou (Eds.), *Data intensive storage services for cloud environments* (pp. 94–108). Hershey, PA: IGI Global. doi:10.4018/978-1-4666-3934-8.ch007

Frank, H., & Mesentean, S. (2012). Efficient communication interfaces for distributed energy resources. In E. Udoh (Ed.), *Evolving developments in grid and cloud computing: Advancing research* (pp. 185–196). Hershey, PA: IGI Global. doi:10.4018/978-1-4666-0056-0.ch013

Gallina, B., & Guelfi, N. (2012). Reusing transaction models for dependable cloud computing. In H. Yang & X. Liu (Eds.), *Software reuse in the emerging cloud computing era* (pp. 248–277). Hershey, PA: IGI Global. doi:10.4018/978-1-4666-0897-9.ch011

Garofalo, D. A. (2013). Empires of the future: Libraries, technology, and the academic environment. In E. Iglesias (Ed.), *Robots in academic libraries: Advancements in library automation* (pp. 180–206). Hershey, PA: IGI Global. doi:10.4018/978-1-4666-3938-6.ch010

Gebremeskel, G. B., He, Z., & Jing, X. (2013). Semantic integrating for intelligent cloud data mining platform and cloud based business intelligence for optimization of mobile social networks. In V. Bhatnagar (Ed.), *Data mining in dynamic social networks and fuzzy systems* (pp. 173–211). Hershey, PA: IGI Global. doi:10.4018/978-1-4666-4213-3.ch009

Gentleman, W. M. (2013). Using the cloud for testing NOT adjunct to development. In S. Tilley & T. Parveen (Eds.), *Software testing in the cloud: Perspectives on an emerging discipline* (pp. 216–230). Hershey, PA: IGI Global. doi:10.4018/978-1-4666-2536-5.ch010

Ghafoor, K. Z., Mohammed, M. A., Abu Bakar, K., Sadiq, A. S., & Lloret, J. (2014). Vehicular cloud computing: Trends and challenges. In J. Rodrigues, K. Lin, & J. Lloret (Eds.), *Mobile networks and cloud computing convergence for progressive services and applications* (pp. 262–274). Hershey, PA: IGI Global. doi:10.4018/978-1-4666-4781-7.ch014

Giannakaki, M. (2012). The "right to be forgotten" in the era of social media and cloud computing. In C. Akrivopoulou & N. Garipidis (Eds.), *Human rights and risks in the digital era: Globalization and the effects of information technologies* (pp. 10–24). Hershey, PA: IGI Global. doi:10.4018/978-1-4666-0891-7.ch002

Gillam, L., Li, B., & O'Loughlin, J. (2012). Teaching clouds: Lessons taught and lessons learnt. In L. Chao (Ed.), *Cloud computing for teaching and learning: Strategies for design and implementation* (pp. 82–94). Hershey, PA: IGI Global. doi:10.4018/978-1-4666-0957-0.ch006

Gonsowski, D. (2013). Compliance in the cloud and the implications on electronic discovery. In K. Ruan (Ed.), *Cybercrime and cloud forensics: Applications for investigation processes* (pp. 230–250). Hershey, PA: IGI Global. doi:10.4018/978-1-4666-2662-1.ch009

Gonzalez-Sanchez, J., Conley, Q., Chavez-Echeagaray, M., & Atkinson, R. K. (2012). Supporting the assembly process by leveraging augmented reality, cloud computing, and mobile devices. *International Journal of Cyber Behavior, Psychology and Learning*, *2*(3), 86–102. doi:10.4018/ijcbpl.2012070107

Gopinath, R., & Geetha, B. (2013). An e-learning system based on secure data storage services in cloud computing. *International Journal of Information Technology and Web Engineering*, *8*(2), 1–17. doi:10.4018/jitwe.2013040101

Gossin, P. C., & LaBrie, R. C. (2013). Data center waste management. In P. Ordóñez de Pablos (Ed.), *Green technologies and business practices: An IT approach* (pp. 226–235). Hershey, PA: IGI Global. doi:10.4018/978-1-4666-1972-2.ch014

Goswami, V., Patra, S. S., & Mund, G. B. (2012). Performance analysis of cloud computing centers for bulk services. *International Journal of Cloud Applications and Computing*, *2*(4), 53–65. doi:10.4018/ijcac.2012100104

Goswami, V., & Sahoo, C. N. (2013). Optimal resource usage in multi-cloud computing environment. *International Journal of Cloud Applications and Computing*, *3*(1), 44–57. doi:10.4018/ijcac.2013010105

Gräuler, M., Teuteberg, F., Mahmoud, T., & Gómez, J. M. (2013). Requirements prioritization and design considerations for the next generation of corporate environmental management information systems: A foundation for innovation. *International Journal of Information Technologies and Systems Approach*, *6*(1), 98–116. doi:10.4018/jitsa.2013010106

Grieve, G. P., & Heston, K. (2012). Finding liquid salvation: Using the cardean ethnographic method to document second life residents and religious cloud communities. In N. Zagalo, L. Morgado, & A. Boa-Ventura (Eds.), *Virtual worlds and metaverse platforms: New communication and identity paradigms* (pp. 288–305). Hershey, PA: IGI Global. doi:10.4018/978-1-60960-854-5.ch019

Grispos, G., Storer, T., & Glisson, W. B. (2012). Calm before the storm: The challenges of cloud computing in digital forensics. *International Journal of Digital Crime and Forensics*, *4*(2), 28–48. doi:10.4018/jdcf.2012040103

Grispos, G., Storer, T., & Glisson, W. B. (2013). Calm before the storm: The challenges of cloud computing in digital forensics. In C. Li (Ed.), *Emerging digital forensics applications for crime detection, prevention, and security* (pp. 211–233). Hershey, PA: IGI Global. doi:10.4018/978-1-4666-4006-1.ch015

Guster, D., & Lee, O. F. (2011). Enhancing the disaster recovery plan through virtualization. *Journal of Information Technology Research*, *4*(4), 18–40. doi:10.4018/jitr.2011100102

Hanawa, T., & Sato, M. (2013). D-Cloud: Software testing environment for dependable distributed systems using cloud computing technology. In S. Tilley & T. Parveen (Eds.), *Software testing in the cloud: Perspectives on an emerging discipline* (pp. 340–355). Hershey, PA: IGI Global. doi:10.4018/978-1-4666-2536-5.ch016

Hardy, J., Liu, L., Lei, C., & Li, J. (2013). Internet-based virtual computing infrastructure for cloud computing. In X. Yang & L. Liu (Eds.), *Principles, methodologies, and service-oriented approaches for cloud computing* (pp. 371–389). Hershey, PA: IGI Global. doi:10.4018/978-1-4666-2854-0.ch016

Hashizume, K., Yoshioka, N., & Fernandez, E. B. (2013). Three misuse patterns for cloud computing. In D. Rosado, D. Mellado, E. Fernandez-Medina, & M. Piattini (Eds.), *Security engineering for cloud computing: Approaches and tools* (pp. 36–53). Hershey, PA: IGI Global. doi:10.4018/978-1-4666-2125-1.ch003

Hassan, Q. F., Riad, A. M., & Hassan, A. E. (2012). Understanding cloud computing. In H. Yang & X. Liu (Eds.), *Software reuse in the emerging cloud computing era* (pp. 204–227). Hershey, PA: IGI Global. doi:10.4018/978-1-4666-0897-9.ch009

Hasselmeyer, P., Katsaros, G., Koller, B., & Wieder, P. (2012). Cloud monitoring. In M. Villari, I. Brandic, & F. Tusa (Eds.), *Achieving federated and self-manageable cloud infrastructures: Theory and practice* (pp. 97–116). Hershey, PA: IGI Global. doi:10.4018/978-1-4666-1631-8.ch006

Hertzler, B. T., Frost, E., Bressler, G. H., & Goehring, C. (2011). Experience report: Using a cloud computing environment during Haiti and Exercise24. *International Journal of Information Systems for Crisis Response and Management, 3*(1), 50–64. doi:10.4018/jiscrm.2011010104

Hertzler, B. T., Frost, E., Bressler, G. H., & Goehring, C. (2013). Experience report: Using a cloud computing environment during Haiti and Exercise24. In M. Jennex (Ed.), *Using social and information technologies for disaster and crisis management* (pp. 52–66). Hershey, PA: IGI Global. doi:10.4018/978-1-4666-2788-8.ch004

Ho, R. (2013). Cloud computing and enterprise migration strategies. In A. Loo (Ed.), *Distributed computing innovations for business, engineering, and science* (pp. 156–175). Hershey, PA: IGI Global. doi:10.4018/978-1-4666-2533-4.ch008

Hobona, G., Jackson, M., & Anand, S. (2012). Implementing geospatial web services for cloud computing. In *Grid and cloud computing: Concepts, methodologies, tools and applications* (pp. 615–636). Hershey, PA: IGI Global. doi:10.4018/978-1-4666-0879-5.ch305

Hochstein, L., Schott, B., & Graybill, R. B. (2011). Computational engineering in the cloud: Benefits and challenges. *Journal of Organizational and End User Computing, 23*(4), 31–50. doi:10.4018/joeuc.2011100103

Hochstein, L., Schott, B., & Graybill, R. B. (2013). Computational engineering in the cloud: Benefits and challenges. In A. Dwivedi & S. Clarke (Eds.), *Innovative strategies and approaches for end-user computing advancements* (pp. 314–332). Hershey, PA: IGI Global. doi:10.4018/978-1-4666-2059-9.ch017

Honarvar, A. R. (2013). Developing an elastic cloud computing application through multi-agent systems. *International Journal of Cloud Applications and Computing, 3*(1), 58–64. doi:10.4018/ijcac.2013010106

Hossain, S. (2013). Cloud computing terms, definitions, and taxonomy. In A. Bento & A. Aggarwal (Eds.), *Cloud computing service and deployment models: Layers and management* (pp. 1–25). Hershey, PA: IGI Global. doi:10.4018/978-1-4666-2187-9.ch001

Hudzia, B., Sinclair, J., & Lindner, M. (2013). Deploying and running enterprise grade applications in a federated cloud. In *Supply chain management: Concepts, methodologies, tools, and applications* (pp. 1350–1370). Hershey, PA: IGI Global. doi:10.4018/978-1-4666-2625-6.ch080

Hung, S., Shieh, J., & Lee, C. (2011). Migrating android applications to the cloud. *International Journal of Grid and High Performance Computing, 3*(2), 14–28. doi:10.4018/jghpc.2011040102

Hung, S., Shieh, J., & Lee, C. (2013). Migrating android applications to the cloud. In E. Udoh (Ed.), *Applications and developments in grid, cloud, and high performance computing* (pp. 307–322). Hershey, PA: IGI Global. doi:10.4018/978-1-4666-2065-0.ch020

Islam, S., Mouratidis, H., & Weippl, E. R. (2013). A goal-driven risk management approach to support security and privacy analysis of cloud-based system. In D. Rosado, D. Mellado, E. Fernandez-Medina, & M. Piattini (Eds.), *Security engineering for cloud computing: Approaches and tools* (pp. 97–122). Hershey, PA: IGI Global. doi:10.4018/978-1-4666-2125-1.ch006

Itani, W., Kayssi, A., & Chehab, A. (2013). Hardware-based security for ensuring data privacy in the cloud. In D. Rosado, D. Mellado, E. Fernandez-Medina, & M. Piattini (Eds.), *Security engineering for cloud computing: Approaches and tools* (pp. 147–170). Hershey, PA: IGI Global. doi:10.4018/978-1-4666-2125-1.ch008

Jackson, A., & Weiland, M. (2013). Cloud computing for scientific simulation and high performance computing. In X. Yang & L. Liu (Eds.), *Principles, methodologies, and service-oriented approaches for cloud computing* (pp. 51–70). Hershey, PA: IGI Global. doi:10.4018/978-1-4666-2854-0.ch003

Jaeger, M. C., & Hohenstein, U. (2013). Content centric storage and current storage systems. In D. Kyriazis, A. Voulodimos, S. Gogouvitis, & T. Varvarigou (Eds.), *Data intensive storage services for cloud environments* (pp. 27–46). Hershey, PA: IGI Global. doi:10.4018/978-1-4666-3934-8.ch003

James, J. I., Shosha, A. F., & Gladyshev, P. (2013). Digital forensic investigation and cloud computing. In K. Ruan (Ed.), *Cybercrime and cloud forensics: Applications for investigation processes* (pp. 1–41). Hershey, PA: IGI Global. doi:10.4018/978-1-4666-2662-1.ch001

Jena, R. K. (2013). Green computing to green business. In P. Ordóñez de Pablos (Ed.), *Green technologies and business practices: An IT approach* (pp. 138–150). Hershey, PA: IGI Global. doi:10.4018/978-1-4666-1972-2.ch007

Jeyarani, R., & Nagaveni, N. (2012). A heuristic meta scheduler for optimal resource utilization and improved QoS in cloud computing environment. *International Journal of Cloud Applications and Computing*, *2*(1), 41–52. doi:10.4018/ijcac.2012010103

Jeyarani, R., Nagaveni, N., & Ram, R. V. (2011). Self adaptive particle swarm optimization for efficient virtual machine provisioning in cloud. *International Journal of Intelligent Information Technologies*, *7*(2), 25–44. doi:10.4018/jiit.2011040102

Jeyarani, R., Nagaveni, N., & Ram, R. V. (2013). Self adaptive particle swarm optimization for efficient virtual machine provisioning in cloud. In V. Sugumaran (Ed.), *Organizational efficiency through intelligent information technologies* (pp. 88–107). Hershey, PA: IGI Global. doi:10.4018/978-1-4666-2047-6.ch006

Jeyarani, R., Nagaveni, N., Sadasivam, S. K., & Rajarathinam, V. R. (2011). Power aware meta scheduler for adaptive VM provisioning in IaaS cloud. *International Journal of Cloud Applications and Computing*, *1*(3), 36–51. doi:10.4018/ijcac.2011070104

Jeyarani, R., Nagaveni, N., Sadasivam, S. K., & Rajarathinam, V. R. (2013). Power aware meta scheduler for adaptive VM provisioning in IaaS cloud. In S. Aljawarneh (Ed.), *Cloud computing advancements in design, implementation, and technologies* (pp. 190–204). Hershey, PA: IGI Global. doi:10.4018/978-1-4666-1879-4.ch014

Jiang, J., Huang, X., Wu, Y., & Yang, G. (2013). Campus cloud storage and preservation: From distributed file system to data sharing service. In X. Yang & L. Liu (Eds.), *Principles, methodologies, and service-oriented approaches for cloud computing* (pp. 284–301). Hershey, PA: IGI Global. doi:10.4018/978-1-4666-2854-0.ch012

Jing, S. (2012). The application exploration of cloud computing in information technology teaching. *International Journal of Advanced Pervasive and Ubiquitous Computing*, *4*(4), 23–27. doi:10.4018/japuc.2012100104

Johansson, D., & Wiberg, M. (2012). Conceptually advancing "application mobility" towards design: Applying a concept-driven approach to the design of mobile IT for home care service groups. *International Journal of Ambient Computing and Intelligence*, *4*(3), 20–32. doi:10.4018/jaci.2012070102

Jorda, J., & M'zoughi, A. (2013). Securing cloud storage. In D. Rosado, D. Mellado, E. Fernandez-Medina, & M. Piattini (Eds.), *Security engineering for cloud computing: Approaches and tools* (pp. 171–190). Hershey, PA: IGI Global. doi:10.4018/978-1-4666-2125-1.ch009

Juiz, C., & Alexander de Pous, V. (2014). Cloud computing: IT governance, legal, and public policy aspects. In I. Portela & F. Almeida (Eds.), *Organizational, legal, and technological dimensions of information system administration* (pp. 139–166). Hershey, PA: IGI Global. doi:10.4018/978-1-4666-4526-4.ch009

Kaisler, S. H., Money, W., & Cohen, S. J. (2013). Cloud computing: A decision framework for small businesses. In A. Bento & A. Aggarwal (Eds.), *Cloud computing service and deployment models: Layers and management* (pp. 151–172). Hershey, PA: IGI Global. doi:10.4018/978-1-4666-2187-9.ch008

Kanamori, Y., & Yen, M. Y. (2013). Cloud computing security and risk management. In A. Bento & A. Aggarwal (Eds.), *Cloud computing service and deployment models: Layers and management* (pp. 222–240). Hershey, PA: IGI Global. doi:10.4018/978-1-4666-2187-9.ch012

Karadsheh, L., & Alhawari, S. (2011). Applying security policies in small business utilizing cloud computing technologies. *International Journal of Cloud Applications and Computing*, *1*(2), 29–40. doi:10.4018/ijcac.2011040103

Karadsheh, L., & Alhawari, S. (2013). Applying security policies in small business utilizing cloud computing technologies. In S. Aljawarneh (Ed.), *Cloud computing advancements in design, implementation, and technologies* (pp. 112–124). Hershey, PA: IGI Global. doi:10.4018/978-1-4666-1879-4.ch008

Kaupins, G. (2012). Laws associated with mobile computing in the cloud. *International Journal of Wireless Networks and Broadband Technologies*, *2*(3), 1–9. doi:10.4018/ijwnbt.2012070101

Kemp, M. L., Robb, S., & Deans, P. C. (2013). The legal implications of cloud computing. In A. Bento & A. Aggarwal (Eds.), *Cloud computing service and deployment models: Layers and management* (pp. 257–272). Hershey, PA: IGI Global. doi:10.4018/978-1-4666-2187-9.ch014

Khan, N., Ahmad, N., Herawan, T., & Inayat, Z. (2012). Cloud computing: Locally sub-clouds instead of globally one cloud. *International Journal of Cloud Applications and Computing*, *2*(3), 68–85. doi:10.4018/ijcac.2012070103

Khan, N., Noraziah, A., Ismail, E. I., Deris, M. M., & Herawan, T. (2012). Cloud computing: Analysis of various platforms. *International Journal of E-Entrepreneurship and Innovation*, *3*(2), 51–59. doi:10.4018/jeei.2012040104

Khansa, L., Forcade, J., Nambari, G., Parasuraman, S., & Cox, P. (2012). Proposing an intelligent cloud-based electronic health record system. *International Journal of Business Data Communications and Networking*, *8*(3), 57–71. doi:10.4018/jbdcn.2012070104

Kierkegaard, S. (2012). Not every cloud brings rain: Legal risks on the horizon. In M. Gupta, J. Walp, & R. Sharman (Eds.), *Strategic and practical approaches for information security governance: Technologies and applied solutions* (pp. 181–194). Hershey, PA: IGI Global. doi:10.4018/978-1-4666-0197-0.ch011

Kifayat, K., Shamsa, T. B., Mackay, M., Merabti, M., & Shi, Q. (2013). Real time risk management in cloud computation. In D. Rosado, D. Mellado, E. Fernandez-Medina, & M. Piattini (Eds.), *Security engineering for cloud computing: Approaches and tools* (pp. 123–145). Hershey, PA: IGI Global. doi:10.4018/978-1-4666-2125-1.ch007

King, T. M., Ganti, A. S., & Froslie, D. (2013). Towards improving the testability of cloud application services. In S. Tilley & T. Parveen (Eds.), *Software testing in the cloud: Perspectives on an emerging discipline* (pp. 322–339). Hershey, PA: IGI Global. doi:10.4018/978-1-4666-2536-5.ch015

Kipp, A., Schneider, R., & Schubert, L. (2013). Encapsulation of complex HPC services. In C. Rückemann (Ed.), *Integrated information and computing systems for natural, spatial, and social sciences* (pp. 153–176). Hershey, PA: IGI Global. doi:10.4018/978-1-4666-2190-9.ch008

Kldiashvili, E. (2012). The cloud computing as the tool for implementation of virtual organization technology for ehealth. *Journal of Information Technology Research*, *5*(1), 18–34. doi:10.4018/jitr.2012010102

Kldiashvili, E. (2013). Implementation of telecytology in georgia for quality assurance programs. *Journal of Information Technology Research*, *6*(2), 24–45. doi:10.4018/jitr.2013040102

Kosmatov, N. (2013). Concolic test generation and the cloud: deployment and verification perspectives. In S. Tilley & T. Parveen (Eds.), *Software testing in the cloud: Perspectives on an emerging discipline* (pp. 231–251). Hershey, PA: IGI Global. doi:10.4018/978-1-4666-2536-5.ch011

Kotamarti, R. M., Thornton, M. A., & Dunham, M. H. (2012). Quantum computing approach for alignment-free sequence search and classification. In S. Ali, N. Abbadeni, & M. Batouche (Eds.), *Multidisciplinary computational intelligence techniques: Applications in business, engineering, and medicine* (pp. 279–300). Hershey, PA: IGI Global. doi:10.4018/978-1-4666-1830-5.ch017

Kremmydas, D., Petsakos, A., & Rozakis, S. (2012). Parametric optimization of linear and non-linear models via parallel computing to enhance web-spatial DSS interactivity. *International Journal of Decision Support System Technology*, *4*(1), 14–29. doi:10.4018/jdsst.2012010102

Krishnadas, N., & Pillai, R. R. (2013). Cloud computing diagnosis: A comprehensive study. In X. Yang & L. Liu (Eds.), *Principles, methodologies, and service-oriented approaches for cloud computing* (pp. 1–18). Hershey, PA: IGI Global. doi:10.4018/978-1-4666-2854-0.ch001

Kübert, R., & Katsaros, G. (2011). Using free software for elastic web hosting on a private cloud. *International Journal of Cloud Applications and Computing*, *1*(2), 14–28. doi:10.4018/ijcac.2011040102

Kübert, R., & Katsaros, G. (2013). Using free software for elastic web hosting on a private cloud. In S. Aljawarneh (Ed.), *Cloud computing advancements in design, implementation, and technologies* (pp. 97–111). Hershey, PA: IGI Global. doi:10.4018/978-1-4666-1879-4.ch007

Kumar, P. S., Ashok, M. S., & Subramanian, R. (2012). A publicly verifiable dynamic secret sharing protocol for secure and dependable data storage in cloud computing. *International Journal of Cloud Applications and Computing*, *2*(3), 1–25. doi:10.4018/ijcac.2012070101

Lasluisa, S., Rodero, I., & Parashar, M. (2013). Software design for passing sarbanes-oxley in cloud computing. In C. Rückemann (Ed.), *Integrated information and computing systems for natural, spatial, and social sciences* (pp. 27–42). Hershey, PA: IGI Global. doi:10.4018/978-1-4666-2190-9.ch002

Lasluisa, S., Rodero, I., & Parashar, M. (2014). Software design for passing sarbanes-oxley in cloud computing. In *Software design and development: Concepts, methodologies, tools, and applications* (pp. 1659–1674). Hershey, PA: IGI Global. doi:10.4018/978-1-4666-4301-7.ch080

Lee, W. N. (2013). An economic analysis of cloud: "Software as a service" (saas) computing and "virtual desktop infrastructure" (VDI) models. In A. Bento & A. Aggarwal (Eds.), *Cloud computing service and deployment models: Layers and management* (pp. 289–295). Hershey, PA: IGI Global. doi:10.4018/978-1-4666-2187-9.ch016

Levine, K., & White, B. A. (2011). A crisis at hafford furniture: Cloud computing case study. *Journal of Cases on Information Technology*, *13*(1), 57–71. doi:10.4018/jcit.2011010104

Levine, K., & White, B. A. (2013). A crisis at Hafford furniture: Cloud computing case study. In M. Khosrow-Pour (Ed.), *Cases on emerging information technology research and applications* (pp. 70–87). Hershey, PA: IGI Global. doi:10.4018/978-1-4666-3619-4.ch004

Li, J., Meng, L., Zhu, Z., Li, X., Huai, J., & Liu, L. (2013). CloudRank: A cloud service ranking method based on both user feedback and service testing. In X. Yang & L. Liu (Eds.), *Principles, methodologies, and service-oriented approaches for cloud computing* (pp. 230–258). Hershey, PA: IGI Global. doi:10.4018/978-1-4666-2854-0.ch010

Liang, T., Lu, F., & Chiu, J. (2012). A hybrid resource reservation method for workflows in clouds. *International Journal of Grid and High Performance Computing*, *4*(4), 1–21. doi:10.4018/jghpc.2012100101

Lorenz, M., Rath-Wiggins, L., Runde, W., Messina, A., Sunna, P., Dimino, G., ... Borgotallo, R. (2013). Media convergence and cloud technologies: Smart storage, better workflows. In D. Kyriazis, A. Voulodimos, S. Gogouvitis, & T. Varvarigou (Eds.), *Data intensive storage services for cloud environments* (pp. 132–144). Hershey, PA: IGI Global. doi:10.4018/978-1-4666-3934-8.ch009

Maharana, S. K., Mali, P. B., Prabhakar, G. J. S., & Kumar, V. (2011). Cloud computing applied for numerical study of thermal characteristics of SIP. *International Journal of Cloud Applications and Computing*, *1*(3), 12–21. doi:10.4018/ijcac.2011070102

Maharana, S. K., Mali, P. B., Prabhakar, G. J. S., & Kumar, V. (2013). Cloud computing applied for numerical study of thermal characteristics of SIP. In S. Aljawarneh (Ed.), *Cloud computing advancements in design, implementation, and technologies* (pp. 166–175). Hershey, PA: IGI Global. doi:10.4018/978-1-4666-1879-4.ch012

Maharana, S. K., P, G. P., & Bhati, A. (2012). A study of cloud computing for retinal image processing through MATLAB. *International Journal of Cloud Applications and Computing*, *2*(2), 59–69. doi:10.4018/ijcac.2012040106

Maharana, S. K., Prabhakar, P. G., & Bhati, A. (2013). A study of cloud computing for retinal image processing through MATLAB. In *Image processing: Concepts, methodologies, tools, and applications* (pp. 101–111). Hershey, PA: IGI Global. doi:10.4018/978-1-4666-3994-2.ch006

Mahesh, S., Landry, B. J., Sridhar, T., & Walsh, K. R. (2011). A decision table for the cloud computing decision in small business. *Information Resources Management Journal*, *24*(3), 9–25. doi:10.4018/irmj.2011070102

Mahesh, S., Landry, B. J., Sridhar, T., & Walsh, K. R. (2013). A decision table for the cloud computing decision in small business. In M. Khosrow-Pour (Ed.), *Managing information resources and technology: Emerging Applications and theories* (pp. 159–176). Hershey, PA: IGI Global. doi:10.4018/978-1-4666-3616-3.ch012

Marquezan, C. C., Metzger, A., Pohl, K., Engen, V., Boniface, M., Phillips, S. C., & Zlatev, Z. (2013). Adaptive future internet applications: Opportunities and challenges for adaptive web services technology. In G. Ortiz & J. Cubo (Eds.), *Adaptive web services for modular and reusable software development: Tactics and solutions* (pp. 333–353). Hershey, PA: IGI Global. doi:10.4018/978-1-4666-2089-6.ch014

Marshall, P. J. (2012). Cloud computing: Next generation education. In L. Chao (Ed.), *Cloud computing for teaching and learning: Strategies for design and implementation* (pp. 180–185). Hershey, PA: IGI Global. doi:10.4018/978-1-4666-0957-0.ch012

Martinez-Ortiz, A. (2012). Open cloud technologies. In L. Vaquero, J. Cáceres, & J. Hierro (Eds.), *Open source cloud computing systems: Practices and paradigms* (pp. 1–17). Hershey, PA: IGI Global. doi:10.4018/978-1-4666-0098-0.ch001

Massonet, P., Michot, A., Naqvi, S., Villari, M., & Latanicki, J. (2013). Securing the external interfaces of a federated infrastructure cloud. In *IT policy and ethics: Concepts, methodologies, tools, and applications* (pp. 1876–1903). Hershey, PA: IGI Global. doi:10.4018/978-1-4666-2919-6.ch082

Mavrogeorgi, N., Gogouvitis, S. V., Voulodimos, A., & Alexandrou, V. (2013). SLA management in storage clouds. In D. Kyriazis, A. Voulodimos, S. Gogouvitis, & T. Varvarigou (Eds.), *Data intensive storage services for cloud environments* (pp. 72–93). Hershey, PA: IGI Global. doi:10.4018/978-1-4666-3934-8.ch006

Mehta, H. K. (2013). Cloud selection for e-business a parameter based solution. In K. Tarnay, S. Imre, & L. Xu (Eds.), *Research and development in e-business through service-oriented solutions* (pp. 199–207). Hershey, PA: IGI Global. doi:10.4018/978-1-4666-4181-5.ch009

Mehta, H. K., & Gupta, E. (2013). Economy based resource allocation in IaaS cloud. *International Journal of Cloud Applications and Computing*, *3*(2), 1–11. doi:10.4018/ijcac.2013040101

Miah, S. J. (2012). Cloud-based intelligent DSS design for emergency professionals. In S. Ali, N. Abbadeni, & M. Batouche (Eds.), *Multidisciplinary computational intelligence techniques: Applications in business, engineering, and medicine* (pp. 47–60). Hershey, PA: IGI Global. doi:10.4018/978-1-4666-1830-5.ch004

Miah, S. J. (2013). Cloud-based intelligent DSS design for emergency professionals. In *Data mining: Concepts, methodologies, tools, and applications* (pp. 991–1003). Hershey, PA: IGI Global. doi:10.4018/978-1-4666-2455-9.ch050

Mikkilineni, R. (2012). Architectural resiliency in distributed computing. *International Journal of Grid and High Performance Computing*, *4*(4), 37–51. doi:10.4018/jghpc.2012100103

Millham, R. (2012). Software asset re-use: Migration of data-intensive legacy system to the cloud computing paradigm. In H. Yang & X. Liu (Eds.), *Software reuse in the emerging cloud computing era* (pp. 1–27). Hershey, PA: IGI Global. doi:10.4018/978-1-4666-0897-9.ch001

Mircea, M. (2011). Building the agile enterprise with service-oriented architecture, business process management and decision management. *International Journal of E-Entrepreneurship and Innovation*, *2*(4), 32–48. doi:10.4018/jeei.2011100103

Modares, H., Lloret, J., Moravejosharieh, A., & Salleh, R. (2014). Security in mobile cloud computing. In J. Rodrigues, K. Lin, & J. Lloret (Eds.), *Mobile networks and cloud computing convergence for progressive services and applications* (pp. 79–91). Hershey, PA: IGI Global. doi:10.4018/978-1-4666-4781-7.ch005

Moedjiono, S., & Mas'at, A. (2012). Cloud computing implementation strategy for information dissemination on meteorology, climatology, air quality, and geophysics (MKKuG). *Journal of Information Technology Research*, *5*(3), 71–84. doi:10.4018/jitr.2012070104

Moiny, J. (2012). Cloud based social network sites: Under whose control? In A. Dudley, J. Braman, & G. Vincenti (Eds.), *Investigating cyber law and cyber ethics: Issues, impacts and practices* (pp. 147–219). Hershey, PA: IGI Global. doi:10.4018/978-1-61350-132-0.ch009

Moreno, I. S., & Xu, J. (2011). Energy-efficiency in cloud computing environments: Towards energy savings without performance degradation. *International Journal of Cloud Applications and Computing*, *1*(1), 17–33. doi:10.4018/ijcac.2011010102

Moreno, I. S., & Xu, J. (2013). Energy-efficiency in cloud computing environments: Towards energy savings without performance degradation. In S. Aljawarneh (Ed.), *Cloud computing advancements in design, implementation, and technologies* (pp. 18–36). Hershey, PA: IGI Global. doi:10.4018/978-1-4666-1879-4.ch002

Muñoz, A., Maña, A., & González, J. (2013). Dynamic security properties monitoring architecture for cloud computing. In D. Rosado, D. Mellado, E. Fernandez-Medina, & M. Piattini (Eds.), *Security engineering for cloud computing: Approaches and tools* (pp. 1–18). Hershey, PA: IGI Global. doi:10.4018/978-1-4666-2125-1.ch001

Mvelase, P., Dlodlo, N., Williams, Q., & Adigun, M. O. (2011). Custom-made cloud enterprise architecture for small medium and micro enterprises. *International Journal of Cloud Applications and Computing*, *1*(3), 52–63. doi:10.4018/ijcac.2011070105

Mvelase, P., Dlodlo, N., Williams, Q., & Adigun, M. O. (2012). Custom-made cloud enterprise architecture for small medium and micro enterprises. In *Grid and cloud computing: Concepts, methodologies, tools and applications* (pp. 589–601). Hershey, PA: IGI Global. doi:10.4018/978-1-4666-0879-5.ch303

Mvelase, P., Dlodlo, N., Williams, Q., & Adigun, M. O. (2013). Custom-made cloud enterprise architecture for small medium and micro enterprises. In S. Aljawarneh (Ed.), *Cloud computing advancements in design, implementation, and technologies* (pp. 205–217). Hershey, PA: IGI Global. doi:10.4018/978-1-4666-1879-4.ch015

Naeem, M. A., Dobbie, G., & Weber, G. (2014). Big data management in the context of real-time data warehousing. In W. Hu & N. Kaabouch (Eds.), *Big data management, technologies, and applications* (pp. 150–176). Hershey, PA: IGI Global. doi:10.4018/978-1-4666-4699-5.ch007

Ofosu, W. K., & Saliah-Hassane, H. (2013). Cloud computing in the education environment for developing nations. *International Journal of Interdisciplinary Telecommunications and Networking*, *5*(3), 54–62. doi:10.4018/jitn.2013070106

Oliveros, E., Cucinotta, T., Phillips, S. C., Yang, X., Middleton, S., & Voith, T. (2012). Monitoring and metering in the cloud. In D. Kyriazis, T. Varvarigou, & K. Konstanteli (Eds.), *Achieving real-time in distributed computing: From grids to clouds* (pp. 94–114). Hershey, PA: IGI Global. doi:10.4018/978-1-60960-827-9.ch006

Orton, I., Alva, A., & Endicott-Popovsky, B. (2013). Legal process and requirements for cloud forensic investigations. In K. Ruan (Ed.), *Cybercrime and cloud forensics: Applications for investigation processes* (pp. 186–229). Hershey, PA: IGI Global. doi:10.4018/978-1-4666-2662-1.ch008

Pakhira, A., & Andras, P. (2013). Leveraging the cloud for large-scale software testing – A case study: Google Chrome on Amazon. In S. Tilley & T. Parveen (Eds.), *Software testing in the cloud: Perspectives on an emerging discipline* (pp. 252–279). Hershey, PA: IGI Global. doi:10.4018/978-1-4666-2536-5.ch012

Pal, K., & Karakostas, B. (2013). The use of cloud computing in shipping logistics. In D. Graham, I. Manikas, & D. Folinas (Eds.), *E-logistics and e-supply chain management: Applications for evolving business* (pp. 104–124). Hershey, PA: IGI Global. doi:10.4018/978-1-4666-3914-0.ch006

Pal, S. (2013). Cloud computing: Security concerns and issues. In A. Bento & A. Aggarwal (Eds.), *Cloud computing service and deployment models: Layers and management* (pp. 191–207). Hershey, PA: IGI Global. doi:10.4018/978-1-4666-2187-9.ch010

Pal, S. (2013). Storage security and technical challenges of cloud computing. In D. Kyriazis, A. Voulodimos, S. Gogouvitis, & T. Varvarigou (Eds.), *Data intensive storage services for cloud environments* (pp. 225–240). Hershey, PA: IGI Global. doi:10.4018/978-1-4666-3934-8.ch014

Palanivel, K., & Kuppuswami, S. (2014). A cloud-oriented reference architecture to digital library systems. In S. Dhamdhere (Ed.), *Cloud computing and virtualization technologies in libraries* (pp. 230–254). Hershey, PA: IGI Global. doi:10.4018/978-1-4666-4631-5.ch014

Paletta, M. (2012). Intelligent clouds: By means of using multi-agent systems environments. In L. Chao (Ed.), *Cloud computing for teaching and learning: Strategies for design and implementation* (pp. 254–279). Hershey, PA: IGI Global. doi:10.4018/978-1-4666-0957-0.ch017

Pallot, M., Le Marc, C., Richir, S., Schmidt, C., & Mathieu, J. (2012). Innovation gaming: An immersive experience environment enabling co-creation. In M. Cruz-Cunha (Ed.), *Handbook of research on serious games as educational, business and research tools* (pp. 1–24). Hershey, PA: IGI Global. doi:10.4018/978-1-4666-0149-9.ch001

Pankowska, M. (2011). Information technology resources virtualization for sustainable development. *International Journal of Applied Logistics*, *2*(2), 35–48. doi:10.4018/jal.2011040103

Pankowska, M. (2013). Information technology resources virtualization for sustainable development. In Z. Luo (Ed.), *Technological solutions for modern logistics and supply chain management* (pp. 248–262). Hershey, PA: IGI Global. doi:10.4018/978-1-4666-2773-4.ch016

Parappallil, J. J., Zarvic, N., & Thomas, O. (2012). A context and content reflection on business-IT alignment research. *International Journal of IT/Business Alignment and Governance*, *3*(2), 21–37. doi:10.4018/jitbag.2012070102

Parashar, V., Vishwakarma, M. L., & Parashar, R. (2014). A new framework for building academic library through cloud computing. In S. Dhamdhere (Ed.), *Cloud computing and virtualization technologies in libraries* (pp. 107–123). Hershey, PA: IGI Global. doi:10.4018/978-1-4666-4631-5.ch007

Pendyala, V. S., & Holliday, J. (2012). Cloud as a computer. In X. Liu & Y. Li (Eds.), *Advanced design approaches to emerging software systems: Principles, methodologies and tools* (pp. 241–249). Hershey, PA: IGI Global. doi:10.4018/978-1-60960-735-7.ch011

Petruch, K., Tamm, G., & Stantchev, V. (2012). Deriving in-depth knowledge from IT-performance data simulations. *International Journal of Knowledge Society Research*, *3*(2), 13–29. doi:10.4018/jksr.2012040102

Philipson, G. (2011). A framework for green computing. *International Journal of Green Computing*, *2*(1), 12–26. doi:10.4018/jgc.2011010102

Philipson, G. (2013). A framework for green computing. In K. Ganesh & S. Anbuudayasankar (Eds.), *International and interdisciplinary studies in green computing* (pp. 12–26). Hershey, PA: IGI Global. doi:10.4018/978-1-4666-2646-1.ch002

Phythian, M. (2013). The 'cloud' of unknowing – What a government cloud may and may not offer: A practitioner perspective. *International Journal of Technoethics*, *4*(1), 1–10. doi:10.4018/jte.2013010101

Pym, D., & Sadler, M. (2012). Information stewardship in cloud computing. In *Grid and cloud computing: Concepts, methodologies, tools and applications* (pp. 185–202). Hershey, PA: IGI Global. doi:10.4018/978-1-4666-0879-5.ch109

Pym, D., & Sadler, M. (2012). Information stewardship in cloud computing. In S. Galup (Ed.), *Technological applications and advancements in service science, management, and engineering* (pp. 52–69). Hershey, PA: IGI Global. doi:10.4018/978-1-4666-1583-0.ch004

Qiu, J., Ekanayake, J., Gunarathne, T., Choi, J. Y., Bae, S., & Ruan, Y. … Tang, H. (2013). Data intensive computing for bioinformatics. In Bioinformatics: Concepts, methodologies, tools, and applications (pp. 287-321). Hershey, PA: IGI Global. doi:10.4018/978-1-4666-3604-0.ch016

Rabaey, M. (2012). A public economics approach to enabling enterprise architecture with the government cloud in Belgium. In P. Saha (Ed.), *Enterprise architecture for connected e-government: Practices and innovations* (pp. 467–493). Hershey, PA: IGI Global. doi:10.4018/978-1-4666-1824-4.ch020

Rabaey, M. (2013). A complex adaptive system thinking approach of government e-procurement in a cloud computing environment. In P. Ordóñez de Pablos, J. Lovelle, J. Gayo, & R. Tennyson (Eds.), *E-procurement management for successful electronic government systems* (pp. 193–219). Hershey, PA: IGI Global. doi:10.4018/978-1-4666-2119-0.ch013

Rabaey, M. (2013). Holistic investment framework for cloud computing: A management-philosophical approach based on complex adaptive systems. In A. Bento & A. Aggarwal (Eds.), *Cloud computing service and deployment models: Layers and management* (pp. 94–122). Hershey, PA: IGI Global. doi:10.4018/978-1-4666-2187-9.ch005

Rak, M., Ficco, M., Luna, J., Ghani, H., Suri, N., Panica, S., & Petcu, D. (2012). Security issues in cloud federations. In M. Villari, I. Brandic, & F. Tusa (Eds.), *Achieving federated and self-manageable cloud infrastructures: Theory and practice* (pp. 176–194). Hershey, PA: IGI Global. doi:10.4018/978-1-4666-1631-8.ch010

Ramanathan, R. (2013). Extending service-driven architectural approaches to the cloud. In R. Ramanathan & K. Raja (Eds.), *Service-driven approaches to architecture and enterprise integration* (pp. 334–359). Hershey, PA: IGI Global. doi:10.4018/978-1-4666-4193-8.ch013

Ramírez, M., Gutiérrez, A., Monguet, J. M., & Muñoz, C. (2012). An internet cost model, assignment of costs based on actual network use. *International Journal of Web Portals*, *4*(4), 19–34. doi:10.4018/jwp.2012100102

Rashid, A., Wang, W. Y., & Tan, F. B. (2013). Value co-creation in cloud services. In A. Lin, J. Foster, & P. Scifleet (Eds.), *Consumer information systems and relationship management: Design, implementation, and use* (pp. 74–91). Hershey, PA: IGI Global. doi:10.4018/978-1-4666-4082-5.ch005

Ratten, V. (2012). Cloud computing services: Theoretical foundations of ethical and entrepreneurial adoption behaviour. *International Journal of Cloud Applications and Computing*, *2*(2), 48–58. doi:10.4018/ijcac.2012040105

Ratten, V. (2013). Exploring behaviors and perceptions affecting the adoption of cloud computing. *International Journal of Innovation in the Digital Economy*, *4*(3), 51–68. doi:10.4018/jide.2013070104

Ravi, V. (2012). Cloud computing paradigm for indian education sector. *International Journal of Cloud Applications and Computing*, *2*(2), 41–47. doi:10.4018/ijcac.2012040104

Rawat, A., Kapoor, P., & Sushil, R. (2014). Application of cloud computing in library information service sector. In S. Dhamdhere (Ed.), *Cloud computing and virtualization technologies in libraries* (pp. 77–89). Hershey, PA: IGI Global. doi:10.4018/978-1-4666-4631-5.ch005

Reich, C., Hübner, S., & Kuijs, H. (2012). Cloud computing for on-demand virtual desktops and labs. In L. Chao (Ed.), *Cloud computing for teaching and learning: strategies for design and implementation* (pp. 111–125). Hershey, PA: IGI Global. doi:10.4018/978-1-4666-0957-0.ch008

Rice, R. W. (2013). Testing in the cloud: Balancing the value and risks of cloud computing. In S. Tilley & T. Parveen (Eds.), *Software testing in the cloud: Perspectives on an emerging discipline* (pp. 404–416). Hershey, PA: IGI Global. doi:10.4018/978-1-4666-2536-5.ch019

Ruan, K. (2013). Designing a forensic-enabling cloud ecosystem. In K. Ruan (Ed.), *Cybercrime and cloud forensics: Applications for investigation processes* (pp. 331–344). Hershey, PA: IGI Global. doi:10.4018/978-1-4666-2662-1.ch014

Sabetzadeh, F., & Tsui, E. (2011). Delivering knowledge services in the cloud. *International Journal of Knowledge and Systems Science*, *2*(4), 14–20. doi:10.4018/jkss.2011100102

Sabetzadeh, F., & Tsui, E. (2013). Delivering knowledge services in the cloud. In G. Yang (Ed.), *Multidisciplinary studies in knowledge and systems science* (pp. 247–254). Hershey, PA: IGI Global. doi:10.4018/978-1-4666-3998-0.ch017

Saedi, A., & Iahad, N. A. (2013). Future research on cloud computing adoption by small and medium-sized enterprises: A critical analysis of relevant theories. *International Journal of Actor-Network Theory and Technological Innovation*, *5*(2), 1–16. doi:10.4018/jantti.2013040101

Saha, D., & Sridhar, V. (2011). Emerging areas of research in business data communications. *International Journal of Business Data Communications and Networking*, *7*(4), 52–59. doi:10.4018/IJBDCN.2011100104

Saha, D., & Sridhar, V. (2013). Platform on platform (PoP) model for meta-networking: A new paradigm for networks of the future. *International Journal of Business Data Communications and Networking*, *9*(1), 1–10. doi:10.4018/jbdcn.2013010101

Sahlin, J. P. (2013). Cloud computing: Past, present, and future. In X. Yang & L. Liu (Eds.), *Principles, methodologies, and service-oriented approaches for cloud computing* (pp. 19–50). Hershey, PA: IGI Global. doi:10.4018/978-1-4666-2854-0.ch002

Salama, M., & Shawish, A. (2012). Libraries: From the classical to cloud-based era. *International Journal of Digital Library Systems*, *3*(3), 14–32. doi:10.4018/jdls.2012070102

Sánchez, C. M., Molina, D., Vozmediano, R. M., Montero, R. S., & Llorente, I. M. (2012). On the use of the hybrid cloud computing paradigm. In M. Villari, I. Brandic, & F. Tusa (Eds.), *Achieving federated and self-manageable cloud infrastructures: Theory and practice* (pp. 196–218). Hershey, PA: IGI Global. doi:10.4018/978-1-4666-1631-8.ch011

Sasikala, P. (2011). Architectural strategies for green cloud computing: Environments, infrastructure and resources. *International Journal of Cloud Applications and Computing*, *1*(4), 1–24. doi:10.4018/ijcac.2011100101

Sasikala, P. (2011). Cloud computing in higher education: Opportunities and issues. *International Journal of Cloud Applications and Computing*, *1*(2), 1–13. doi:10.4018/ijcac.2011040101

Sasikala, P. (2011). Cloud computing towards technological convergence. *International Journal of Cloud Applications and Computing*, *1*(4), 44–59. doi:10.4018/ijcac.2011100104

Sasikala, P. (2012). Cloud computing and e-governance: Advances, opportunities and challenges. *International Journal of Cloud Applications and Computing*, *2*(4), 32–52. doi:10.4018/ijcac.2012100103

Sasikala, P. (2012). Cloud computing in higher education: Opportunities and issues. In *Grid and cloud computing: Concepts, methodologies, tools and applications* (pp. 1672–1685). Hershey, PA: IGI Global. doi:10.4018/978-1-4666-0879-5.ch709

Sasikala, P. (2012). Cloud computing towards technological convergence. In *Grid and cloud computing: Concepts, methodologies, tools and applications* (pp. 1576–1592). Hershey, PA: IGI Global. doi:10.4018/978-1-4666-0879-5.ch703

Sasikala, P. (2013). Architectural strategies for green cloud computing: Environments, infrastructure and resources. In S. Aljawarneh (Ed.), *Cloud computing advancements in design, implementation, and technologies* (pp. 218–242). Hershey, PA: IGI Global. doi:10.4018/978-1-4666-1879-4.ch016

Sasikala, P. (2013). Cloud computing in higher education: Opportunities and issues. In S. Aljawarneh (Ed.), *Cloud computing advancements in design, implementation, and technologies* (pp. 83–96). Hershey, PA: IGI Global. doi:10.4018/978-1-4666-1879-4.ch006

Sasikala, P. (2013). Cloud computing towards technological convergence. In S. Aljawarneh (Ed.), *Cloud computing advancements in design, implementation, and technologies* (pp. 263–279). Hershey, PA: IGI Global. doi:10.4018/978-1-4666-1879-4.ch019

Sasikala, P. (2013). New media cloud computing: Opportunities and challenges. *International Journal of Cloud Applications and Computing*, *3*(2), 61–72. doi:10.4018/ijcac.2013040106

Schrödl, H., & Wind, S. (2013). Requirements engineering for cloud application development. In A. Bento & A. Aggarwal (Eds.), *Cloud computing service and deployment models: Layers and management* (pp. 137–150). Hershey, PA: IGI Global. doi:10.4018/978-1-4666-2187-9.ch007

Sclater, N. (2012). Legal and contractual issues of cloud computing for educational institutions. In L. Chao (Ed.), *Cloud computing for teaching and learning: Strategies for design and implementation* (pp. 186–199). Hershey, PA: IGI Global. doi:10.4018/978-1-4666-0957-0.ch013

Sen, J. (2014). Security and privacy issues in cloud computing. In A. Ruiz-Martinez, R. Marin-Lopez, & F. Pereniguez-Garcia (Eds.), *Architectures and protocols for secure information technology infrastructures* (pp. 1–45). Hershey, PA: IGI Global. doi:10.4018/978-1-4666-4514-1.ch001

Shah, B. (2013). Cloud environment controls assessment framework. In *IT policy and ethics: Concepts, methodologies, tools, and applications* (pp. 1822–1847). Hershey, PA: IGI Global. doi:10.4018/978-1-4666-2919-6.ch080

Shah, B. (2013). Cloud environment controls assessment framework. In S. Tilley & T. Parveen (Eds.), *Software testing in the cloud: Perspectives on an emerging discipline* (pp. 28–53). Hershey, PA: IGI Global. doi:10.4018/978-1-4666-2536-5.ch002

Shang, X., Zhang, R., & Chen, Y. (2012). Internet of things (IoT) service architecture and its application in e-commerce. *Journal of Electronic Commerce in Organizations, 10*(3), 44–55. doi:10.4018/jeco.2012070104

Shankararaman, V., & Kit, L. E. (2013). Integrating the cloud scenarios and solutions. In A. Bento & A. Aggarwal (Eds.), *Cloud computing service and deployment models: Layers and management* (pp. 173–189). Hershey, PA: IGI Global. doi:10.4018/978-1-4666-2187-9.ch009

Sharma, A., & Maurer, F. (2013). A roadmap for software engineering for the cloud: Results of a systematic review. In X. Wang, N. Ali, I. Ramos, & R. Vidgen (Eds.), *Agile and lean service-oriented development: Foundations, theory, and practice* (pp. 48–63). Hershey, PA: IGI Global. doi:10.4018/978-1-4666-2503-7.ch003

Sharma, A., & Maurer, F. (2014). A roadmap for software engineering for the cloud: Results of a systematic review. In *Software design and development: Concepts, methodologies, tools, and applications* (pp. 1–16). Hershey, PA: IGI Global. doi:10.4018/978-1-4666-4301-7.ch001

Sharma, S. C., & Bagoria, H. (2014). Libraries and cloud computing models: A changing paradigm. In S. Dhamdhere (Ed.), *Cloud computing and virtualization technologies in libraries* (pp. 124–149). Hershey, PA: IGI Global. doi:10.4018/978-1-4666-4631-5.ch008

Shawish, A., & Salama, M. (2013). Cloud computing in academia, governments, and industry. In X. Yang & L. Liu (Eds.), *Principles, methodologies, and service-oriented approaches for cloud computing* (pp. 71–114). Hershey, PA: IGI Global. doi:10.4018/978-1-4666-2854-0.ch004

Shebanow, A., Perez, R., & Howard, C. (2012). The effect of firewall testing types on cloud security policies. *International Journal of Strategic Information Technology and Applications, 3*(3), 60–68. doi:10.4018/jsita.2012070105

Sheikhalishahi, M., Devare, M., Grandinetti, L., & Incutti, M. C. (2012). A complementary approach to grid and cloud distributed computing paradigms. In *Grid and cloud computing: Concepts, methodologies, tools and applications* (pp. 1929–1942). Hershey, PA: IGI Global. doi:10.4018/978-1-4666-0879-5.ch811

Sheikhalishahi, M., Devare, M., Grandinetti, L., & Incutti, M. C. (2012). A complementary approach to grid and cloud distributed computing paradigms. In N. Preve (Ed.), *Computational and data grids: Principles, applications and design* (pp. 31–44). Hershey, PA: IGI Global. doi:10.4018/978-1-61350-113-9.ch002

Shen, Y., Li, Y., Wu, L., Liu, S., & Wen, Q. (2014). Cloud computing overview. In Y. Shen, Y. Li, L. Wu, S. Liu, & Q. Wen (Eds.), *Enabling the new era of cloud computing: Data security, transfer, and management* (pp. 1–24). Hershey, PA: IGI Global. doi:10.4018/978-1-4666-4801-2.ch001

Shen, Y., Li, Y., Wu, L., Liu, S., & Wen, Q. (2014). Main components of cloud computing. In Y. Shen, Y. Li, L. Wu, S. Liu, & Q. Wen (Eds.), *Enabling the new era of cloud computing: Data security, transfer, and management* (pp. 25–50). Hershey, PA: IGI Global. doi:10.4018/978-1-4666-4801-2.ch002

Shen, Y., Yang, J., & Keskin, T. (2014). Impact of cultural differences on the cloud computing ecosystems in the USA and China. In Y. Shen, Y. Li, L. Wu, S. Liu, & Q. Wen (Eds.), *Enabling the new era of cloud computing: Data security, transfer, and management* (pp. 269–283). Hershey, PA: IGI Global. doi:10.4018/978-1-4666-4801-2.ch014

Shetty, S., & Rawat, D. B. (2013). Cloud computing based cognitive radio networking. In N. Meghanathan & Y. Reddy (Eds.), *Cognitive radio technology applications for wireless and mobile ad hoc networks* (pp. 153–164). Hershey, PA: IGI Global. doi:10.4018/978-1-4666-4221-8.ch008

Shi, Z., & Beard, C. (2014). QoS in the mobile cloud computing environment. In J. Rodrigues, K. Lin, & J. Lloret (Eds.), *Mobile networks and cloud computing convergence for progressive services and applications* (pp. 200–217). Hershey, PA: IGI Global. doi:10.4018/978-1-4666-4781-7.ch011

Shuster, L. (2013). Enterprise integration: Challenges and solution architecture. In R. Ramanathan & K. Raja (Eds.), *Service-driven approaches to architecture and enterprise integration* (pp. 43–66). Hershey, PA: IGI Global. doi:10.4018/978-1-4666-4193-8.ch002

Siahos, Y., Papanagiotou, I., Georgopoulos, A., Tsamis, F., & Papaioannou, I. (2012). An architecture paradigm for providing cloud services in school labs based on open source software to enhance ICT in education. *International Journal of Cyber Ethics in Education*, *2*(1), 44–57. doi:10.4018/ijcee.2012010105

Simon, E., & Estublier, J. (2013). Model driven integration of heterogeneous software artifacts in service oriented computing. In A. Ionita, M. Litoiu, & G. Lewis (Eds.), *Migrating legacy applications: Challenges in service oriented architecture and cloud computing environments* (pp. 332–360). Hershey, PA: IGI Global. doi:10.4018/978-1-4666-2488-7.ch014

Singh, J., & Kumar, V. (2013). Compliance and regulatory standards for cloud computing. In R. Khurana & R. Aggarwal (Eds.), *Interdisciplinary perspectives on business convergence, computing, and legality* (pp. 54–64). Hershey, PA: IGI Global. doi:10.4018/978-1-4666-4209-6.ch006

Singh, V. V. (2012). Software development using service syndication based on API handshake approach between cloud-based and SOA-based reusable services. In H. Yang & X. Liu (Eds.), *Software reuse in the emerging cloud computing era* (pp. 136–157). Hershey, PA: IGI Global. doi:10.4018/978-1-4666-0897-9.ch006

Smeitink, M., & Spruit, M. (2013). Maturity for sustainability in IT: Introducing the MITS. *International Journal of Information Technologies and Systems Approach*, *6*(1), 39–56. doi:10.4018/jitsa.2013010103

Smith, P. A., & Cockburn, T. (2013). Socio-digital technologies. In *Dynamic leadership models for global business: Enhancing digitally connected environments* (pp. 142–168). Hershey, PA: IGI Global. doi:10.4018/978-1-4666-2836-6.ch006

Sneed, H. M. (2013). Testing web services in the cloud. In S. Tilley & T. Parveen (Eds.), *Software testing in the cloud: Perspectives on an emerging discipline* (pp. 136–173). Hershey, PA: IGI Global. doi:10.4018/978-1-4666-2536-5.ch007

Solomon, B., Ionescu, D., Gadea, C., & Litoiu, M. (2013). Geographically distributed cloud-based collaborative application. In A. Ionita, M. Litoiu, & G. Lewis (Eds.), *Migrating legacy applications: Challenges in service oriented architecture and cloud computing environments* (pp. 248–274). Hershey, PA: IGI Global. doi:10.4018/978-1-4666-2488-7.ch011

Song, W., & Xiao, Z. (2013). An infrastructure-as-a-service cloud: On-demand resource provisioning. In X. Yang & L. Liu (Eds.), *Principles, methodologies, and service-oriented approaches for cloud computing* (pp. 302–324). Hershey, PA: IGI Global. doi:10.4018/978-1-4666-2854-0.ch013

Sood, S. K. (2013). A value based dynamic resource provisioning model in cloud. *International Journal of Cloud Applications and Computing*, *3*(1), 1–12. doi:10.4018/ijcac.2013010101

Sotiriadis, S., Bessis, N., & Antonopoulos, N. (2012). Exploring inter-cloud load balancing by utilizing historical service submission records. *International Journal of Distributed Systems and Technologies*, *3*(3), 72–81. doi:10.4018/jdst.2012070106

Soyata, T., Ba, H., Heinzelman, W., Kwon, M., & Shi, J. (2014). Accelerating mobile-cloud computing: A survey. In H. Mouftah & B. Kantarci (Eds.), *Communication infrastructures for cloud computing* (pp. 175–197). Hershey, PA: IGI Global. doi:10.4018/978-1-4666-4522-6.ch008

Spyridopoulos, T., & Katos, V. (2011). Requirements for a forensically ready cloud storage service. *International Journal of Digital Crime and Forensics*, *3*(3), 19–36. doi:10.4018/jdcf.2011070102

Spyridopoulos, T., & Katos, V. (2013). Data recovery strategies for cloud environments. In K. Ruan (Ed.), *Cybercrime and cloud forensics: Applications for investigation processes* (pp. 251–265). Hershey, PA: IGI Global. doi:10.4018/978-1-4666-2662-1.ch010

Srinivasa, K. G., S, H. R. C., H, M. K. S., & Venkatesh, N. (2012). MeghaOS: A framework for scalable, interoperable cloud based operating system. *International Journal of Cloud Applications and Computing*, *2*(1), 53–70. doi:10.4018/ijcac.2012010104

Stantchev, V., & Stantcheva, L. (2012). Extending traditional IT-governance knowledge towards SOA and cloud governance. *International Journal of Knowledge Society Research*, *3*(2), 30–43. doi:10.4018/jksr.2012040103

Stantchev, V., & Tamm, G. (2012). Reducing information asymmetry in cloud marketplaces. *International Journal of Human Capital and Information Technology Professionals*, *3*(4), 1–10. doi:10.4018/jhcitp.2012100101

Steinbuß, S., & Weißenberg, N. (2013). Service design and process design for the logistics mall cloud. In X. Yang & L. Liu (Eds.), *Principles, methodologies, and service-oriented approaches for cloud computing* (pp. 186–206). Hershey, PA: IGI Global. doi:10.4018/978-1-4666-2854-0.ch008

Stender, J., Berlin, M., & Reinefeld, A. (2013). XtreemFS: A file system for the cloud. In D. Kyriazis, A. Voulodimos, S. Gogouvitis, & T. Varvarigou (Eds.), *Data intensive storage services for cloud environments* (pp. 267–285). Hershey, PA: IGI Global. doi:10.4018/978-1-4666-3934-8.ch016

Sticklen, D. J., & Issa, T. (2011). An initial examination of free and proprietary software-selection in organizations. *International Journal of Web Portals*, *3*(4), 27–43. doi:10.4018/jwp.2011100103

Sun, Y., White, J., Gray, J., & Gokhale, A. (2012). Model-driven automated error recovery in cloud computing. In *Grid and cloud computing: Concepts, methodologies, tools and applications* (pp. 680–700). Hershey, PA: IGI Global. doi:10.4018/978-1-4666-0879-5.ch308

Sun, Z., Yang, Y., Zhou, Y., & Cruickshank, H. (2014). Agent-based resource management for mobile cloud. In J. Rodrigues, K. Lin, & J. Lloret (Eds.), *Mobile networks and cloud computing convergence for progressive services and applications* (pp. 118–134). Hershey, PA: IGI Global. doi:10.4018/978-1-4666-4781-7.ch007

Sutherland, S. (2013). Convergence of interoperability of cloud computing, service oriented architecture and enterprise architecture. *International Journal of E-Entrepreneurship and Innovation*, *4*(1), 43–51. doi:10.4018/jeei.2013010104

Takabi, H., & Joshi, J. B. (2013). Policy management in cloud: Challenges and approaches. In D. Rosado, D. Mellado, E. Fernandez-Medina, & M. Piattini (Eds.), *Security engineering for cloud computing: Approaches and tools* (pp. 191–211). Hershey, PA: IGI Global. doi:10.4018/978-1-4666-2125-1.ch010

Takabi, H., & Joshi, J. B. (2013). Policy management in cloud: Challenges and approaches. In *IT policy and ethics: Concepts, methodologies, tools, and applications* (pp. 814–834). Hershey, PA: IGI Global. doi:10.4018/978-1-4666-2919-6.ch037

Takabi, H., Joshi, J. B., & Ahn, G. (2013). Security and privacy in cloud computing: Towards a comprehensive framework. In X. Yang & L. Liu (Eds.), *Principles, methodologies, and service-oriented approaches for cloud computing* (pp. 164–184). Hershey, PA: IGI Global. doi:10.4018/978-1-4666-2854-0.ch007

Takabi, H., Zargar, S. T., & Joshi, J. B. (2014). Mobile cloud computing and its security and privacy challenges. In D. Rawat, B. Bista, & G. Yan (Eds.), *Security, privacy, trust, and resource management in mobile and wireless communications* (pp. 384–407). Hershey, PA: IGI Global. doi:10.4018/978-1-4666-4691-9.ch016

Teixeira, C., Pinto, J. S., Ferreira, F., Oliveira, A., Teixeira, A., & Pereira, C. (2013). Cloud computing enhanced service development architecture for the living usability lab. In R. Martinho, R. Rijo, M. Cruz-Cunha, & J. Varajão (Eds.), *Information systems and technologies for enhancing health and social care* (pp. 33–53). Hershey, PA: IGI Global. doi:10.4018/978-1-4666-3667-5.ch003

Thimm, H. (2012). Cloud-based collaborative decision making: Design considerations and architecture of the GRUPO-MOD system. *International Journal of Decision Support System Technology*, *4*(4), 39–59. doi:10.4018/jdsst.2012100103

Thomas, P. (2012). Harnessing the potential of cloud computing to transform higher education. In L. Chao (Ed.), *Cloud computing for teaching and learning: Strategies for design and implementation* (pp. 147–158). Hershey, PA: IGI Global. doi:10.4018/978-1-4666-0957-0.ch010

Toka, A., Aivazidou, E., Antoniou, A., & Arvanitopoulos-Darginis, K. (2013). Cloud computing in supply chain management: An overview. In D. Graham, I. Manikas, & D. Folinas (Eds.), *E-logistics and e-supply chain management: Applications for evolving business* (pp. 218–231). Hershey, PA: IGI Global. doi:10.4018/978-1-4666-3914-0.ch012

Torrealba, S. M., Morales, P. M., Campos, J. M., & Meza, S. M. (2013). A software tool to support risks analysis about what should or should not go to the cloud. In D. Rosado, D. Mellado, E. Fernandez-Medina, & M. Piattini (Eds.), *Security engineering for cloud computing: Approaches and tools* (pp. 72–96). Hershey, PA: IGI Global. doi:10.4018/978-1-4666-2125-1.ch005

Trivedi, M., & Suthar, V. (2013). Cloud computing: A feasible platform for ICT enabled health science libraries in India. *International Journal of User-Driven Healthcare*, *3*(2), 69–77. doi:10.4018/ijudh.2013040108

Truong, H., Pham, T., Thoai, N., & Dustdar, S. (2012). Cloud computing for education and research in developing countries. In L. Chao (Ed.), *Cloud computing for teaching and learning: Strategies for design and implementation* (pp. 64–80). Hershey, PA: IGI Global. doi:10.4018/978-1-4666-0957-0.ch005

Tsirmpas, C., Giokas, K., Iliopoulou, D., & Koutsouris, D. (2012). Magnetic resonance imaging and magnetic resonance spectroscopy cloud computing framework. *International Journal of Reliable and Quality E-Healthcare*, *1*(4), 1–12. doi:10.4018/ijrqeh.2012100101

Turner, H., White, J., Reed, J., Galindo, J., Porter, A., Marathe, M., ... Gokhale, A. (2013). Building a cloud-based mobile application testbed. In *IT policy and ethics: Concepts, methodologies, tools, and applications* (pp. 879–899). Hershey, PA: IGI Global. doi:10.4018/978-1-4666-2919-6.ch040

Turner, H., White, J., Reed, J., Galindo, J., Porter, A., Marathe, M., ... Gokhale, A. (2013). Building a cloud-based mobile application testbed. In S. Tilley & T. Parveen (Eds.), *Software testing in the cloud: Perspectives on an emerging discipline* (pp. 382–403). Hershey, PA: IGI Global. doi:10.4018/978-1-4666-2536-5.ch018

Tusa, F., Paone, M., & Villari, M. (2012). CLEVER: A cloud middleware beyond the federation. In M. Villari, I. Brandic, & F. Tusa (Eds.), *Achieving federated and self-manageable cloud infrastructures: Theory and practice* (pp. 219–241). Hershey, PA: IGI Global. doi:10.4018/978-1-4666-1631-8.ch012

Udoh, E. (2012). Technology acceptance model applied to the adoption of grid and cloud technology. *International Journal of Grid and High Performance Computing*, *4*(1), 1–20. doi:10.4018/jghpc.2012010101

Vannoy, S. A. (2011). A structured content analytic assessment of business services advertisements in the cloud-based web services marketplace. *International Journal of Dependable and Trustworthy Information Systems, 2*(1), 18–49. doi:10.4018/jdtis.2011010102

Vaquero, L. M., Cáceres, J., & Morán, D. (2011). The challenge of service level scalability for the cloud. *International Journal of Cloud Applications and Computing, 1*(1), 34–44. doi:10.4018/ijcac.2011010103

Vaquero, L. M., Cáceres, J., & Morán, D. (2013). The challenge of service level scalability for the cloud. In S. Aljawarneh (Ed.), *Cloud computing advancements in design, implementation, and technologies* (pp. 37–48). Hershey, PA: IGI Global. doi:10.4018/978-1-4666-1879-4.ch003

Venkatraman, R., Venkatraman, S., & Asaithambi, S. P. (2013). A practical cloud services implementation framework for e-businesses. In K. Tarnay, S. Imre, & L. Xu (Eds.), *Research and development in e-business through service-oriented solutions* (pp. 167–198). Hershey, PA: IGI Global. doi:10.4018/978-1-4666-4181-5.ch008

Venkatraman, S. (2013). Software engineering research gaps in the cloud. *Journal of Information Technology Research, 6*(1), 1–19. doi:10.4018/jitr.2013010101

Vijaykumar, S., Rajkarthick, K. S., & Priya, J. (2012). Innovative business opportunities and smart business management techniques from green cloud TPS. *International Journal of Asian Business and Information Management, 3*(4), 62–72. doi:10.4018/jabim.2012100107

Wang, C., Lam, K. T., & Ma, K. R. K. (2012). A computation migration approach to elasticity of cloud computing. In J. Abawajy, M. Pathan, M. Rahman, A. Pathan, & M. Deris (Eds.) Network and traffic engineering in emerging distributed computing applications (pp. 145-178). Hershey, PA: IGI Global. doi:10.4018/978-1-4666-1888-6.ch007

Wang, D., & Wu, J. (2014). Carrier-grade distributed cloud computing: Demands, challenges, designs, and future perspectives. In H. Mouftah & B. Kantarci (Eds.), *Communication infrastructures for cloud computing* (pp. 264–281). Hershey, PA: IGI Global. doi:10.4018/978-1-4666-4522-6.ch012

Wang, H., & Philips, D. (2012). Implement virtual programming lab with cloud computing for web-based distance education. In L. Chao (Ed.), *Cloud computing for teaching and learning: Strategies for design and implementation* (pp. 95–110). Hershey, PA: IGI Global. doi:10.4018/978-1-4666-0957-0.ch007

Warneke, D. (2013). Ad-hoc parallel data processing on pay-as-you-go clouds with nephele. In A. Loo (Ed.), *Distributed computing innovations for business, engineering, and science* (pp. 191–218). Hershey, PA: IGI Global. doi:10.4018/978-1-4666-2533-4.ch010

Wei, Y., & Blake, M. B. (2013). Adaptive web services monitoring in cloud environments. *International Journal of Web Portals, 5*(1), 15–27. doi:10.4018/jwp.2013010102

White, S. C., Sedigh, S., & Hurson, A. R. (2013). Security concepts for cloud computing. In X. Yang & L. Liu (Eds.), *Principles, methodologies, and service-oriented approaches for cloud computing* (pp. 116–142). Hershey, PA: IGI Global. doi:10.4018/978-1-4666-2854-0.ch005

Williams, A. J. (2013). The role of emerging technologies in developing and sustaining diverse suppliers in competitive markets. In *Enterprise resource planning: Concepts, methodologies, tools, and applications* (pp. 1550–1560). Hershey, PA: IGI Global. doi:10.4018/978-1-4666-4153-2.ch082

Williams, A. J. (2013). The role of emerging technologies in developing and sustaining diverse suppliers in competitive markets. In J. Lewis, A. Green, & D. Surry (Eds.), *Technology as a tool for diversity leadership: Implementation and future implications* (pp. 95–105). Hershey, PA: IGI Global. doi:10.4018/978-1-4666-2668-3.ch007

Wilson, L., Goh, T. T., & Wang, W. Y. (2012). Big data management challenges in a meteorological organisation. *International Journal of E-Adoption*, *4*(2), 1–14. doi:10.4018/jea.2012040101

Wu, R., Ahn, G., & Hu, H. (2012). Towards HIPAA-compliant healthcare systems in cloud computing. *International Journal of Computational Models and Algorithms in Medicine*, *3*(2), 1–22. doi:10.4018/jcmam.2012040101

Xiao, J., Wang, M., Wang, L., & Zhu, X. (2013). Design and implementation of C-iLearning: A cloud-based intelligent learning system. *International Journal of Distance Education Technologies*, *11*(3), 79–97. doi:10.4018/jdet.2013070106

Xing, R., Wang, Z., & Peterson, R. L. (2011). Redefining the information technology in the 21st century. *International Journal of Strategic Information Technology and Applications*, *2*(1), 1–10. doi:10.4018/jsita.2011010101

Xu, L., Huang, D., Tsai, W., & Atkinson, R. K. (2012). V-lab: A mobile, cloud-based virtual laboratory platform for hands-on networking courses. *International Journal of Cyber Behavior, Psychology and Learning*, *2*(3), 73–85. doi:10.4018/ijcbpl.2012070106

Xu, Y., & Mao, S. (2014). Mobile cloud media: State of the art and outlook. In J. Rodrigues, K. Lin, & J. Lloret (Eds.), *Mobile networks and cloud computing convergence for progressive services and applications* (pp. 18–38). Hershey, PA: IGI Global. doi:10.4018/978-1-4666-4781-7.ch002

Xu, Z., Yan, B., & Zou, Y. (2013). Beyond hadoop: Recent directions in data computing for internet services. In S. Aljawarneh (Ed.), *Cloud computing advancements in design, implementation, and technologies* (pp. 49–66). Hershey, PA: IGI Global. doi:10.4018/978-1-4666-1879-4.ch004

Yan, Z. (2014). Trust management in mobile cloud computing. In *Trust management in mobile environments: Autonomic and usable models* (pp. 54–93). Hershey, PA: IGI Global. doi:10.4018/978-1-4666-4765-7.ch004

Yang, D. X. (2012). QoS-oriented service computing: Bringing SOA into cloud environment. In X. Liu & Y. Li (Eds.), *Advanced design approaches to emerging software systems: Principles, methodologies and tools* (pp. 274–296). Hershey, PA: IGI Global. doi:10.4018/978-1-60960-735-7.ch013

Yang, H., Huff, S. L., & Tate, M. (2013). Managing the cloud for information systems agility. In A. Bento & A. Aggarwal (Eds.), *Cloud computing service and deployment models: Layers and management* (pp. 70–93). Hershey, PA: IGI Global. doi:10.4018/978-1-4666-2187-9.ch004

Yang, M., Kuo, C., & Yeh, Y. (2011). Dynamic rightsizing with quality-controlled algorithms in virtualization environments. *International Journal of Grid and High Performance Computing*, *3*(2), 29–43. doi:10.4018/jghpc.2011040103

Yang, X. (2012). QoS-oriented service computing: Bringing SOA into cloud environment. In *Grid and cloud computing: Concepts, methodologies, tools and applications* (pp. 1621–1643). Hershey, PA: IGI Global. doi:10.4018/978-1-4666-0879-5.ch706

Yang, Y., Chen, J., & Hu, H. (2012). The convergence between cloud computing and cable TV. *International Journal of Technology Diffusion*, *3*(2), 1–11. doi:10.4018/jtd.2012040101

Yassein, M. O., Khamayseh, Y. M., & Hatamleh, A. M. (2013). Intelligent randomize round robin for cloud computing. *International Journal of Cloud Applications and Computing*, *3*(1), 27–33. doi:10.4018/ijcac.2013010103

Yau, S. S., An, H. G., & Buduru, A. B. (2012). An approach to data confidentiality protection in cloud environments. *International Journal of Web Services Research*, *9*(3), 67–83. doi:10.4018/jwsr.2012070104

Yu, W. D., Adiga, A. S., Rao, S., & Panakkel, M. J. (2012). A SOA based system development methodology for cloud computing environment: Using uhealthcare as practice. *International Journal of E-Health and Medical Communications*, *3*(4), 42–63. doi:10.4018/jehmc.2012100104

Yu, W. D., & Bhagwat, R. (2011). Modeling emergency and telemedicine heath support system: A service oriented architecture approach using cloud computing. *International Journal of E-Health and Medical Communications*, *2*(3), 63–88. doi:10.4018/jehmc.2011070104

Yu, W. D., & Bhagwat, R. (2013). Modeling emergency and telemedicine health support system: A service oriented architecture approach using cloud computing. In J. Rodrigues (Ed.), *Digital advances in medicine, e-health, and communication technologies* (pp. 187–213). Hershey, PA: IGI Global. doi:10.4018/978-1-4666-2794-9.ch011

Yuan, D., Lewandowski, C., & Zhong, J. (2012). Developing a private cloud based IP telephony laboratory and curriculum. In L. Chao (Ed.), *Cloud computing for teaching and learning: Strategies for design and implementation* (pp. 126–145). Hershey, PA: IGI Global. doi:10.4018/978-1-4666-0957-0.ch009

Yuvaraj, M. (2014). Cloud libraries: Issues and challenges. In S. Dhamdhere (Ed.), *Cloud computing and virtualization technologies in libraries* (pp. 316–338). Hershey, PA: IGI Global. doi:10.4018/978-1-4666-4631-5.ch018

Zaman, M., Simmers, C. A., & Anandarajan, M. (2013). Using an ethical framework to examine linkages between "going green" in research practices and information and communication technologies. In B. Medlin (Ed.), *Integrations of technology utilization and social dynamics in organizations* (pp. 243–262). Hershey, PA: IGI Global. doi:10.4018/978-1-4666-1948-7.ch015

Zapata, B. C., & Alemán, J. L. (2013). Security risks in cloud computing: An analysis of the main vulnerabilities. In D. Rosado, D. Mellado, E. Fernandez-Medina, & M. Piattini (Eds.), *Security engineering for cloud computing: Approaches and tools* (pp. 55–71). Hershey, PA: IGI Global. doi:10.4018/978-1-4666-2125-1.ch004

Zapata, B. C., & Alemán, J. L. (2014). Security risks in cloud computing: An analysis of the main vulnerabilities. In *Software design and development: Concepts, methodologies, tools, and applications* (pp. 936–952). Hershey, PA: IGI Global. doi:10.4018/978-1-4666-4301-7.ch045

Zardari, S., Faniyi, F., & Bahsoon, R. (2013). Using obstacles for systematically modeling, analysing, and mitigating risks in cloud adoption. In I. Mistrik, A. Tang, R. Bahsoon, & J. Stafford (Eds.), *Aligning enterprise, system, and software architectures* (pp. 275–296). Hershey, PA: IGI Global. doi:10.4018/978-1-4666-2199-2.ch014

Zech, P., Kalb, P., Felderer, M., & Breu, R. (2013). Threatening the cloud: Securing services and data by continuous, model-driven negative security testing. In S. Tilley & T. Parveen (Eds.), *Software testing in the cloud: Perspectives on an emerging discipline* (pp. 280–304). Hershey, PA: IGI Global. doi:10.4018/978-1-4666-2536-5.ch013

Zhang, F., Cao, J., Cai, H., & Wu, C. (2011). Provisioning virtual resources adaptively in elastic compute cloud platforms. *International Journal of Web Services Research, 8*(3), 54–69. doi:10.4018/jwsr.2011070103

Zhang, G., Li, C., Xue, S., Liu, Y., Zhang, Y., & Xing, C. (2012). A new electronic commerce architecture in the cloud. *Journal of Electronic Commerce in Organizations, 10*(4), 42–56. doi:10.4018/jeco.2012100104

Zhang, J., Yao, J., Chen, S., & Levy, D. (2011). Facilitating biodefense research with mobile-cloud computing. *International Journal of Systems and Service-Oriented Engineering, 2*(3), 18–31. doi:10.4018/jssoe.2011070102

Zhang, J., Yao, J., Chen, S., & Levy, D. (2013). Facilitating biodefense research with mobile-cloud computing. In D. Chiu (Ed.), *Mobile and web innovations in systems and service-oriented engineering* (pp. 318–332). Hershey, PA: IGI Global. doi:10.4018/978-1-4666-2470-2.ch017

Zheng, S., Chen, F., Yang, H., & Li, J. (2013). An approach to evolving legacy software system into cloud computing environment. In X. Yang & L. Liu (Eds.), *Principles, methodologies, and service-oriented approaches for cloud computing* (pp. 207–229). Hershey, PA: IGI Global. doi:10.4018/978-1-4666-2854-0.ch009

Zhou, J., Athukorala, K., Gilman, E., Riekki, J., & Ylianttila, M. (2012). Cloud architecture for dynamic service composition. *International Journal of Grid and High Performance Computing, 4*(2), 17–31. doi:10.4018/jghpc.2012040102

Compilation of References

Abdelhamid, S., Hassanein, H., & Takahara, G. (2015). Vehicle as a resource (VaaR). *IEEE Network, 29*(1), 12–17. doi:10.1109/MNET.2015.7018198

Abolfazli, S., Sanaei, Z., Ahmed, E., Gani, A., & Buyya, R. (2013). Cloud-based augmentation for mobile devices: motivation, taxonomies, and open challenges. IEEE Communications Surveys and Tutorials. doi:10.1109/SURV

Abolhasan, M., Wysocki, T., & Dutkiewicz, E. (2004). A review of routing protocols for mobile ad hoc networks. *Ad Hoc Networks, 2*(1), 1–22. doi:10.1016/S1570-8705(03)00043-X

Abuelela, M., & Olariu, S. (2010). Taking VANET to the Clouds. *Proc. ACM MoMM.*

Abuelela, M., & Olariu, S. (2010). Taking VANET to the clouds. *Proceedings of the 8th international conference on advances in mobile computing and multimedia (MoMM'10)*, 6–13. 10.1145/1971519.1971522

Ahmad, F., Kazim, M., & Adnane, A. (2015). Vehicular Cloud Networks: Architecture and Security. In Guide to Security Assurance for Cloud Computing (pp. 211-226). Springer International Publishing.

Ahmad, F., Kazim, M., Adnane, A., & Awad, A. (2015, December). Vehicular cloud networks: Architecture, applications and security issues. In *Utility and Cloud Computing (UCC), 2015 IEEE/ACM 8th International Conference on* (pp. 571-576). IEEE.

Ahmad, F., Kazim, M., Adnane, A., & Awad, A. (2015, December). Vehicular cloud networks: architecture, applications, and security issues. In *Utility and Cloud Computing (UCC), 2015 IEEE/ACM 8th International Conference on* (pp. 571-576). IEEE.

Ahmad, I., Noor, R. M., Ali, I., & Qureshi, M. A. (2016, September). The Role of Vehicular Cloud Computing in Road Traffic Management: A Survey. In *International Conference on Future Intelligent Vehicular Technologies* (pp. 123-131). Springer.

Alam, K. M., Saini, M., & El Saddik, A. (2015). Toward social internet of vehicles: Concept, architecture, and applications. *IEEE Access: Practical Innovations, Open Solutions, 3*, 343–357. doi:10.1109/ACCESS.2015.2416657

Alonso, J. D., Vidal, E. R., Rotter, A., & Mühlenberg, M. (2008). Lane-Change Decision Aid System Based on Motion-Driven Vehicle Tracking. *IEEE Transactions on Vehicular Technology, 57*(5), •••.

Aloqaily, M., Kantarci, B., & Mouftah, H. T. (2015, November). Vehicular clouds: State of the art, challenges and future directions. In *Applied Electrical Engineering and Computing Technologies (AEECT), 2015 IEEE Jordan Conference on* (pp. 1-6). IEEE.

Aloqaily, M., Kantarci, B., & Mouftah, H. T. (2015, November). Vehicular Clouds: State of the art, challenges and future directions. In *Applied Electrical Engineering and Computing Technologies (AEECT), 2015 IEEE Jordan Conference on* (pp. 1-6). IEEE.

Aloqaily, M., Kantarci, B., & Mouftah, H. T. (2016). Multiagent/multiobjective interaction game system for service provisioning in vehicular cloud. *IEEE Access: Practical Innovations, Open Solutions*, *4*, 3153–3168. doi:10.1109/ACCESS.2016.2575038

Alsaqour, O. (2015). *A Comparative Study of Simulation Based Performance Evaluation of Routing Protocol for Ad-Hoc Networks*. Springer International Publishing.

Arif, S., Olariu, S., Wang, J., Yan, G., Yang, W., & Khalil, I. (2012). Datacenter at the airport: Reasoning about time-dependent parking lot occupancy. *IEEE Transactions on Parallel and Distributed Systems*, *23*(11), 2067–2080. doi:10.1109/TPDS.2012.47

Arkian, H. R., Atani, R. E., & Kamali, S. (2014). FcVcA: A fuzzy clustering-based vehicular cloud architecture. *2014 7th International Workshop on Communication Technologies for Vehicles (Nets4Cars-Fall)*, 24-28. 10.1109/Nets4CarsFall.2014.7000907

Armbrust, M., Fox, A., Griffith, R., Joseph, A. D., Katz, R., & Konwinski, A. (2010). A view of cloud computing. *Communications of the ACM*, *53*(4), 50–58. doi:10.1145/1721654.1721672

Artimy, M. M., Robertson, W., & Phillips, W. J. (2005), Assignment of dynamic transmission range based on estimation of vehicle density. *Proceedings of the Second International Workshop on Vehicular Ad Hoc Networks*. 10.1145/1080754.1080761

Azizian, M., Cherkaoui, S., & Hafid, A. (2016). An Optimized Flow Allocation in Vehicular Cloud. *IEEE Access: Practical Innovations, Open Solutions*, *4*, 6766–6779. doi:10.1109/ACCESS.2016.2615323

Azouqa, M., Jamoos, A., & Bali, S. (2013). *Investigation of vehicle-to-vehicle communication in a road intersection. Department of Electronic Engineering Al-Quds University.*

Baby, D., Sabareesh, R. D., Saravanaguru, R. A. K., & Thangavelu, A. (2013). VCR: vehicular cloud for road side scenarios. In Advances in Computing and Information Technology (pp. 541-552). Springer Berlin Heidelberg. doi:10.1007/978-3-642-31600-5_53

Bai, F., Krishnan, H., Sadekar, V., Holl, G., & Elbatt, T. (2006). Towards characterizing and classifying communication-based automotive applications from a wire- less networking perspective. *IEEE Workshop on Automotive Networking and Applications (AutoNet)*. doi:10.1371/journal.pone.0094602

Bajpai, A., & Singh, S. (2016). A survey on Security Analysis in Cloud computing. In *Proceedings of the International Conference on Recent Cognizance in Wireless Communication & Image Processing* (pp. 249-262). Springer India.

Balakrishnan, K., Deng, J., & Varshney, P. K. (2005). TWOACK: Preventing Selfishness in Mobile Ad Hoc Networks. *Proceedings WCNC*, 2137-2142.

Balon, N. (2006). *Introduction to vehicular ad hoc networks and the broadcast storm problem*. Academic Press.

Barlow, R. F., & Proschan, F. (1967). Mathematical Theory of Reliability. John Wiley and Sons.

Baron, B., Campista, M., Spathis, P., Costa, L. H., de Amonim, M. D., Duarte, O. C., ... Viniotis, Y. (2016). Virtualizing vehicular node resources: Feasibility study of virtual machine migration. *Vehicular Communications*, *4*, 39–46. doi:10.1016/j.vehcom.2016.04.001

Baron, B., Spathis, P., de Amorim, M. D., & Ammar, M. (2016). Cloud storage for mobile users using pre-positioned storage facilities. *Proc. 2nd Workshop on Experiences in the Design and Implementation of Smart Objects*, 11–16. 10.1145/2980147.2980151

Barroso, L. A., Calderas, J., & Holzle, U. (2013). *The datacenter as a computer: An introduction to the design of warehouse-scale machines* (2nd ed.). San Rafael, CA: Morgan & Claypool.

Beijar, N. (2002). Zone routing protocol (ZRP). Networking Laboratory, Helsinki University of Technology.

Belding-Royer, E., & Das, S. (2003). *Ad hoc on-demand distance vector (AODV) routing. draft-ietf-manet-aodv-13. txt.* INTERNET DRAFT.

Beloglazov, A., Abawajy, J., & Buyya, R. (2012). Energy-efficient resource allocation heuristics for efficient management of datacenters for cloud computing. *Future Generation Computer Systems*, *28*(5), 755–768. doi:10.1016/j.future.2011.04.017

Benslimane, A., Taleb, T., & Sivaraj, R. (2012, March). Dynamic clustering-based adaptive mobile gateway management in integrated VANET - 3G heterogeneous wireless networks. *IEEE Journal on Selected Areas in Communications*, *29*(3), 559–570. doi:10.1109/JSAC.2011.110306

Bento, L. C., Parafita, R., & Nunes, U. (2012). Intelligent traffic management at intersections supported by V2V and V2I communications. *Intelligent Transportation Systems (ITSC)15th International IEEE Conference on.*

Bevish Jinila, Y. (2015). *Anonymization based location privacy preservation in Vehicular ad hoc networks.* Academic Press.

Bevish Jinila, Y., & Komathy, K. (2014). Distributed and secured dynamic pseudo ID generation for privacy preservation in Vehicular ad hoc networks. *Journal of Applied and Theoretical Information Technology, 66*(1), 126 – 134.

Bevish Jinila, Y., & Komathy, K. (2013). A privacy preserving authentication framework for safety messages in vanet. *4th International Conference on Sustainable Energy and Intelligent System (SEISCON 2013)*, 456-461.

Bhagavathi, D., Looges, P. J., Olariu, S., Schwing, J. L., & Zhang, J. (1994). A fast selection algorithms on meshes with multiple broadcasting. *IEEE Transactions on Parallel and Distributed Systems*, *5*(7), 772–778. doi:10.1109/71.296326

Bhoi, S. K., & Khilar, P. M. (2014). Vehicular communication: A survey. *IET Networks*, *3*(3), 204–217. doi:10.1049/iet-net.2013.0065

Bilal, M., & Umar, W. (2010, June). *Issues of Routing in VANET* (Master Computer Science Thesis no: MCS-2010-20). Blekinge Institute of Technology, Sweden.

Biswas, S., & Misic, J. (2013). A Cross-Layer Approach to Privacy Preserving Authentication in WAVE-Enabled VANETs. *IEEE Transactions on Vehicular Technology*, *62*(5), 2182–2192. doi:10.1109/TVT.2013.2238566

Biswas, S., Tatchikou, R., & Dion, F. (2006). Vehicle-to-Vehicle wireless communication pro-tocols for enhancing highway traffic safety. *IEEE Communications Magazine*, *44*(1), 74–82. doi:10.1109/MCOM.2006.1580935

Bitam, S., Mellouk, A., & Zeadally, S. (2015, February). VANET-cloud: A generic cloud computing model for vehicular Ad Hoc networks. *IEEE Wireless Communications*, *22*(1), 96–102. doi:10.1109/MWC.2015.7054724

Blum, J. J., Eskandarian, A., & Hoffman, L. J. (2004, December). Challenges of inter vehicle ad hoc networks. *IEEE Transactions on Intelligent Transportation Systems*, *5*(4), 347–351. doi:10.1109/TITS.2004.838218

Blum, J., & Eskandarian, A. (2006). Fast, robust message forwarding for inter-vehicle communication netw. *2006 IEEE Intelligent Transportation Systems Conference*. 10.1109/ITSC.2006.1707422

Bonomi, F. Milito, R. & Zhu, Z. and Sateesh Addepalli. (2012). Fog computing and its role in the internet of things. In Proceedings of the first edition of the MCC workshop on Mobile cloud computing (MCC '12). ACM, New York, NY, USA, 13-16.

Bonomi, F. (2011). Connected Vehicles, Internet of things and fog computing *Proceedings of 8th ACM international Workshop on Vehicular Inter-Networking*.

Bordley, L., Cherry, C. R., Stephens, D., Zimmer, R., & Petrolino, J. (2012). Commercial motor vehicle wireless roadside inspection pilot test, Part B: Stakeholder perceptions. *91st annual meeting of the transportation research board.*

Brik, B., Lagraa, N., Ghamri-Doudane, Y., & Lakas, A. (2016). Finding the most adequate public bus in vehicular clouds. *Proc. 2016 International Conference on Wireless Networks and Mobile Communications*, 67-74. 10.1109/WINCOM.2016.7777193

Buttyan, L., Holczer, T., Weimerskirch, A., & Whyte, W. (2009). Slow: A practical pseudonym changing scheme for location privacy in vanets. *IEEE Vehicular Networking Conference (VNC)*, 1-8. 10.1109/VNC.2009.5416380

Buyya, R., Vecchiola, C., & Thamarai Selvi, S. (2013). *Mastering Cloud Computing: Foundations and Applications Programming*. Elsevier.

Calandriello, G., Papadimitratos, P., Hubaux, J. P., & Lioy, A. (2007). Efficient and Robust Pseudonymous Authentication in VANET. *Proc. 4th ACM Int. Workshop VANET*, 19-28. 10.1145/1287748.1287752

CANU Research Group (Stuttgart University). (n.d.). Retrieved from http://canu.informatik.uni-stuttgart.de/mobisim/

CANU Research Group. CanuMobiSim (Mobility Simulation Environment). (2001). Institute of Parallel and Distributed Systems (IPVS). University of Stuttgart. Retrieved from http://canu.informatik.uni-stuttgart.de/

Cardosa, M., Singh, A., Pucha, H., & Chandra, A. (2012). Exploiting spatio-temporal tradeoffs for energy-aware MapReduce in the cloud. *IEEE Transactions on Computers*, *61*(12), 1737–1751. doi:10.1109/TC.2012.166

Cardoso, A., Moreira, F., & Simões, P. (2014). A Survey of Cloud Computing Migration Issues and Frameworks. In *New Perspectives in Information Systems and Technologies* (Vol. 1, pp. 161–170). Springer International Publishing. doi:10.1007/978-3-319-05951-8_16

Chakeres, I. (2008). Dynamic MANET on-demand (DYMO) routing. *Internet Draft (draft-ietf-manet-dymo-14).*

Chakeres, I. D., & Belding-Royer, E. M. (2004, March). AODV routing protocol implementation design. In *Distributed Computing Systems Workshops, 2004. Proceedings. 24th International Conference on* (pp. 698-703). IEEE. 10.1109/ICDCSW.2004.1284108

Chang, Y. C., Chen, J. L., Ma, Y. W., & Chiu, P. S. (2015, December). Vehicular Cloud Serving Systems with Software-Defined Networking. In *International Conference on Internet of Vehicles* (pp. 58-67). Springer International Publishing 10.1007/978-3-319-27293-1_6

Chaqfeh, M., Mohamed, N., Jawhar, I., & Wu, J. (2016). Vehicular cloud data collection for intelligent transportation systems. Smart cloud networks & systems (scns), 1–6. doi:10.1109/SCNS.2016.7870555

Chen, H., & Tian, J. (2009). Research on the Controller Area Network. *Networking and Digital Society, 2009. ICNDS '09. International Conference on, 2,* 251-254. 10.1109/ICNDS.2009.142

Chen, J., Long, H., Zheng, Q., Xing, M., & Wang, W. (2015). An SMDP-Based Resource Management Scheme for Distributed Cloud Systems. *2015 IEEE 81st Vehicular Technology Conference (VTC Spring).* doi:10.1109/vtcspring.2015.7145692

Cho, H., & Kim, B. (2014). Cooperative Intersection Collision-Warning System Based on Vehicle-to-Vehicle Communication. *Contemporary Engineering Sciences, 7*(22), 1147-1154.

Choi, J. Y., Jakobsson, M., & Wetzel, S. (2005). Balancing auditability and privacy in vehicular networks. *Proceedings of the 1st ACM international workshop on Quality of service & security in wireless and mobile networks*, 79-87.

CintraM. O. (2014). Retrieved from http://blogchicao.tripod.com/custocarro.html

Clausen, T., & Jacquet, P. (2003). *Optimized link state routing protocol (OLSR)* (No. RFC 3626).

Cordeschi, N., Amendola, D., & Baccarelli, E. (2015). Reliable Adaptive Resource Management for Cognitive Cloud Vehicular Networks. *IEEE Transactions on Vehicular Technology*, *64*(6), 2528–2537. doi:10.1109/tvt.2014.2345767

Cushman, I. J., Rawat, D. B., Chen, L., & Yang, Q. (2016, August). Performance Evaluation of Vehicular Ad Hoc Networks for Rapid Response Traffic Information Delivery. In *International Conference on Wireless Algorithms, Systems, and Applications* (pp. 571-579). Springer International Publishing 10.1007/978-3-319-42836-9_50

Darus, M. Y. B., & Bakar, K. A. (2011). Congestion control framework for disseminating safety messages in Vehicular Ad-Hoc Networks (VANETs). *International Journal of Digital Content Technology and Its Applications, 5*(2), 173-180.

Darus, M. Y., & Bakar, K. A. (2011). A review of congestion control algorithm for event-driven safety messages in vehicular networks. *Channels, 10*, 13.

Datta, S. K. & Costa, D. C., & Härri, J. (2016). *Christian Bonnet Communication Systems Department, France Integrating Connected Vehicles in Internet of Things Ecosystems: Challenges and Solutions.* French research project and the HIGHTS project funded by the European Commission (636537-H2020).

De Fuentes, J. M., González-Manzano, L., González-Tablas, A. I., & Blasco, J. (2014). Security models in vehicular ad-hoc networks: A survey. *IETE Technical Review, 31*(1), 47–64. doi:10.1080/02564602.2014.890844

De Oliveira Barros, M. T. A., de Morais Gomes, R. C., & da Costa, A. F. B. F. (2013). A Top-down Multi-layer Routing Architecture for Vehicular Ad-Hoc Networks. *IEEE Latin America Transactions, 11*(6), 1344–1352. doi:10.1109/TLA.2013.6710382

Decker, L. (2016, October). private communication.

Deng, J., Huang, S.-H., Han, Y., & Deng, J. (2010, December). Fault tolerant and reliable computation in cloud computing. In *Globecom workshops (gc wkshps), 2010* (pp. 1601–1605). IEEE. doi:10.1109/GLOCOMW.2010.5700210

Devi, Y. U., & Rukmini, M. S. S. (2016). IoT in Connected Vehicles: Challenges and Issues- A Review. *International conference on Signal Processing,Communication, Power and Embedded System.*

Ding, D., Gebel, K., Phongsavan, P., Bauman, A. E., & Merom, D. (2014). Driving: A Road to Unhealthy Lifestyles and Poor Health Outcomes. *PLoS One, 9*(6).

Dinh, H. T., Lee, C., Niyato, D., & Wang, P. (2013). A survey of mobile cloud computing: Architecture, applications, and approaches. *Wireless Communications and Mobile Computing, 13*(18), 1587–1611. doi:10.1002/wcm.1203

Djenouri, D. & Khelladi, L. & Badache, A. N. (2005). A survey of Security Issues in mobile Ad Hoc Networks. *IEEE Communication Surveys Tutorials, 7*(4), 2-28.

Dobashi, K. (2005). Adaptive MAC Protocol for high-load Inter-Vehicle Communication. *Wireless and Mobile Computing, Networking and Communications, 2005. (WiMob'2005), IEEE International Conference on, 3.* 10.1109/WIMOB.2005.1512896

Dong, X., Wei, L., Zhu, H., Cao, Z., & Wang, L. (2011). An Efficient Privacy-Preserving Data-Forwarding Scheme for Service-Oriented Vehicular Ad Hoc Networks. Vehicular Technology. *IEEE Transactions on, 60*(2), 580–591.

Dressler, F., Handle, P., & Sommer, C. (2014, August). Towards a vehicular cloud-using parked vehicles as a temporary network and storage infrastructure. In *Proceedings of the 2014 ACM international workshop on Wireless and mobile technologies for smart cities* (pp. 11-18). ACM. 10.1145/2633661.2633671

Egwutuoha, I. P., Chen, S., Levy, D., & Selic, B. (2012, May). A fault tolerance framework for high performance computing in cloud, *Proc. 12th IEEE/ACM International Symposium on Cluster, Cloud and Grid Computing*, 709-710. 10.1109/CCGrid.2012.80

El Zarki, Mehrotra, Tsudik, & Venkatasubramanian. (2002). Security issues in a future vehicular network. *European Wireless*.

Eltahir & Saeed. (2015). Performance Evaluation of an Enhanced Hybrid Wireless Mesh Protocol (E-HWMP) Protocol for VANET. *International Conference On Computing, Control, Networking, Electronics And Embedded Systems Engineering (ICCNEEE 2015)*, 95 - 100.

Eltahir, A. A., & Saeed, R. A. (2015, September). Performance Evaluation of an Enhanced Hybrid Wireless Mesh Protocol (E-HWMP) Protocol for VANET. In *Computing, Control, Networking, Electronics and Embedded Systems Engineering (ICCNEEE), 2015 International Conference on* (pp. 95-100). IEEE.

Eltahir, Saeed, & Alawi. (2013). An Enhanced Hybrid Wireless Mesh Protocol (E-HWMP) Protocol for Multi-hop Vehicular Communications. *International Conference On Computing, Electrical And Electronic Engineering (ICCEEE 2013)*, 1 - 8.

Eltahir, Saeed, & Mokhtar. (2014). Vehicular Communication and Cellular Network Integration: Gateway Selection Perspective. *International Conference On Computer & Communication Engineering (ICCCE 2014)*, 64 - 67.

Eltahir, A. A., Saeed, R. A., Mukherjee, A., & Hasan, M. K. (2016). Evaluation and analysis of an enhanced hybrid wireless mesh protocol for vehicular ad hoc network. *EURASIP Journal on Wireless Communications and Networking*, (1): 1–11.

Eltahir, A. A., Saeed, R. A., Mukherjee, A., & Hasan, M. K. (2016). Evaluation and analysis of an enhanced hybrid wireless mesh protocol for vehicular ad hoc network. *EURASIP Journal on Wireless Communications and Networking*, *2016*(169), 1–11. doi:10.118613638-016-0666-5

Eltoweissy, M., Olariu, S., & Younis, M. (2010). *Towards autonomous Vehicular clouds Ad Hoc Networks*. Berlin, Germany: Springer; doi:10.15439/2014F503

Eltoweissy, M., Olariu, S., & Younis, M. (2010). Towards autonomous vehicular clouds. *Ad Hoc Networks*, 1–16.

Eltoweissy, M., Olariu, S., & Younis, M. (2010). Towards autonomous vehicular clouds. *Proc. of AdHocNets'2010*.

Eltoweissy, M., Olariu, S., & Younis, M. (2010b). Towards autonomous vehicular clouds. In J. Zheng, D. Simplot-Ryl, & V. M. Leung (Eds.), *Ad hoc networks* (Vol. 49, pp. 1–16). Berlin: Springer. doi:10.1007/978-3-642-17994-5_1

Falchetti, A., Azurdia-Meza, C., & Cespedes, S. (2015, October). Vehicular cloud computing in the dawn of 5G. In *Electrical, Electronics Engineering, Information and Communication Technologies (CHILECON), 2015 CHILEAN Conference on* (pp. 301-305). IEEE.

Falchetti, A., Azurdia-Meza, C., & Cespedes, S. (2015). Vehicular cloud computing in the dawn of 5G. *2015 CHILEAN Conference on Electrical, Electronics Engineering, Information and Communication Technologies (CHILECON)*, 301-305. 10.1109/Chilecon.2015.7400392

Falko, D., Hannes, H., Onur, A., & Ozan, T. (2014, June). InterVehicle Communication: Quo Vadis. *IEEE Communications Magazine*, *52*(6), 170–177. doi:10.1109/MCOM.2014.6829960

Fatima, B., & Shoad, A. K. (2005). Traffic Estimation and Real Time Predication Using Ad Hoc Networks. *Proceedings of IEEE International Conference on Emerging Technologies 2005.*

Feiri, M., Pielage, R., Petit, J., Zannone, N., & Kargl, F. (2015). Pre-distribution of certificates for pseudonymous broadcast authentication in VANET. *IEEE 81st Vehicular Technology Conference (VTC Spring),* 1-5. 10.1109/VTCSpring.2015.7146029

Fernandes, P., & Nunes, U. (2007). *Vehicle communications: A short survey.* Institute of Systems and Robotics Department of Electrical and Computer Engineering, University of Coimbra.

Fernando, N., Loke, S. W., & Rahayu, W. (2013). Mobile cloud computing: A survey. *Future Generation Computer Systems*, *29*(1), 84–106. doi:10.1016/j.future.2012.05.023

Firdhous, M., Ghazali, O., & Hassan, S. (2014). Fog computing: will it be the future of cloud computing? *Proceedings of the 3rd International Conference on Informatics and Applica-tions (ICIA'14)*, 8–15.

Florin, R., Abolghasemi, S., Ghazi Zadeh, A., & Olariu, S. (2017). Big Data in the parking lot. In A. Z. Kuan-Ching Li Hai Jiang (Ed.), *Big Data Management and Processing* (pp. 425–449). Boca Raton, FL: Taylor and Francis.

Florin, R., Ghazi Zadeh, A., Ghazizadeh, P., & Olariu, S. (2017). Promoting reliability and availability in vehicular clouds. *IEEE Transactions on Cloud Computing*.

Florin, R., Ghazizadeh, P., Ghazi Zadeh, A., & Olariu, A. (2015, Oct.). Enhancing dependability through redundancy in military vehicular clouds. *Proc. IEEE MILCOM'*2015. 10.1109/MILCOM.2015.7357586

Florin, R., Ghazizadeh, P., Ghazi Zadeh, A., & Olariu, S. (2017). Reasoning about job completion time in vehicular clouds. *IEEE Transactions on Intelligent Transportation Systems*, *18*(7), 1762–1771. doi:10.1109/TITS.2016.2620434

Ford-Comp. (2010). *University of Michigan Reveal Students' Vision for Future of In-Car Cloud Computing Apps*. Ford Online.

Frodigh, M., Johansson, P., & Larsson, P. (2000). Wireless ad hoc networking: the art of networking without a network. *Ericsson Review, 4*(4), 249.

Fujise, M., Kato, A., Sato, K., & Harada, H. (2002). Intelligent transport systems. In *Wireless Communication Technologies: New Multimedia Systems* (pp. 171–200). Boston, MA: Springer. doi:10.1007/0-306-47326-7_8

Garai, M., Rekhis, S., & Boudriga, N. (2015, July). Communication as a service for cloud vanets. In *Computers and Communication (ISCC), 2015 IEEE Symposium on* (pp. 371-377). IEEE 10.1109/ISCC.2015.7405543

Gayomali, C. (2011). *Toyota and microsoft team up to bring cloud computing on the road*. Time Tech.

Gerla, M. (2012). Vehicular Cloud Computing. *2012 The 11th Annual Mediterranean Ad Hoc Networking Workshop (Med-Hoc-Net).* doi:10.1109/medhocnet.2012.6257116

Gerla, M. (2012). Vehicular cloud computing. *Proceedings of the 11th annual Mediterranean ad hoc networking workshop (Med-Hoc-Net)*, 152–5. 10.1109/MedHocNet.2012.6257116

Gerla, M., Hong, X., & Pei, G. (2002). *Fisheye state routing protocol (FSR) for ad hod networks*. draft-ietf-manet-fsr-03. txt.

Gerla, M., Lee, E. K., Pau, G., & Lee, U. (2014, March). Internet of vehicles: From intelligent grid to autonomous cars and vehicular clouds. In *Internet of Things (WF-IoT), 2014 IEEE World Forum on* (pp. 241-246). IEEE.

Gerlach, M., & Guttler, F. (2007). Privacy in VANETs using changing pseudonyms-ideal and real. In *Vehicular Technology Conference, 2007. VTC2007-Spring*. IEEE. 10.1109/VETECS.2007.519

Gerla, M. (2012). Vehicular Cloud Computing. *Workshop on Vehicular Communications and Applications*.

Gerla, M., Pei, G., & Lee, S. J. (1999). Wireless, mobile ad-hoc network routing. *Proceedings of IEEE/ACM FOCUS'99*, 1112-1124.

Gerla, M., Weng, J.-T., & Pau, G. (2013). Pics-on-wheels: Photo Surveillance in the Vehicular Cloud. In *Proceedings of the International Conference on Computing, Networking and Communications (ICNC)*. IEEE. 10.1109/ICCNC.2013.6504250

Ghafoor, K. Z., Mohammed, M. A., & Lloret, J. L. (2016). Vehicular Cloud Computing: Trends and Challenges. *ZANCO Journal of Pure and Applied Sciences*, *28*(3), 67–77.

Ghazizadeh, P., Florin, R., Ghazi Zadeh, A., & Olariu, S. (2016). Reasoning about the Mean-Time-to-Failure in vehicular clouds. *IEEE Transactions on Intelligent Transportation Systems*, *17*(3), 751–761. doi:10.1109/TITS.2015.2486523

Giang, N. K., Victor, C. M., & Rodger, L. (2016). On Developing Smart Transportation Appli-cations in Fog Computing Paradigm. *Proceedings of the 6th ACM Symposium on develop-ment and Analysis of Intellegent Vehicular Networks and Applications (DIVANet'16)*, 91–98.

Gollan & Meinel. (2002). Digital signatures for automobiles. In *Systemics, Cybernetics and Informatics*. SCI.

Gong, H., Yu, L., Liu, N., & Zhang, X. (2016). Mobile content distribution with vehicular cloud in urban VANETS. *China Communications*, *13*(8), 84–96. doi:10.1109/CC.2016.7563691

Gu, L., Zeng, D., & Guo, S. (2013). Vehicular cloud computing: A survey. *2013 IEEE Globecom Workshops (GC Wkshps)*, 403-407. doi: 10.1109/GLOCOMW.2013.6825021

Gu, L., Zeng, D., & Guo, S. (2013, December). Vehicular cloud computing: A survey. In Globecom Workshops (GC Wkshps), 2013 IEEE (pp. 403-407). IEEE.

Gu, L., Zeng, D., &Guo, S. (2013). Vehicular cloud computing: A survey. *2013 IEEE Globecom Workshops (GC Wkshps).* doi:10.1109/glocomw.2013.6825021

Guerrero-ibanez, J. A., Zeadally, S., & Contreras-Castillo, J. (2015). Integration challenges of intelligent transportation systems with connected vehicle, cloud computing, and internet of things technologies. *IEEE Wireless Communications*, *22*(6), 122–128. doi:10.1109/MWC.2015.7368833

Haas, Z. J., Pearlman, M. R., & Samar, P. (2002). *The zone routing protocol (ZRP) for ad hoc networks.* Academic Press.

Habib, S. M., Hauke, S., Ries, S., & Mühlhäuser, M. (2012). Trust as a facilitator in cloud computing: A survey. *Journal of Cloud Computing: Advances, Systems, and Applications*, *1*(1), 19. doi:10.1186/2192-113X-1-19

Haerri, J., Fiore, M., Fethi, F., & Bonnet, C. (2006). *VanetMobiSim: generating realistic mobility patterns for VANETs.* Institut Eurécom and Politecnico Di Torino.

Hamidian. (2003). *A Study of Internet Connectivity for Mobile Ad Hoc Networks in NS 2.* Department of Communication Systems Lund Institute of Technology, Lund University.

Harri, J., Filali, F., & Bonnet, C. (2009). Mobility models for vehicular ad hoc networks: A survey and taxonomy. *IEEE Communications Surveys and Tutorials*, *11*(4), 19–41. doi:10.1109/SURV.2009.090403

Harris, C. (2013). *General motors connects cars to 4 G LTE Tech page one.* Academic Press.

Hartenstein, H., & Laberteaux, K. (2010). *VANET Vehicular Applications and Inter-Networking Technologies.* Academic Press.

Hartenstein, H., & Laberteaux, K. P. (2008). A tutorial survey on vehicular ad hoc networks. *Communications Magazine, IEEE*, *46*(6), 164–171. doi:10.1109/MCOM.2008.4539481

Hassanein, H. S., Abdelhamid, S., & Elgazzar, K. (2015, October). A framework for vehicular cloud computing. In *Connected Vehicles and Expo (ICCVE), 2015 International Conference on* (pp. 238-239). IEEE. 10.1109/ICCVE.2015.51

Hatch, D. (2013, November). *Singapore strives to become 'The Smartest City': Singapore is using data to redefine what it means to be a 21st-century metropolis.* Academic Press.

Hellbruck, H., Cao, J., & Zuo, T. (2009). Fast prototyping for VANET applications with PDAs Wireless Communication, Vehicular Technology, Information Theory and Aerospace & Electronics Systems Technology, 2009. *Wireless VITAE 2009.1st International Conference on.*

Hennessy, J. L., & Patterson, D. A. (2012). Computer Architecture: A Quantitative Approach (5th ed.). Elsevier.

He, W., Yan, G., & Da Xu, L. (2014). Developing vehicular data cloud services in the IoT environment. *IEEE Transactions on Industrial Informatics*, *10*(2), 1587–1595. doi:10.1109/TII.2014.2299233

Hindia, M. N., Reza, A. W., Dakkak, O., Awang Nor, S., & Noordin, K. A. (2014). Cloud computing applications and platforms: a survey (pp. 515–524). Academic Press.

Hoover, J.N., & Martin, R. (2008). Demystifying the Cloud. *Information Week Research & Reports*, 30–37.

Hu, X., Wang, L., & Sheng, Z. & TalebiFard, P. & Zhou, L. & Liu, J. & r C.M. Leung, V. (2014). Towards a service centric contextualized vehicular cloud. In Proceedings of the fourth ACM international symposium on Development and analysis of intelligent vehicular networks and applications (DIVANet '14). ACM, New York, NY, USA, 73-80.

Huang, C. Y., & Xu, K. (2016, July). Reliable realtime streaming in vehicular cloud-fog computing networks. *Proc. 2016 IEEE/CIC International Conference on Communications in China*, 1-6. 10.1109/ICCChina.2016.7636838

Huang, C., Lu, R., Zhu, H., Hu, H., & Lin, X. (2016, December). PTVC: *Achieving Privacy-Preserving Trust-Based Verifiable Vehicular Cloud Computing. In Global Communications Conference (GLOBECOM)*, 2016 *IEEE* (pp. 1-6). IEEE.

Huang, D., Xing, T., & Fang, S. (2013). Mobile Cloud Computing Service Models: A User Centric Approach. *IEEE Network*, *27*(5), 6–11.

Huang, S. C., Chen, B. H., Chou, S. K., Hwang, J. N., & Lee, K. H. (2016). Smart Car. *IEEE Computational Intelligence Magazine*.

Hubaux, J.-P., Capkun, S., & Luo, J. (2004). The security and privacy of smart vehicles. *IEEE Security and Privacy Magazine*, *2*(3), 49–55. doi:10.1109/MSP.2004.26

Hussain, R., & Oh, H. (2014). A secure and privacy-aware route tracing and revocation mechanism in vanet-based clouds. 정보보호학회논문지, *24*(5), 795-807.

Hussain, R., Son, J., Eun, H., Kim, S., & Oh, H. (2012, December). Rethinking vehicular communications: Merging VANET with cloud computing. In *Cloud Computing Technology and Science (CloudCom), 2012 IEEE 4th International Conference on* (pp. 606-609). IEEE.

Hussain, R., Son, J., Eun, H., Kim, S., & Oh, H. (2012). Rethinking Vehicular Communications: Merging VANET with cloud computing. *4th IEEE International Conference on Cloud Computing Technology and Science Proceedings*. doi:10.1109/CloudCom.2012.6427481

Iwai, A., & Aoyama, M. (2011, July). Automotive cloud service systems based on service-oriented architecture and its evaluation. In *Cloud Computing (CLOUD), 2011 IEEE International Conference on* (pp. 638-645). IEEE. 10.1109/CLOUD.2011.119

Jacquet, P., Muhlethaler, P., Clausen, T., Laouiti, A., Qayyum, A., & Viennot, L. (2001). Optimized link state routing protocol for ad hoc networks. In *Multi Topic Conference, 2001. IEEE INMIC 2001. Technology for the 21st Century. Proceedings. IEEE International* (pp. 62-68). IEEE. 10.1109/INMIC.2001.995315

Jeong, J., Jeong, H., Lee, E., Oh, T., & Du, D. H. C. (2016, June). Saint: Self-adaptive interactive navigation tool for cloud-based vehicular traffic optimization. *IEEE Transactions on Vehicular Technology*, *65*(6), 4053–4067. doi:10.1109/TVT.2015.2476958

Jiang, D., & Delgrossi, L. (2008). IEEE 802.11p: Towards an International Standard for Wireless Access in Vehicular Environments. *Vehicular Technology Conference, VTC Spring 2008, IEEE*, 2036–2040.

Jiang, H., & Garcia-Luna-Aceves, J. J. (2001). Performance comparison of three routing protocols for ad hoc networks. In *Computer Communications and Networks, 2001. Proceedings. Tenth International Conference on* (pp. 547-554). IEEE. 10.1109/ICCCN.2001.956320

Jiang, D., Taliwal, V., Meier, A., Holfelder, W., & Herrtwich, R. (2006). Design of 5.9 GHz DSRC-based vehicular safety communication. *IEEE Wireless Communications*, *13*(5), 36–43. doi:10.1109/WC-M.2006.250356

Jiau, M.-K. (2015). Multimedia Services in Cloud-Based Vehicular Networks. *IEEE Intell. Transport. Syst. Mag.*, *7*(3), 62–79.

Johnson, D., Hu, Y. C., & Maltz, D. (2007). *The dynamic source routing protocol (DSR) for mobile ad hoc networks for IPv4* (No. RFC 4728).

Johnson, D. B., Maltz, D. A., & Broch, J. (2001). DSR: The dynamic source routing protocol for multi-hop wireless ad hoc networks. *Ad Hoc Networks*, *5*, 139–172.

Kai, K., Cong, W., & Tao, L. (2016). Fog computing for Vehicular Ad hoc Networks: Paradigms, Scenarios and Issues. *Journal of China Universities of Posts and Telecommunications*, *23*(2), 56–96.

Karagiannis, G., Altintas, O., Ekici, E., Heijenk, G., Jarupan, B., Lin, K., & Weil, T. (2011). Vehicular networking: A survey and tutorial on requirements, architectures, challenges, standards, and solutions. *IEEE Communications Surveys and Tutorials*, *13*(4), 584–616. doi:10.1109/SURV.2011.061411.00019

Kathirvel & Loyd. (2014). Optimized Hybrid Wireless Mesh Protocol using Estimation of Paket Loss Rate Algorithm for VANET. *International Journal of Computer Science and Mobile Applications, 2*(3), 117-123.

Khabazian, M., Aissa, S., & Mehmet, M. (2011). Performance Modeling of message dissemination in vehicular ad hoc networks with priority. *IEEE Journal on Selected Areas in Communications*, *29*(1), 61–71.

Khan, A. A. (2015). *An Evolutionary Game Theoretic Approach for Stable Clustering in Vehicular Ad hoc Networks (VANETs)* (Master of Engineering thesis). University of Technology.

Kim, S.H., Kim, B.H., Kim, Y.K., & Lee, D.H. (n.d.). *Auditable and privacy preserving authentication in vehicular networks*. IEEE.

Kim, Seo & Kim. (2008). A gateway system for an automotive system: LIN, CAN, and FlexRay Industrial Informatics. *6th IEEE International Conference.*

Kim, J., Jo, M. J., & Jung, Y. (2009), Efficient Routing Scheme Using Circular Node for VANET Ubiquitous Information Technologies & Applications, 2009. *ICUT '09. Proceedings of the 4th International Conference on.*

Kim, O. T. T., Dang Tri, N., Nguyen, V. D., Tran, N. H., & Hong, C. S. (2015). A Shared Parking Model in Vehicular Network Using Fog and Cloud environment. *7th Asia Pacific Network Operations and Management Symposium (APNOMS)*, 321–326. 10.1109/APNOMS.2015.7275447

Kim, R., Lim, H., & Krishnamachari, B. (2016, January). Prefetching-based data dissemination in vehicular cloud systems. *IEEE Transactions on Vehicular Technology, 65*(1), 292–306. doi:10.1109/TVT.2015.2388851

Kleberger, P., Olovsson, T., & Jonsson, E. (2011). Security Aspects of the in vehicle network in the connected car. *Proceedings of IEEE Intelligent Vehicle Symposium*, 528-533. 10.1109/IVS.2011.5940525

Ko, S. Y., Hoque, I., Cho, B., & Gupta, I. (2010). Making cloud intermediate data fault-tolerant. *Proc. 1-st ACM Symposium on Cloud Computing*, 181–192. 10.1145/1807128.1807160

Kumar, N., Singh, J. P., Bali, R. S., Misra, S., & Ullah, S. (2015). An intelligent clustering scheme for distributed intrusion detection in vehicular cloud computing. *Cluster Computing*, 1263-1283.

Kumar, A., & Singla, A. K. (2011). Performance evaluation of MANET routing protocols on the basis of TCP traffic pattern. *International Journal of Information Technology Convergence and Services, 1*(5), 41.

Kwak, D., Liu, R., Kim, D., Nath, B., & Iftode, L. (2016). Seeing is believing: Sharing real-time visual traffic information via vehicular clouds. *IEEE Access: Practical Innovations, Open Solutions, 4*, 3617–3631. doi:10.1109/ACCESS.2016.2569585

Lakakis, K., & Kyriakou, K. (2015). Creating and intelligent transportation system for smart cities: performance evaluation of spatial-temporal algorithms for traffic prediction. *Proc. 14-th International Conference on Environmental Science and Technology.*

Le, L., Festag, A., Baldesari, R., & Zhang, W. (2009). CAR-2-X Communications in Europe. In Vehicular Networks: From Theory to Practice (pp. 4.1–b8.32). Taylor and Francis.

Lee, E., Gerla, M., & Oh, S. (2014). Vehicular cloud networking: architecture and design principles. *IEEE Communications Magazine, 52*(2), 148-155. doi:10.1109/mcom.2014.6736756

Lee, S. J., Gerla, M., & Toh, C. K. (1999). A simulation study of table-driven and on-demand routing protocols for mobile ad hoc networks. *IEEE Network, 13*(4), 48-54.

Lee, E., Lee, E. K., Gerla, M., & Oh, S. Y. (2014, February). Vehicular cloud networking: Architecture and design principles. *IEEE Communications Magazine, 52*(2), 148–155. doi:10.1109/MCOM.2014.6736756

Lee, U., Cheung, R., & Gerla, M. (2009). Emerging Vehicular Applications. In S. Olariu & M. C. Weigle (Eds.), *Vehicular Networks: From Theory to Practice* (pp. 6.1–6.30). Boca Raton, FL: Taylor and Francis. doi:10.1201/9781420085891-c6

Li, B., Mirhashemi, M. S., Laurent, X., & Gao, J. (2010). Wireless access for vehicular environments. *Telematics Communication Technologies and Vehicular Networks: Wireless Architectures and Applications*, 90-107.

Li, F., & Wang, L. & Can. (2008). Bus Communication System Based on Matlab/Simulink Wireless Communications, Networking and Mobile Computing. *WiCOM '08. 4th International Conference on.*

Li, F., & Wang, Y. (2007). Routing in vehicular ad hoc networks: A survey. *IEEE Vehicular Technology Magazine*, *2*(2), 12–22. doi:10.1109/MVT.2007.912927

Li, M. W., Wu, T. H., Lin, W. Y., Lan, K. C., Chou, C. M., & Hsu, C. H. (2011). On the feasibility of using 802.11 p for communication of electronic toll collection systems. *ISRN Communications and Networking*.

Lind, R. (1999). The Network Vehicle-a glimpse into the future of mobile multi-media. *IEEE Aerospace and Electronic Systems Magazine, 14*(9), 27-32.

Lin, X., Sun, X., Ho, P. H., & Shen, X. (2007). GSIS: A Secure and Privacy preserving protocol for vehicular communications. *IEEE Transactions on Vehicular Technology*, *56*(6), 3442–3456. doi:10.1109/TVT.2007.906878

Li, P., Guo, S., Yu, S., & Vasilakos, A. V. (2014, December). Reliable Multicast with Pipelined Network Coding Using Opportunistic Feeding and Routing. *IEEE Transactions on Parallel and Distributed Systems*, *25*(12), 3264–3273. doi:10.1109/TPDS.2013.2297105

Li, T., Zhao, M., Liu, A., & Huang, C. (2017). On Selecting Vehicles as Recommenders for Vehicular Social Networks. *IEEE Access: Practical Innovations, Open Solutions.*

Litman, T. (2015, December). *Autonomous vehicle implementation predictions: Implications for transport planning*. Presented at the *2015 Transportation Research Board Annual Meeting*.

Liu, Jiang, Nishiyama, Kato, & Shen. (2012). End-to-End Delay in Mobile Ad Hoc Networks with Generalized Transmission Range and Limited Packet Redundancy. *IEEE Wireless Communications and Networ king Conference (WCNC 2012)*, 1731-1736.

Liu, W., Wen, X. Z., Duan, B., Yuan, H., & Wang, N. (2007). Rear Vehicle Detection and Tracking for Lane Change Assist, Member. *IEEE Proceedings of the 2007 IEEE Intelligent Vehicles Symposium.*

Liu, B., Jia, D., Wang, J., Lu, K., & Wu, L. (2015). Cloud-Assisted Safety Message Dissemination in VANET-Cellular Heterogeneous Wireless Network. *IEEE System Journal*, 1–12.

Liu, N., Liu, M., Lou, W., Chen, G., & Cao, G. (2011). *PVA in VANETs: Stopped cars are not silent* (pp. 431–435). Shanghai: IEEE INFOCOM.

Loo, J., Mauri, J. L., & Ortiz, J. H. (Eds.). (2016). *Mobile ad hoc networks: current status and future trends*. CRC Press.

Lorido-Botran, T., Miguel-Alonso, J., & Lozano, J. A. (2014). A review of auto-scaling techniques for elastic applications in cloud environments. *Journal of Grid Computing*, *12*(4), 559–592. doi:10.100710723-014-9314-7

Lo, S.-C., Lin, Y.-J., & Gao, J.-S. (2013). A multi-head clustering algorithm in vehicular ad hoc networks. *International Journal of Computer Theory and Engineering*, *5*(2), 242–247. doi:10.7763/IJCTE.2013.V5.686

Lozano, J. G., Montero, M. I., & Martínez, M. A. (2011). Power Electrical and Electronic Systems (PE&ES). *School of Industrial Engineering (University of Extremadura) Compatibility and Power Electronics (CPE), 7th International Conference-Workshop.*

Luan, T. H., Gao, L. X., & Xiang, Y. (2015). *Fog computing: focusing on mobile users at the edge.* CoRR, arXiv:1502.0181v1

Lu, D., Li, Z., Huang, D., Lu, X., Deng, Y., Chowdhary, A., & Li, B. (2016). VC-bots: a vehicular cloud computing testbed with mobile robots. *Proc. First International Workshop on Internet of Vehicles and Vehicles of Internet*, 31–36. 10.1145/2938681.2938683

Lu, R. (2012). Pseudonym changing at social spots: An effective strategy for location privacy in vanets. *IEEE Transactions on Vehicular Technology*, *61*(1), 86–96. doi:10.1109/TVT.2011.2162864

Ma, M., Huang, Y., Chu, C. H., & Wang, P. (2012, December). User-driven cloud transportation system for smart driving. In *Cloud Computing Technology and Science (CloudCom), 2012 IEEE 4th International Conference on* (pp. 658-665). IEEE. 10.1109/CloudCom.2012.6427600

Malandrino, F., Casetti, C., & Chiasserini, C. F & Sommer. & C. Dressler, F. (2014). The Role of Parked Cars in Content Downloading for Vehicular Networks. *IEEE Transactions on Vehicular Technology*, *63*(9), 4606–4617.

Malik, S., & Huet, F. (2011). Adaptive fault tolerance in realtime cloud computing. *Proc. IEEE World Congress on Services*, 280-287.

Marinescu, D. C. (2013). *Cloud Computing, Theory and Applications.* Elsevier.

Marinescu, D. C., Florin, R., & Olariu, S. (2017). Towards platoon-aware retiming of traffic lights in smart cities. In F. Hu (Ed.), *Vehicle-to-Vehicle and Vehicle-to-Infrastructure Communications: A Technical Approach.* New York, NY: Taylor and Francis.

Marinescu, D. C., Paya, A., Morrison, J. P., & Olariu, S. (2017). An approach for scaling cloud resource management. *Cluster Computing*, *20*(1), 909–924. doi:10.100710586-016-0700-8

Mario, G. (2012). Vehicular Cloud Computing. *Vehicular Communications and Applications Workshop*, 152-155.

Md AliAl, M., Khairul, A., Md Fakhrul Alam, O., & Esfar- E- Alam, A. (2012). *Proceedings of the World Congress on Engineering and Computer Science 2012,* 1.

Md, W., Mehdi, S., Abdullah, G., & Rajkumar, B. (2014). A survey on vehicular cloud computing. *Journal of Network and Computer Applications*, *40*, 325–344. doi:10.1016/j.jnca.2013.08.004

Mei, X., Nagasaka, N., Okumura, B., & Prokhorov, D. (2015, June). Detection and motion planning for roadside parked vehicles at long distance. In *Intelligent Vehicles Symposium (IV)*, 2015 *IEEE* (pp. 412-418). IEEE. 10.1109/IVS.2015.7225720

Mekki, T., Jabri, I., Rachedi, A., & Ben Jemaa, M. (2016). Vehicular cloud networks: Challenges, architectures, and future directions. *Vehicular Communications*.

Mell, P., & Grace, T. (2011). *NIST definition of cloud computing*. National Institute of Standards and Technology. doi:10.6028/NIST.SP.800-145

Meneguette, R. I. (2016). A Vehicular Cloud-Based Framework for the Intelligent Transport Management of Big Cities. *International Journal of Distributed Sensor Networks*, *12*(5), 8198597. doi:10.1155/2016/8198597

Meneguette, R. I., Bittencourt, L. F., & Madeira, E. R. (2013). A seamless flow mobility management architecture for vehicular communication networks. *Journal of Communications and Networks (Seoul)*, *15*(2), 207–216. doi:10.1109/JCN.2013.000034

Meneguette, R. I., Boukerche, A., & de Grande, R. (2016, Dec). Smart: An efficient resource search and management scheme for vehicular cloud-connected system. *Proc. IEEE Global Communications Conference*, 1-6. 10.1109/GLOCOM.2016.7842271

Meng, T. (2016). Spatial Reusability-Aware Routing in Multi-Hop Wireless Networks. IEEE TC, 65(1), 244-255.

Mershad, K., & Artail, H. (2013). Finding a STAR in a Vehicular Cloud. *IEEE Intelligent Transportation Systems Magazine, 5*(2), 55-68.

Mohamed, E., Stephan, O., & Mohamed, Y. (2010). Towards Autonomous Vehicular Clouds - A Position Paper. ADHOCNETS 2010. *LNICST*, *49*, 1–16.

Mohammad, P., Mohammad, U. F., Mahboob, U. A., & Khaleel, U. R. (2015). Vehicular Cloud Computing: leading towards tomorrow's Internet of Vehicles. *Journal of Wireless Sensor Networks, 1*, 1-0006.

Morgan, Y. L. (2010). Managing DSRC and WAVE standards operations in a V2V scenario. *International Journal of Vehicular Technology.*

Mousannif, H., Khalil, I., & Al Moatassime, H. (2011). Cooperation as a Service in VANETs. *J. UCS*, *17*(8), 1202–1218.

Murthy, S., & Garcia-Luna-Aceves, J. J. (1996). An efficient routing protocol for wireless networks. *Mobile Networks and Applications*, *1*(2), 183–197. doi:10.1007/BF01193336

Naboulsi, D., & Fiore, M. (2013). On the instantaneous topology of a large scale urban vehicular network: The Cologne case. *Proceedings of ACM MobiHoc*, 167-176. 10.1145/2491288.2491312

Nagaraj, U., & Dhamal, P. (2012). Broadcasting routing protocols in VANET. *Network and Complex Systems*, *1*(2), 13–19.

Nageswara Rao, S., & Reddy, B. R. M. (2015). Developing Data Cloud Services in Various Environments. International Journal & Magazine of Engineering, Technology, Management and Research, 2(8).

Nakano, K., & Olariu, S. (2000). Randomized leader election protocols in radio networks with no collision detection. *Proc. International Symposium on Algorithms and Computation*, 362-373. 10.1007/3-540-40996-3_31

Namboodiri & Gao. (2007). Prediction-based routingfor vehicular ad hoc networks. *Vehicular Technology, IEEE Transactions on, 56*(4), 2332–2345.

Nassar, L., & Karray, F. (2016) Fuzzy Logic in VANET Context Aware Congested Road & Automatic Crash Notification Electrical & Computer Engineering University of Waterloo Waterloo, ON, Canada. *IEEE International Conference on Fuzzy Systems (FUZZ).*

Neha Roy, Y., & Jinila, B. (2015). A survey on security challenges and malicious vehicle detection in Vehicular ad hoc networks. *Contemporary Engineering Sciences, 8*(5-8), 235-240.

ns. (n.d.). *ns2*. Retrieved from http://www.isi.edu/nsnam/ns/

nsnam. (2014). *Publications*. Retrieved from http://www.nsnam.org/overview/publications/

Okrouhly, M., & Waraus, D. (2011). Anti-Lock Braking System Based on FlexRay Protocol, Czech Technical University in Prague. *The 6th IEEE International Conference on Intelligent Data Acquisition and Advanced Computing Systems: Technology and Applications.*

Olariu, S., & Weigle, M. C. (Eds.). (2009). Vehicular Networks: From Theory to Practice. Taylor & Francis.

Olariu, S., Eltoweissy, M., & Younis, M. (2007). ANSWER: AutoNomouS netWorked sEnsoR system. *Journal of Parallel and Distributed Computing*, *67*(1), 114–126. doi:10.1016/j.jpdc.2006.08.013

Olariu, S., Eltoweissy, M., & Younis, M. (2011). Towards autonomous vehicular clouds. EAI Endorsed Trans. *Mobile Communications Applications*, *1*(1), e2. doi:10.4108/icst.trans.mca.2011.e2

Olariu, S., Eltoweissy, M., & Younis, M. (2011). Towards autonomous vehicular clouds. *ICST Transactions on Mobile Communications and Applications*, *11*(7-9), 1–11.

Olariu, S., Hristov, T., & Yan, G. (2013). The next paradigm shift: from vehicular networks to vehicular clouds. In S. Basagni & ... (Eds.), *Mobile Ad Hoc Networking Cutting Edge Directions* (pp. 645–700). New York: Wiley and Sons. doi:10.1002/9781118511305.ch19

Olariu, S., Khalil, I., & Abuelela, M. (2011, February). Taking VANET to the clouds. *International Journal of Pervasive Computing and Communications*, *7*(1), 7–21. doi:10.1108/17427371111123577

Olariu, S., Schwing, J. L., & Zhang, J. (1991). Fundamental algorithms on reconfigurable meshes. *Proc. 29-th Allerton Conference on Communications, Control and Computing*, 811–820.

Olariu, S., Schwing, J. L., & Zhang, J. (1992). Optimal parallel algorithms for problems modeled by a family of intervals. *IEEE Transactions on Parallel and Distributed Systems*, *3*(3), 364–374. doi:10.1109/71.139209

Oliver, N., & Pentland, A. P. (2000). Driver behavior recognition and prediction in a smartcar. Proceedings of the SPIE-The International Society for Optical Engineering, 280-90.

Othman, M., Madani, S. A., & Khan, S. U. (2014). A survey of mobile cloud computing application models. *IEEE Communications Surveys and Tutorials*, *16*(1), 393–413. doi:10.1109/SURV.2013.062613.00160

Papadimitratos, P., Fortelle, A. L., Evenssen, K., Brignolo, R., & Cosenza, S. (2009). Vehicular communication systems: Enabling technologies, applications, and future outlook on intelligent transportation. *IEEE Communications Magazine*, *47*(11), 84–95. doi:10.1109/MCOM.2009.5307471

Park, V. (2000). *Temporally-ordered routing algorithm (TORA) version 1 functional specification*. draft-ietf-manet-tora-spec-03. txt.

Park, J. S., Yu, H. C., Chung, K. S., & Lee, E. (2011, March). Markov chain based monitoring service for fault tolerance in mobile cloud computing. *Proc. IEEE Workshop on Advanced Information Networking and Applications*, 520-525. 10.1109/WAINA.2011.10

Pathak, P. H., Feng, X., Hu, P., & Mohapatra, P. (2015). Visible Light Communication, Networking and Sensing: A Survey, Potential and Challenges. Computer Science Department, Electrical and Computer Engineering Department, University of California, Davis, CA.

Perkins, C. E., Royer, E. M., Das, S. R., & Marina, M. K. (2001). Performance comparison of two on-demand routing protocols for ad hoc networks. *IEEE Personal communications, 8*(1), 16-28.

Perkins, C., Belding-Royer, E., & Das, S. (2003). *Ad hoc on-demand distance vector (AODV) routing* (No. RFC 3561).

Perkins, C. E., & Bhagwat, P. (1994, October). Highly dynamic destination-sequenced distance-vector routing (DSDV) for mobile computers. *Computer Communication Review*, *24*(4), 234–244. doi:10.1145/190809.190336

Pesel, R., & Maslouh, O. (2011). *Vehicular Ad Hoc Networks (VANET) applied to Intelligent Transportation Systems*. Universite de Limoges, France: ITS.

Prasad, K. V., Hodges, S., & Noble. (2016). Vehicle Spaces. *IEEE Pervasive Computing, 15*(1).

Qin, E., Long, Y., Zhang, C., & Huang, L. (2013). Cloud Computing and the Internet of Things: Technology Innovation in Automobile Service. Human Interface and the Management of Information. Information and Interaction for Health, Safety, Mobility and Complex Environments Lecture Notes in Computer Science, 173-180. doi:10.1007/978-3-642-39215-3_21

Qiu, H., Ahmad, F., Govindan, R., Gruteser, M., Bai, F., & Kar, G. (2017). Augmented vehicular reality: Enabling extended vision for future vehicles. *Proc. 18th ACM International Workshop on Mobile Computing Systems and Applications*, 67–72. 10.1145/3032970.3032976

Racehorse, V. J., Zhang, G., Sussman, A., Jian, A., & Parker, T. (2014). Bus rapid transit system deployment for high quality and cost-effective transit service: A comprehensive review and comparative analysis. *IET Intelligent Transport Systems*, *9*(2), 175–183. doi:10.1049/iet-its.2013.0176

Rajput, U., Abbas, F., & Oh, H. (2016). A Hierarchical Privacy Preserving Pseudonymous Authentication Protocol for VANET. *IEEE Access: Practical Innovations, Open Solutions*, *4*, 7770–7784. doi:10.1109/ACCESS.2016.2620999

Rasheed, H. (2015). A Paradigm Shift from Vehicular Ad Hoc Networks to VANET-Based Clouds. *Wireless Personal Communications*, *83*(2), 1131–1158. doi:10.100711277-015-2442-y

Ravishankar, S., & Theetharappan, N. (2017). Cloud Connected Smart Car Park. *International conference on I-SMAC- IoT in Social, Mobile, Analytics and Cloud.*

Rawat, R., Yan, G., Popescu, D. C., Weigle, M. C., & Olariu, S. (2009). Dynamic Adaptation of Joint Transmission Power and Contention Window in VANET. *Proc. of the IEEE Vehicular Technology Conference.*

Raya, M., & Hubaux, J.-P. (2005). A security of vehicular ad hoc networks. *Proc. SASN.* 10.1145/1102219.1102223

Raya, M., & Hubaux, J.-P. (2007). Securing vehicular ad hoc networks. *Vehicular Technology. IEEE Transactions on*, *15*(1), 39–68.

Refaat, T. K., Kantarci, B., & Mouftah, H. T. (2016). Virtual machine migration and management for vehicular clouds. *Vehicular Communications*, *4*, 47–56. doi:10.1109/VETECF.2009.5378793

Riaz, F., & Niazi, M. A. (2016). Road collisions avoidance using vehicular cyber-physical systems: A taxonomy and review. *Complex Adaptive Systems Modeling*, *4*(1), 15. doi:10.118640294-016-0025-8

Rodolfo, I. M. (2016). A Vehicular Cloud-Based Framework for the Intelligent Transport Management of Big Cities. *International Journal of Distributed Sensor Networks*, *12*(5).

Roess, R. P., & Prassas, E. S. (2011). *Traffic Engineering* (4th ed.). Pearson Prentice Hall.

Royer, E. M., & Perkins, C. E. (2000). An implementation study of the AODV routing protocol. In Wireless Communications and Networking Confernce, 2000. WCNC. 2000 IEEE (Vol. 3, pp. 1003-1008). IEEE. doi:10.1109/WCNC.2000.904764

Saeed, Alawi, & Hassan. (2011). Simplified Gateway Selection in Integrated VANET-UMTS Heterogeneous Wireless Network. IST-Africa 2012 Conference & Exhibition.

Safda, H., & Sasase, I. (2009). Multiple End- to-End QoS Metrics Gateway Selection Scheme in Mobile Ad hoc Networks. *International Conference on Emerging Technologies.*

Salahuddin, M. A., Al-Fuqaha, A., & Guizani, M. (2016). Reinforcement learning for resource provisioning in the vehicular cloud. *IEEE Wireless Communications*, *23*(4), 128–135. doi:10.1109/MWC.2016.7553036

Salian, P. P., Prabhu, S., Amin, P., Naik, S.K., & Parashuram, M. K. (2013). Visible Light Communication. *Institute of Technology, Nitte, Karkala, Udupi Texas Instruments India Educators' Conference.*

Salim, B., & Abdelhamid, M. (2012). ITS-Cloud: Cloud Computing for Intelligent Transportation System. *Globecom 2012 – IEEE, Communications Software, Services and Multimedia Symposium*, 2054-2059.

Sallam, G., & Mahmoud, A. (2015, April). Performance Evaluation of OLSR and AODV in VANET Cloud Computing Using Fading Model with SUMO and NS3. In *Cloud Computing (ICCC), 2015 International Conference on* (pp. 1-5). IEEE. 10.1109/CLOUDCOMP.2015.7149649

Sam Mathews, Y., & Jinila, B. (2014). An effective strategy for pseudonym generation and changing scheme with privacy preservation for VANET. *International Conference on Electronics and Communication Systems (ICECS) 2014*, 1-6.

Sampigethaya, K. M. Y., Huang, L. P., & Poovendran, R. (2005). Amoeba: Robust Location Privacy Scheme for VANET. *IEEE Journal on Selected Areas in Communications*, *25*(8), 1569–1589. doi:10.1109/JSAC.2007.071007

Sampigethaya, K., Huang, L., Li, M., Poovendran, R., Matsuura, K., & Sezaki, K. (2005a). *CARAVAN: Providing location privacy for VANET*. Washington Univ Seattle Dept of Electrical Engineering.

Sanaei, Z., Abolfazli, S., Gani, A., & Buyya, R. (2013). Heterogeneity in mobile cloud computing: Taxonomy and open challenges. *IEEE Communications Surveys and Tutorials*. doi:10.1109/SURV.2013.050113.00090

Sanzgiri, K., Dahill, B., Levine, B. N., Shields, C., & Belding-Royer, E. M. (2002, November). A secure routing protocol for ad hoc networks. In *Network Protocols, 2002. Proceedings. 10th IEEE International Conference on* (pp. 78-87). IEEE. 10.1109/ICNP.2002.1181388

Sarkar, S., & Misra, S. (2016). Theoretical modelling of fog computing: A Green Computing to support IoT Applications. *IET Networks*, *5*(2), 23–29. doi:10.1049/iet-net.2015.0034

Sasikala, G., & Deepti, A.R. (2013). Real Time Services for Cloud Computing Enabled Vehicle Networks. *IOSR Journal of Computer Engineering, 11*(1), 8-14.

Sehgal, V. K., Mehrotra, S., & Marwah, H. (2016). Car Security using Internet of Things. *1st IEEE International Conference on Power Electronics, Intelligent Control and Energy Systems.*

F. P. Setiawan, S. H. Bouk, & I. Sasase (Eds.). (2008). An optimum multiple metrics gateway selection mechanism in MANET and infrastructured networks integration. *IEEE Wireless Communications and Networking Conference (WCNC 2008)*. 10.1109/WCNC.2008.394

Sha, K., Xi, Y., Shi, W., Schwiebert, L., & Zhang, T. (2006). Adaptive Privacy-Preserving Authentication in Vehicular Networks. *Proceedings of theFirst International Conference on Communications and Networking in China,* 1–8. 10.1109/CHINACOM.2006.344746

Sharma, M. K., & Kaur, A. (2015, September). A survey on vehicular cloud computing and its security. In *Next Generation Computing Technologies (NGCT), 2015 1st International Conference on* (pp. 67-71). IEEE

Sharma, M. K., & Kaur, A. (2015, September). A survey on vehicular cloud computing and its security. In *Next Generation Computing Technologies (NGCT), 2015 1st International Conference on* (pp. 67-71). IEEE.

Sharma, M. K., Bali, R. S., & Kaur, A. (2015, October). Dyanimc key based authentication scheme for Vehicular Cloud Computing. In *Green Computing and Internet of Things (ICGCIoT), 2015 International Conference on* (pp. 1059-1064). IEEE. 10.1109/ICGCIoT.2015.7380620

Shastri, A., Dadhich, R., & Poonia, R. C. (2011). Performance analysis of on-demand Routing protocols for vehicular ad-hoc Networks. *International Journal of Wireless & Mobile Networks*, *3*(4), 103–111. doi:10.5121/ijwmn.2011.3407

Shim, K. A. (2012). An Efficient Conditional Privacy-Preserving Authentication Scheme for Vehicular Sensor Networks. *Vehicular Technology. IEEE Transactions on*, *61*(4), 1874–1883.

Shiraz, M., Gani, A., Khokhar, R., & Buyya, R. (2012). A review on distributed application processing frameworks in smart mobile devices for mobile cloud computing. *IEEE Communications Surveys and Tutorials*, *15*, 1–20.

Shobin, (2012). Dynamic Clustering and Gateway Management in Vehicular Ad-Hoc Networks. *International Journal of Communications and Engineering, 3*(1).

Shravanthi, C., & Guruprasad, H. S. (2014, May). *Mobile Cloud Computing as future for mobile applications. International Journal of Research in Engineering and Technology.*

Sibai, R. E., Atechian, T., Abdo, J. B., Demerjian, J., & Tawil, R. (2015). A new software-based service provision approach for vehicular cloud. *2015 Global Summit on Computer & Information Technology (GSCIT)*. doi:.10.1109/gscit.2015.7353333

Sibai, R. E., Atechian, T., Abdo, J. B., Tawil, R., & Demerjian, J. (2015). Connectivity-aware service provision in vehicular cloud. *2015 International Conference on Cloud Technologies and Applications (CloudTech)*. 10.1109/CloudTech.2015.7337017

Singh, G., Chakrabarty, N., & Gupta, K. (2014). Traffic Congestion Detection and Management Using Vehicular Ad-Hoc Networks (VANETs) In India, Traffic Engineering and Safety Division,CSIR-CRRI. *International Journal of Advanced Computer Technology.*

Singh, P., Kaur, S., & Sandhu, S. S. (2015). An Overview of Mobile Cloud Computing. *International Journal of Computer and Communication System Engineering*, *2*(2), 255–259.

Singh, R., Singh, D. K., & Kumar, L. (2011). Performance Evaluation of DSR and DSDV Routing Protocols for Wireless Ad Hoc Networks. *Int. J. Advanced Networking and Applications*, *2*(04), 732–737.

Song, W. Z. X., Chen, Q., & Luo, H. (2013). Adaptive resource provisioning for the cloud using online bin packing. *IEEE Transactions on Computers*, *73*(11), 2647–2660.

Squatriglia, C. (2011). *Toyota, microsoft to bring the cloud to cars*. Wired.

Steenstrup, M. (1995). Routing in communications networks. Prentice Hall International (UK) Ltd.

Stojmenovic, I., & Wen, S. (2014). The Fog Computing Paradigm: Scenario and security Issues. *Proceedings of the 2014 Federated Conference on Computer Science and Information Systems (FedCSIS'14)*, 8–16.

Stubing. (2013). *Multilayered security and privacy protection in car-to-X networks: Solution from application down to physical layer*. Springer.

Subir, B., Raymond, T., & Francois, D. (2006). Vehicle-to-Vehicle Wireless Communication Protocols for Enhancing Highway Traffic Safety. *IEEE Communications Magazine*, 74–82.

Sugumar, R., Rengarajan, A., & Jayakumar, C. (2016). Trust based authentication technique for cluster based vehicular ad hoc networks (VANET). *Wireless Networks*, 1–10.

Sun, Y., Lu, R., Lin, X., Shen, X., & Su, J. (2010). An efficient pseudonymous authentication scheme with strong privacy preservation for vehicular communications. *Vehicular Technology. IEEE Transactions on*, *59*(7), 3589–3603.

Sur, C., Park, Y., & Rhee, K. H. (2016). An efficient and secure navigation protocol based on vehicular cloud. *International Journal of Computer Mathematics*, *93*(2), 325–344. doi:10.1080/00207160.2014.934685

Tachi, S., Inami, M., & Uema, Y. (2014). Augmented Reality helps drivers see around blind spots. *IEEE Spectrum*, 31.

Tang, L., Hong, X., & Bradford, P. (2006). Secure relative location determination in vehicular network. *Mobile Ad hoc and sensor networks*, 543-554.

Tang, M., & Pan, S. (2014). A hybrid genetic algorithm for the energy-efficient virtual machine placement problem in datacenters. *Neural Processing Letters*, *1*(5), 1–11.

Tekbiyik, N., & Uysal-Biyikoglu, E. (2011). Energy efficient wireless unicast routing alternatives for machine-to-machine networks. *Journal of Network and Computer Applications*, *34*(5), 1587–1614. doi:10.1016/j.jnca.2011.02.005

Teng, F., Yu, L., Li, T., Deng, D., & Magoules, F. (2017). Energy efficiency of VM consolidation in IaaS clouds. *The Journal of Supercomputing*, *73*(2), 782–809. doi:10.100711227-016-1797-5

Towards Cloud-based Vehicular Networks - TSConIT. (n.d.). Retrieved May 17, 2017, from http://www.bing.com/cr?IG=917999F337DC48CF92519EF2CE98A5A6&CID=129E6989BD476F831B5B630CBCD76E04&rd=1&h=8vvJF-7_Y73ixMatQK8KK37JFmgw5bHLc3alqlNW1yE&v=1&r=http%3a%2f%2freferaat.cs.utwente.nl%2fconference%2f25%2fpaper%2f7577%2ftowards-cloud-based-vehicular-networks.pdf&p=DevEx,5061.1

Townsend, A. M. (2013). *Smart cities: Big data, civic hackers, and the quest for a new utopia*. New York, NY: W. W. Norton.

Tsirigos, A., & Haas, Z. J. (2001). Multipath routing in the presence of frequent topological changes. *IEEE Communications Magazine*, *39*(11), 132–138. doi:10.1109/35.965371

Tuli, A., Hasteer, N., Sharma, M., & Bansal, A. (2013). Exploring Challenges in Mobile Cloud Computing: An Overview. *Proceedings of 4th International Conference on Next Generation Information Technical Summit*, Noida, India, 496-501.

Tyagi, A. K., & Niladhuri, S. (2016, August). Providing trust enabled services in vehicular cloud computing. In *Proceedings of the International Conference on Informatics and Analytics* (p. 3). ACM.

Ucar, Ergen, & Ozkasap. (2013). VMaSC: Vehicular multi-hop algorithm for stable clustering in Vehicular Ad Hoc Networks. Wireless Communications and Networking Conference (WCNC).

USDOT. (2015). *2015 -2019 Strategic Plan Intelligent Transportation Systems (ITS)*. Retrieved from http://www.its.dot.gov/strategicplan.pdf

Uzcategui, R. A., & Acosta-Marum, G. (2009, May). WAVE: A Tutorial. *IEEE Communications Magazine*, *47*(5), 126–133. doi:10.1109/MCOM.2009.4939288

Varaiya, P. (1993). Smart Cars on Smart Roads: Problems of Control. IEEE Transactions on Automatic Control, 38(2).

Vauero, L. M., & Rodero Merino, L. (2014). Finding your Way in the Fog: Towards a Comprehensive Definition of Fog computing. *Computer Communication Review*, *44*(5), 27–32. doi:10.1145/2677046.2677052

Verma, A., & Kaushal, S. (2011). Cloud computing security issues and challenges: a survey. *Advances in Computing and Communications*, 445-454.

Vidhale, B., & Dorle, S. S. (2011, August). Performance analysis of routing protocols in realistic environment for vehicular Ad Hoc networks. In *Systems Engineering (ICSEng), 2011 21st International Conference on* (pp. 267-272). IEEE. 10.1109/ICSEng.2011.55

Vigneri, L., Spyropoulos, T., & Barakat, C. (2016). Storage on wheels: Offloading popular contents through a vehicular cloud. *Proc. 17-th IEEE International Symposium on a World of Wireless, Mobile and Multimedia Networks*, 1-9. 10.1109/WoWMoM.2016.7523506

Viriyasitavat, W., Boban, M., Tsai, H. M., & Vasilakos, A. (2015). Vehicular communications: Survey and challenges of channel and propagation models. *IEEE Vehicular Technology Magazine*, *10*(2), 55–66. doi:10.1109/MVT.2015.2410341

Viriyasitavat, W., Tonguz, O. K., & Bai, F. (2009). Network Connectivity of VANET in Urban Areas. *Proceedings IEEE SECON*, 1-9. 10.1109/SAHCN.2009.5168949

Wahyono, W., & Jo, K. H. (2017). Cumulative Dual Foreground Differences For Illegally Parked Vehicles Detection. *IEEE Transactions on Industrial Informatics*, *13*(5), 2464–2473. doi:10.1109/TII.2017.2665584

Wang, T., Song, L., & Han, Z. (2013). Coalitional Graph Games for Popular Content Distribution in Cognitive Radio VANETs. *IEEE Transactions on Vehicular Technologies, 62*(8), 4010 – 4019. DOI: 10.1109/TVT.2013.2245353

Wang, W. (2002). A digital-driving system for smart vehicles. *IEEE Intelligent Systems, 17*(5).

Wang, S. Y., & Lin, C. C. (2008, September). NCTUns 5.0: A network simulator for IEEE 802.11 (p) and 1609 wireless vehicular network researches. In *Vehicular Technology Conference, 2008. VTC 2008-Fall. IEEE 68th* (pp. 1-2). IEEE. 10.1109/VETECF.2008.464

Wang, T. (n.d.). *Popular Content Distribution in CR-VANETs with Joint Spectrum Sensing and Channel Access*. IEEE.

Wan, J., Liu, J., Shao, Z., Vasilakos, A., Imran, M., & Zhou, K. (2016). Mobile Crowd Sensing for Traffic Prediction in Internet of Vehicles. *Sensors (Basel), 16*(1), 88. doi:10.339016010088 PMID:26761013

Wan, J., Zhang, D., Zhao, S., Yang, L., & Lloret, J. (2014). Context-aware vehicular cyber-physical systems with cloud support: Architecture, challenges, and solutions. *IEEE Communications Magazine, 52*(8), 106–113. doi:10.1109/MCOM.2014.6871677

Wan, S., Tang, J., & Wolff, R. (2008). Reliable routing for road side to vehicle Communications in rural areas. *Proceedings of IEEE International Conference on Communications*, 3017-3021.

Wasef, A., & Shen, X. (2013). EMAP: Expedite message authentication protocol for vehicular ad hoc networks. *IEEE Transactions on Mobile Computing, 12*(1), 78–89. doi:10.1109/TMC.2011.246

Whaiduzzaman, Sookhak, Gani, & Buyya. (2013). A Survey on Vehicular cloud computing. *Journal of Network and Computer Applications*.

Whaiduzzaman, M., Gani, A., & Naveed, A. (2014). PEFC: Performance Enhancement Framework for Cloudlet in mobile cloud computing. *2014 IEEE International Symposium on Robotics and Manufacturing Automation (ROMA)*. 10.1109/ROMA.2014.7295892

Xiao-bo, W. A. N. G., Yu-liang, Y. A. N. G., & Jian-wei, A. N. (2009). *Multi-Metric Routing Decisions in VANET*. Beijing, China: Department of Communication Engineering, University of Science and Technology Beijing.

Xie, S., Liu, Y., Zhang, Y., & Yu. (2010). A Parallel Cooperative Spectrum Sensing in Cognitive Radio Networks. *IEEE Transactions on Vehicular Technology, 59*(8), 4079-4092.

Xi, Y., Sha, K., Shi, W., Schwiebert, L., & Zhang, T. (2007). Enforcing Privacy UsingSymmetric Random Key-Set in Vehicular Networks. *Proceedings of EighthInternational Symposium on Autonomous Decentralized Systems, ISADS'07*, 344–351. 10.1109/ISADS.2007.37

Yan, G., Rawat, D. B., & Olariu, S. (2011). Smart Parking: A secure and intelligent parking system. In IEEE Intelligent Transportation Systems Magazine 3(1):18–30.

Yan, G., Olariu, S., & Weigle, M. (2009). Providing location security in Vehicular ad hoc networks. *IEEE Wireless Communications, 16*(6), 48–55. doi:10.1109/MWC.2009.5361178

Yan, G., Olariu, S., & Weigle, M. C. (2008, July). Providing VANET security through active position detection. *Computer Communications, 31*(12), 2883–2897. doi:10.1016/j.comcom.2008.01.009

Yan, G., Wen, D., Olariu, S., & Weigle, M. C. (2013). Security challenges in vehicular cloud computing. *IEEE Transactions on Intelligent Transportation Systems, 14*(1), 284–294. doi:10.1109/TITS.2012.2211870

Yang, Y., & Bagrodia, R. (2009). Evaluation of VANET-based Advanced Intelligent Transportation Systems. *Proc. ACM VANET*. 10.1145/1614269.1614273

Yi, S. H., Li, C., & Li, Q. (2015). A Survey of fog computing, concepts, applications and Issues. *Proceedings of the 2015 workshop on mobile big data (MobiData'15)*, 37–42. 10.1145/2757384.2757397

Yousefi, S., & Fathy, M. (2006). Vehicular Ad-hoc NETworks (VAN ETs): Challenges and Per-spectives. *6th International Conference on ITS Telecommunications, IEEE*, 761–766.

Yu, R., Zhang, Y., Gjessing, S., Xia, W., & Yang, K. (2013). Toward cloud-based vehicular networks with efficient resource management. *IEEE Network, 27*(5), 48-55. doi: 10.1109/MNET.2013.6616115

Yu, R. Y., Zhang, Y., Yi, L., Xie, S., Song, L., & Guizani, M. (2012, April). Secondary Users Cooperation in Cognitive Radio Networks: Balancing Sensing Accuracy and Efficiency. *IEEE Wireless Communications Magazine, 19*(2), 2–9. doi:10.1109/MWC.2012.6189410

Zadeh, A. G. (2018). *Virtual machine managent in vehicular clouds* (PhD Thesis). Department of Computer Science, Old Dominion University.

Zardosht, B., Beauchemin, S., & Bauer, M. A. (2013). A Decision Making Module for Cooperative Collision Warning Systems Using Vehicular Ad-Hoc Networks. *Proceedings of the 16th International IEEE Annual Conference on Intelligent Transportation Systems*. 10.1109/ITSC.2013.6728481

Zhang, Lin, Lu, & Ho. (2008). An efficient message authentication scheme for vehicular communications. *IEEE Transactions on Vehicular Technology, 57*, 3357–3368.

Zhang, Lin, & Lu, Ho, & Xuemin. (2008, November). An efficient message authentication scheme for vehicular communications. *IEEE Transactions on Vehicular Technology, 57*(6).

Zhang, Q., Cheng, L., & Boutaba, R. (2010). Cloud computing: State-of-the-art and research challenges. *Journal of Internet Services and Applications, 1*(1), 7–18. doi:10.100713174-010-0007-6

Zheng, K., Meng, H., Chatzimisios, P., Lei, L., & Shen, X. (2015, December). An smdp-based resource allocation in vehicular cloud computing systems. *IEEE Transactions on Industrial Electronics, 62*(12), 7920–7928. doi:10.1109/TIE.2015.2482119

Zhioua, G. (2014). A fuzzy multi-metric qos-balancing gateway selection algorithmin a clustered vanet to lte advanced hybrid cellular network. *Vehicular Technology. IEEE Transactions on, 64*(2), 804–817.

Zhong, S., Chen, J., & Yang, Y. R. (2003). Sprite: A Simple, Cheat Proof, Credit based System for Mobile Ad Hoc Networks. *Proceedings INFOCOM*, 1987-1997. 10.1109/INFCOM.2003.1209220

Zhou, Choudhury, Ning, & Chakrabarty. (2011). P2DAP – Sybil Attacks Detection in Vehicular Ad Hoc Networks. *IEEE Journal on Selected Areas in Communications, 29*(3), 582 – 594.

Zhou, J. (2015). Seure and privacy preserving protocol for cloud-based vehicular DTNs. *IEEE Transactions on Information Forensics and Security*, *10*(6), 1299–1314. doi:10.1109/TIFS.2015.2407326

Zhu, H., Lu, R., Huang, C., Chen, L., & Li, H. (2016). An Efficient Privacy-Preserving Location-Based Services Query Scheme in Outsourced Cloud. *IEEE Transactions on Vehicular Technology*, *65*(9), 7729–7739. doi:10.1109/TVT.2015.2499791

Zhu, K., Niyato, D., Wang, P., Hossain, E., & In Kim, D. (2011). Mobility and handoff management in vehicular networks: A survey. *Wireless Communications and Mobile Computing*, *11*(4), 459–476. doi:10.1002/wcm.853

Zingirian, N., & Valenti, C. (2012). Sensor clouds for Intelligent Truck Monitoring. *2012 IEEE Intelligent Vehicles Symposium*. 10.1109/IVS.2012.6232192

Zingirian, N., & Valenti, C. (2012). Sensor clouds for intelligent truck monitoring. *Proceedings of the IEEE Intelligent Vehicles Symposium (IV '12)*, 999–1004.

Zingirian, N., & Valenti, C. (2012). Sensor clouds for intelligent truck monitoring. *Proceedings of the IEEE Intelligent Vehicles Symposium.*

Zomaya, A., Clements, M., & Olariu, S. (1998). A framework for reinforcement-based scheduling in parallel processor systems. *IEEE Transactions on Parallel and Distributed Systems*, *9*(3), 249–260. doi:10.1109/71.674317

About the Contributors

Jyoti Grover received a PhD degree in Computer Engineering from Malaviya National Institute of Technology, Jaipur (India) in 2013. Prior to this, she completed M.Tech degree in Computer Science and Engineering from G.J. University in 2004 and B.E in Computer Science and Engineering from M.D. University, Rohtak, India in 2002. Currently, she is an Associate Professor in Computer Science and Engineering department at Manipal University Jaipur (India). Her areas of interest are Ad hoc networks security, VANET, cloud and mobile computing.

Vinod P. is Professor in Department of Computer Science & Engineering at SCMS School of Engineering & Technology, Cochin, Kerala. He holds his Ph.D in Computer Engineering from Malaviya National Institute of Technology, Jaipur, India. He has also served as programme committee member in the International Security Conference like Security of Information and Networks, ICACCI, ICISSP etc. He also a reviewer of Inderscience Journal of Information and Computer Security, Elsevier Journal of Information Security and Application. Dr. Vinod P's research uses feature selection and dimensionality reduction methods which is major principle in Machine Learning Domain in the detection of malicious application. He has more than 70 research articles published in peer reviewed Journals and International Conferences. His current research is involved in the development of offline malware scanner for mobile application using machine learning techniques. Vinod's area of interest is Malware Analysis, Data Mining, Ethical Hacking and Natural Language Processing.

* * *

Zeinab E. Ahmed received her MSc in Computer and Network Engineering, UofG, Sudan. She has been a lecturer since 2011 in UofG. Her areas of research interest include computer network, cognitive computing, computer engineering, wireless broadband, Internet of Things, cloud computing, vehicular cloud computing.

Amal A. Eltahir received her PhD in Computer Engineering, SUST, Sudan. She is A lecturer since 2013 in Tabuk University, Saudia Arabia. Dr. Amal has been published more than 5 research papers on wireless communications and networking in peer-reviewed academic journals and conferences. Her areas of research interest include computer network, cognitive computing, computer engineering, wireless broadband, WiMAX Femtocell.

Ryan Florin is currently PhD Candidate in Computer Science at Old Dominion University. In 2005, he received the Bachelors Degree from Old Dominion with a Major in Computer Science and Minors in both Math and Physics. In 2011, he received the Masters Degree from Old Dominion in Computer Science with a thesis on clustering in mobile ad hoc networks. His main focus is vehicular clouds and using vehicles as distributed sensors to measure traffic conditions in real time; however, his interests are far reaching. He is passionate about real world problems of today, such as mitigating road traffic congestion, and tomorrow, such as virtual machine migration in a data centers built on cars in a parking lot. Additionally he spends many a night dreaming of smart cities, driverless vehicles, and the problems that can be solved with tomorrow's technology.

Komathy K. graduated from Madras University, India in 1982 and obtained her master and doctoral degrees in Computer Science and Engineering from Anna University, India. She served in Industry, academic and research for about 30+ years. She has published about 70+ research papers that are based on wired and wireless networking, network security, social networking, data mining and artificial intelligence. She contributes her research guidance and support to many PhD scholars registered under various universities. She chaired many National and International Workshops and Conferences. She serves as a member of the Editorial board of Journals and Technical Reviewer of many reputed National and International Journals. With the support of research funding from AICTE India, she submitted a project report on monitoring air pollution of a metro city using Vehicular Networking. She also received Sir CV Raman research Award from IET Chennai Chapter. She shares her experience and knowledge through professional participation in CSI, IEEE, ACM and ISTE societies.

T. Kavitha completed her Ph.D in the Faculty of Information and Communication Engineering, Anna University Chennai, India in the year 2014. She received her M.E. degree in Systems Engineering and Operations Research from Anna University, Chennai India in the year 2006. B.E. in Electronics and Communication Engineering from Bharathidasan University, India in the year 2000. Presently she is working as an Associate Professor in BNM Institute of Technology. She has 15+ years of experience in teaching and research engineering colleges. Her fields of interests are Wireless Networks, Wireless Sensor Network and information security, etc. She has published 20 papers in national/International conferences and 12 in International Journals. She is a life member of ISTE, 2011.

Riti Kushwaha is a Research Scholar in Department of Computer Engineering at Malaviya National Institute of Technology Jaipur, India. She obtained his M.Tech. from MNIT Jaipur. Her research interests include Image Processing, Pattern Recognition.

Stephan Olairu received his PhD from McGill University, Montreal, Canada. Over the years, he has held many different roles and responsibilities as a member of numerous organizations and teams. Much of his experience has been with the design and implementation of robust protocols for wireless networks and their applications. Professor Olariu is applying mathematical modeling and analytical frameworks to the resolution of problems ranging from securing communications, to predicting the behavior of complex systems, to evaluating performance of wireless networks and transportation systems. His most recent research interests are in the area of vehicular clouds and the design of protocols in support of efficient transportation in smart cities.

Ramesh C. Poonia has rich experience of 15+ years as an academician. At present he is an Associate Professor of Computer Science at Amity Institute of Information Technology, Amity University Rajasthan. He has received his PhD in Computer Science from the Banasthali University - Rajasthan. Dr. Poonia has endorsed with the prestigious 'Faculty Appreciation Award' in 2013 for commendable services. He has been engaged as faculty guide for more than 150 undergraduate and graduate students. He has published more than 50 papers and books; some of them have been published in referred journals/ conference proceedings as: IEEE Trans. on Computers, IEEE International Conference on Signal Processing and Communication Engineering Systems, ACM - International Conference Proceedings Series (ICPS), International Journal of Wireless & Mobile Networks – AIRCC, Springer and Inderscience. Dr. Poonia is serving as senior member, technical program committee member, member of editorial board/ reviewer board of various renowned national and international conferences/journals.

Linesh Raja is serving as an Academician from past 7+ years. Currently he is working as an Assistant Professor at Amity Institute of Information Technology, Amity University Rajasthan. He was awarded PhD in Computer Science from the Jaipur National University – Rajasthan. His research area is Vehicular Networks, Wireless & Mobile Networks, Network Protocol Evaluation, Network Simulation & Modeling, and Internet Congestion Control. At present he is supervising three PhD research scholars in the field of Vehicular Wireless Networks & Mobile Computing. He also acted as project supervisor of Master degree in Computer Science and guided undergraduate and graduate students. He has published several research papers in referred journals/conference proceedings He is the Guest Editor of various reputed journals of Taylor & Francis, Bentham Science and Indersceince etc. He is also acting as reviewer and member of various renowned national and international conferences/journals. Few of them are Elsevier, IGI Global, Taylor & Francis, Inderscience and Bentham Science etc.

Gaurav Singal is an Assistant Professor in Computer Science Engineering Department at Bennett University Greater Noida, India. He obtained his PhD from CSE department of Malaviya National Institute of Technology Jaipur, India. He obtained his M.Tech. from MNIT Jaipur. His research interests include Wireless and Adhoc Networking, QoS-aware algorithms, network attacks and simulation.

Ashish Thomas obtained his M.Tech. from Rajasthan University Jaipur from Rajasthan Institute of Engineering and Technology Jaipur. His research interests include Adhoc Networking, VANET, network attacks and simulation.

Bevish Jinila Y. graduated from Manonmaniam Sundaranar University, India in 2003 and obtained her master degree in Computer Science and Engineering from Anna University, India in 2005. She obtained her doctorate from Sathyabama University, India in 2017. She served in academia for 12 years. She has published about 30+research papers based on privacy preservation in vehicular networks, network security, Internet of things and wireless networks. She serves as a reviewer in reputed national and international journals. She attended and organized workshops and seminars. Her research interests include privacy and security in networks, Internet of Things, Machine Learning, Big data analytics and cloud computing.

Index

O

P

R

S

U

V

W

www.ingramcontent.com/pod-product-compliance
Lightning Source LLC
LaVergne TN
LVHW062154300926
849161LV00010B/2664
* 9 7 8 1 5 2 2 5 3 9 8 1 0 *